Lecture Notes in Chemis

Edited by:

Prof. Dr. Gaston Berthier
Université de Paris

Prof. Dr. Hanns Fischer
Universität Zürich

Prof. Dr. Kenichi Fukui
Kyoto University

Prof. Dr. George G. Hall
University of Nottingham

Prof. Dr. Jürgen Hinze
Universität Bielefeld

Prof. Dr. Joshua Jortner
Tel-Aviv University

Prof. Dr. Werner Kutzelnigg
Universität Bochum

Prof. Dr. Klaus Ruedenberg
Iowa State University

Prof Dr. Jacopo Tomasi
Università di Pisa

Springer
Berlin
Heidelberg
New York
Barcelona
Budapest
Hong Kong
London
Milan
Paris
Tokyo

S. Fraga   J. M. R. Parker   J. M. Pocock

Computer Simulations of Protein Structures and Interactions

Springer

Authors

S. Fraga
Department of Chemistry, University of Alberta, Edmonton, AB, Canada T6G 2G2
and
Departamento de Química Física Aplicada, Universidad Autónoma de Madrid
28049 Canto Blanco (Madrid), Spain

J. M. R. Parker
Alberta Peptide Institute and Department of Biochemistry, University of Alberta
Edmontom, AB, Canada T6G 2S2

J. M. Pocock
Department of Biochemistry, University of Dundee, Dundee, Scotland DD1 4HN
United Kingdom

Cataloging-in-Publication Data applied for

Die Deutsche Bibliothek - CIP-Einheitsaufnahme

Fraga, Serafin:
Computer simulations of protein structures and interactions /
S. Fraga ; J. M. R. Parker ; J. M. Pocock. - Berlin ; Heidelberg
; New York ; London ; Paris ; Tokyo ; Hong Kong ; Barcelona
; Budapest : Springer, 1995
  (Lecture notes in chemistry ; 66)
  ISBN 3-540-60133-3 (Berlin)
  ISBN 0-387-60133-3 (New York)
NE: Parker, J. M. Robert:; Pocock, Jennifer M.:; GT

ISBN 3-540-60133-3 Springer-Verlag Berlin Heidelberg New York

Typesetting: Camera ready by author
SPIN: 10500620   51/3142 - 543210 - Printed on acid-free paper

Preface

The *de novo* prediction of the tertiary structure of peptides and proteins and, ultimately, the development of molecular switches, synthetic vaccines, pepzymes, peptidomimetics, ..., are the goals of research in biotechnology. This field relies on a cooperative effort between many branches of the life and natural sciences and this work strives to present the corresponding picture, from quantum mechanics to rational drug design, through computer science, synthetic chemistry, immunology, pharmacology, ... We hope that it will serve its dual purpose, as a learning tool and as reference.

In this type of work it is advisable to try and profit from the knowledge of experts in the various fields and we have been fortunate in counting with the comments of G. Arteca (Laurentian University), A. Cameron (Synphar Laboratories Inc.), J.M. Garcia de la Vega (Universidad Autonoma de Madrid), M. Klobukowski (University of Alberta), G. Kotovych (University of Alberta), G.R. Loppnow (University of Alberta), I. Rozas (Instituto de Química Médica, C.S.I.C.), D.S. Wishart (University of Alberta), H. Yamamoto (Osaka National Research Institute), and S. Yoshikawa (Osaka National Research Institute).

And, as always, we acknowledge the efficiency, patience, and initiative of J. Jorgensen (Department of Chemistry, University of Alberta).

S. Fraga,   J.M.R. Parker,   J.M. Pocock
Dundee Edmonton Madrid

Table of Contents

Acknowledgments

In this work, data, methods and formulations, and results from many sources are examined, discussed, and compared, with reproduction in some instances of copyright material.

We would like to acknowledge here our indebtedness to those organizations from whose publications information and/or copyright material has been used:

Academic Press Limited, Adenine Press, Alan R. Liss, Inc., American Association for the Advancement of Science, Annual Reviews Inc., Bailliére Tindall, Biophysical Society, Butterworth-Heineman, Cambridge University Press, Consejo Superior de Investigaciones Científicas (Spain), Current Biology Ltd., Elsevier Biomedical Press, Elsevier Science Publishers B.V., ESCOM Science Publishers B.V., European Peptide Society, International League Against Epilepsy, International Union of Crystallography, Interscience Publishers, IRL Press, John Wiley & Sons, Inc., MacMillan Journals Ltd., Munksgaard Ltd., National Academy of Sciences (U.S.A.), National Research Council (Canada), New Forum Press, Inc., North-Holland Publishing Company, Oxford University Press, Pergamon Press, Inc., Pharmacotherapy Pub. Inc., Plenum Publishing Corporation, Portland Press Ltd., Protein Research Foundation (Japan), Raven Press, Ltd., Real Sociedad Española de Química, Societa Chimica Italiana, Springer-Verlag, The American Chemical Society, The American Institute of Physics, The American Physical Society, The American Society for Biochemistry and Molecular Biology, Inc., The Chemical Society (U.K.), The National Academy of Sciences (U.S.A.), The Physical Society (Japan), The Protein Society, VCH Verlagsgesellschaft mbH, Wiley-Liss, Inc.

Specific mention must be made, in particular, of the following works from which copyright material has been taken:

A.D. Buckingham. *Permanent and Induced Molecular Moments and Long-Range Intermolecular Forces,* in *Intermolecular Forces,* edited by J.O. Hirschfelder. Advances in Chemical Physics, vol. 12, pp.107-142. Interscience Publishers, New York (1967).

D.A. Clark, G.J. Barton, and C.J. Rawlings. *A Knowledge-Based Architecture for Protein Sequence Analysis and Structure Predictions.* Journal of Molecular Graphics 8, 94-107 (1990).

A. Godzik, A. Kolinski, and J. Skolnick. *De Novo and Inverse Folding Predictions of Protein Structure and Dynamics.* Journal of Computer-Aided Molecular Design 7, 397-438 (1993).

R.D. King and M.J.E. Sternberg. *Machine Learning Approach for the Prediction of Protein Secondary Structure.* Journal of Molecular Biology *216*, 441-457 (1990).

G.H. Loew, H.O. Villar, and I. Alkorta. *Strategies for Indirect Computer-Aided Drug Design.* Pharmaceutical Research *10*, 475-486 (1993).

M.A. Navia and D.A. Peattie. *Structure-based Drug Design: Applications in Immunopharmacology and Immunosuppression.* Immunology Today *14*, 296-302 (1993).

W.R. Taylor. *Protein Structure Prediction*, in *Nucleic Acid and Protein Sequence Analysis. A Practical Approach*, edited by M.J. Bishop and C.J. Rawlings, pp.285-322. IRL Press, Oxford (1987).

Introduction

The modeling and analysis of protein structures and interactions has developed into a major field of research, diverse, difficult, and of exceptional scientific and practical relevance.

Those factors – diversity, difficulty, and relevance – were the motivation for a non-mathematical review (Fraga and Parker 1994), prepared for the *3rd* International Congress on Amino Acids, Peptides, and Analogues, held in Vienna, August 23-27, 1993 [*Amino Acids 5*, 103-216 (1993)]. The interest of many colleagues for that review prompted us to expand it into the present work, maintaining the same objective: to acquaint the reader with the perils and rewards and emphasize the need for a collaborative interaction between experimental and theoretical researchers.

The mixed feelings among researchers were highlighted in that review through a selected sampling of opinions, ranging from an early optimism to a realistic appreciation of the many problems to be faced (Bradley 1970, Wilson and Klausner 1984, King 1989, Wilson and Doniach 1989, Holm and Sander 1992, Ngo and Marks 1992, Honig *et al*. 1993, Eisen *et al*. 1994, Gronbech-Jensen and Doniach 1994, Kolinski and Skolnick 1994, van Gelder *et al*. 1994), with the conclusion that information from as many sources as possible should be used (Thornton 1988), with close collaboration between experimental and theoretical/computational researchers (Daggett and Levitt 1993).

It is this last point that we would like to explore here in more detail. The practitioner in this field might be, in some very special cases, knowledgeable in the life sciences (from biology and biochemistry through pharmacology and immunology), chemistry, quantum mechanics, and computational techniques. In general, however, it will not be so. On one hand, for example, it may be that a worker specialized in computational chemistry will feel tempted to apply his expertise to problems in the life sciences. Two avenues are possible in such a situation: either to dedicate a considerable amount of time in order to become knowledgeable in that new field or to try and engage in a collaboration with an experimental researcher.

Equally serious are the difficulties to be encountered by an experimental worker, who decides to complement/expand the experimental information with computer simulations. Such a researcher would have to decide whether to try and become an expert in theoretical methods and computational techniques, interact with a theoretician, or rely on the use of existing software.

The difficulties facing both types of researchers, when working on their own, as well as the communication gap existing between them may be brought into focus with the following example (*not* a recommendation) of a possible strategy for the development of a new drug:

> Preliminary work on the target protein (assuming that its tertiary structure is not known) might be carried out using a *symbolic network*, incorporating a *bi-level non-symbolic network* for the prediction of the secondary structure. Actual calculations might then be started with an initial build-up *from fragments*, through an *inverse folding* procedure, or by means of a *genetic algorithm*, followed by a *brute force energy minimization*, with refinement and prediction of additional quantities by either a *Monte Carlo with minimization*, *simulated annealing*, or *molecular dynamics* (in an *essential subspace*) procedure, using an appropriate *potential energy function (AMBER/CHARMm/ECEPP/GROMOS/MM3/?*), with a *generalized force shifted potential truncation* method for the *non-bonded interactions*. The interaction of prospective *ligands* with the protein could then be studied first within the framework of only the *electrostatic interactions* (from the solutions of the corresponding *linearized Poisson-Boltzmann* equation) and then refined through a *docking* procedure, either making use (with care) of *shape/chemical complementarities* or in a complete treatment (perhaps with *atomic mass weighting*) in either an *explicit* or *non-explicit-solvent* approach. The prediction of *agonists/antagonists/inhibitors* would then be completed with an appropriate *3D-QSAR* methodology (again with care) and the final determination and refinement of the corresponding *peptidomimetics* (if such were the purpose of the study) through quantum-chemical calculations, using the appropriate software (*GAMESS/GAUSSIAN 92/HONDO/MELDF/MOLCAS/?*). These last steps in the study would benefit, naturally, from the information obtained from *combinatorial libraries*.

It is our hope that the present work will be of help in bridging the communication gap between experimental and theoretical researchers, thus encouraging fruitful collaborations. The coverage is wide enough and with sufficient detail to provide a fair picture, with a selected sampling of the literature. Its limitations, which should be evident to the reader, stem from the staggering amount of material appearing in an endless flow, such that each chapter in this work could be easily expanded into an independent monograph.

Lest the reader is given a false sense of optimism, the difficulties and deficiencies have been pointed out again and again, but we would like to stress that there are also success stories (Boyd 1990, Gund 1994), which add further interest to this very exciting field.

Protein Folding

This work would be incomplete without a summary of the basic concepts regarding peptides and proteins to acquaint the reader, who ventures into this field for the first time, with the terminology in use.

The building blocks for peptides and proteins are the L- and D-amino acids, whether natural or non-natural. The chemical formulas of the twenty natural L-amino acids are given in the following chapter and their Cartesian coordinates are presented in Appendix 2.

The peptide chain, formed by amide (peptide) bonds between successive residues is characterized by the appropriate torsion angles (ϕ, ψ, ω, χ), mentioned repeatedly throughout the text, and their definitions may be found in the following chapter.

Designations such as primary, secondary, tertiary, and quaternary structure; α-helices, β-sheets, β-turns, γ-turns; domains; structural motifs; ... are also defined in the following chapter, with graphic representations where appropriate.

In any case, the reader might wish to consult the well-known work of Schulz and Schirmer (1979), Richardson (1981), Ghelis and Yon (1982), Creighton (1983), Chothia (1984), Jaenicke (1987), Valencia Herrera *et al.* (1988), Chothia (1990), and Chothia and Finkelstein (1990), describing in detail the properties of amino acids, the chemical and structural properties of proteins, the energetics of protein conformations, folding patterns, protein biosynthesis, origins and evolution of proteins, ...

2 Amino Acids, Peptides, and Proteins

2.1 Introduction

Proteins play a central role in most biological processes, interacting with *DNA*, *RNA*, other proteins, carbohydrates, lipids, and other organic and inorganic molecules. Proteins transmit chemical and physical signals between molecules in the cell, act as receptors on the cell surface, control the activity of other proteins as well as of *DNA*, transport oxygen, lipids and metals in the blood, act as storage proteins, and control the flow of ions and other molecules across the cell membrane and participate in the transfer of electrons in photosynthesis. There are many proteins that play a major protective role in the immune system, and others that act as important structural and functional components in the cell.

Most of these proteins are the end product of the nucleic acid genetic code. The universality of the genetic code in bacteria and eukaryotes suggests that this relationship between proteins and nucleic acids was established early in evolution. It is this relationship that has led to a long standing debate on the origins of proteins, nucleic acids, and life on earth. Explanations for the origins of amino acids include the prebiotic formation of amino acids by electrical discharge in mixtures of methane, ammonia, hydrogen, and water (Miller 1953), and the formation of peptide bonds by cyanamide (Oro 1963) and reductive acetylation of amino acids associated with pyrite formation (Keller *et al*. 1994); it has also been proposed that adenine was formed from aqueous solutions of ammonia and cyanide (Oro 1960). It has been particularly difficult, however, to explain the formation of the sugar portion of *DNA* or *RNA* and their subsequent polymerization, having been proposed that early sugars were made from glycoaldehyde phosphates (Eschenmoser and Loewenthal 1992) to give phosphorylated ribonucleotides. The appearance of membranes to contain *RNA* or *DNA* is proposed as an important evolutionary step which provided ways to contain and improve the genetic information (Ourisson and Nakatani 1994).

The first step in protein expression is transcription of the *DNA* sequence to the messenger *RNA* (*mRNA*), which requires several regulatory proteins and *RNA* polymerase. In some cases the *mRNA* is then modified to remove non-protein coding nucleotide sequences (*introns*). Mature *mRNA* is then translated to the amino acid code by interaction with transfer *RNA (tRNA)*, which carry the amino acids that are incorporated into the growing protein chain, and with a complex of ribosomes. Ribosomes are complexes of several proteins and ribosomal *RNA (rRNA)* (Lewin 1990). Proteins are synthesized from the *N*- to the *C*-terminus, which corresponds to the sequence of *DNA* read from the 5' to the 3' end.

It is thought that double stranded *DNA* conserves the genetic code fidelity. It is interesting to note that the codes (Table 2.1) with the most stable nucleoside base pair, *G-C*, code for the single amino acid residues glycine, alanine, proline and arginine. This code is universal but there are some variations in mycoplasma, protozoa, and especially mitochondria. While one strand of the *DNA* code, called the *sense* strand, is used almost exclusively to generate the coded protein, little is known about the purpose of the *antisense* strand of *DNA*. An interesting proposal is that structural information is retained in both sense and anti-sense strands of *DNA* (Table 2.2) (Zull and Smith 1990). It is thought that the protein coding portion of genes is only 3% of the total *DNA* in the 4,000 genes that have been identified of the assumed 50,000-100,000 total human genes. Several points of view exist on whether exons, which encode protein sequences, and introns, which are non-protein coding *DNA* segments inserted between exons, provide evidence to the evolution of protein sequences (Stoltzfus *et al.* 1994). The sequences of proteins have been used to determine evolutionary ancestors and to provide information on their structural and functional properties (Doolittle 1992). A database of 551 ancient conserved regions in proteins (Green, 1994) has been proposed.

Today it is generally thought that *RNA*, which plays an intermediate role in protein synthesis, carried the original genetic information for protein synthesis (Eigen *et al.* 1981, Orgel 1994, Schimmel and Henderson 1994). It is proposed that, because of the stability of the *G-C* base pair, the early triplet codes may have been *GGC, GCC, GAC, GUC* or *GGG, GCG, GAG, GUG*. This means that the original amino acids probably were glycine, alanine, aspartic acid, glutamic acid, and valine, but it might as well be that the evolution of the genetic code is directly related to the precursor-product relationship in the biosynthetic pathway of amino acids where all amino acids originated from Ala, Asp, Glu, Gly, Phe and Val (Wong 1975). Recent information has shown that *RNA* can replicate in the absence of proteins (Kruger *et al.* 1982, Guerrier-Takada *et al.* 1983, Bartel and Szostak 1993) and that it is the *RNA* in ribosomes that catalyzes peptide bond formation (Noller 1993).

In another question related to the origins of proteins, it is not known how the two types of aminoacyl-*tRNA* synthetases, that specify which amino acid is coupled to *tRNA*, coevolved with protein synthesis and the genetic code (Steitz 1991, Moras 1992, Delarue 1995). Type II synthetases may have evolved first, since they are

Table 2.1. Genetic code for *mRNA* translation to amino acid residue.[a]

	G	C	U	A	
G	Gly	Ala	Val	Glu	G
G	Gly	Ala	Val	Asp	C
G	Gly	Ala	Val	Asp	U
G	Gly	Ala	Val	Glu	A
C	Arg	Pro	Leu	Gln	G
C	Arg	Pro	Leu	His	C
C	Arg	Pro	Leu	His	U
C	Arg	Pro	Leu	Gln	A
U	Trp	Ser	Leu	STOP	G
U	Cys	Ser	Phe	Tyr	C
U	Cys	Ser	Phe	Tyr	U
U	STOP	Ser	Leu	STOP	A
A	Arg	Thr	Met	Lys	G
A	Ser	Thr	Ile	Asn	C
A	Ser	Thr	Ile	Asn	U
A	Arg	Thr	Ile	Lys	A

[a]$G, C, U,$ and A stand for guanine, cytosine, uracil, and adenine, respectively. The first column (G, C, U, A) represents the nucleoside code for the first letter in the *mRNA* nucleoside code, the top row (G, C, U, A) is the second letter code and the last column (G, C, U, A) is the third letter code. Note that the strong hydrogen bonded pairs of G and C for *mRNA* to *tRNA* (*GGX, GCX, CGX* and *CCX*) code for the single amino acids Gly, Ala, Arg and Pro. Codes with G and C as the first and last letters (*GXG, GXC, CXG, CXC*) translate to amino acids which can be considered as the basic elements for a stable peptide or protein, including small residues (Gly, Pro, Ala), charged residues (Arg, Asp, Glu), nonpolar residues (Val, Leu) and residues frequently observed in enzymatic sites (His, Gln, Asp, Glu). Tryptophan and Methionine are unique, since they are the only amino acids coded by single triplet *mRNA* codes. The codes *UAA, UAG* and *UGA* code for protein chain termination.

Table 2.2. Sense-antisense exchange for codons.[a]

Class   Anticodon Exchange

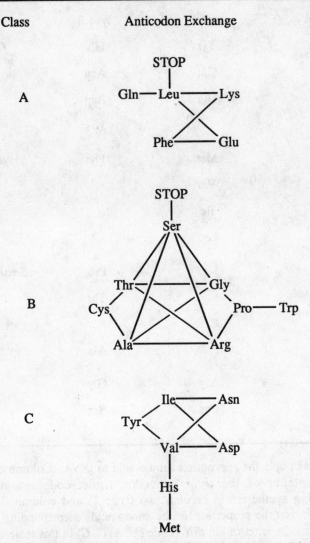

A

B

C

[a]For example, in class A, a sense code for Leu (*CUG, CUC, CUU, CUA*, see Table 2.1) would correspond to codes in the antisense strand for Gln (*GAC*, translated as *CAG* = Gln), Glu (*GAG*, translated as *GAG* = Glu), Lys (*GAA*, translated as *AAG* = Lys) and STOP (*GAU*, translated as *UAG* = STOP). Classes A, B, and C group the α-helical, small and polar (except Arg and Trp), and β-sheet residues, respectively.

Table 2.3. Recognition of *tRNA* by amino acyl *tRNA* synthetase.[a]

tRNA	type 1	type 2	Property
U	Tyr	His	Charged
U	Gln	Asn	
U	Glu	Asp	
U		Lys	
A	Met	Phe	Hydrophobic
A	Val		
A	Ile		
A	Leu		
G		Pro	Small and Polar
G		Thr	
G		Ser	
G		Ala	
C	Arg	Gly	
C	Cys	Ser	
C	Trp		

[a]Synthetases couple the appropriate amino acid to *tRNA*. Column one represents the middle letter of the *tRNA* anticodon triplet code recognized by the corresponding synthetase in column two (type 1) and column three (type 2). Column four lists the properties for the amino acids corresponding to each *tRNA* code. The middle letter of the *tRNA* code (*U, A, G, C*) in this table corresponds to the base pair of the middle letter of the *mRNA* code (top row, *A, U, C, G*) in Table 2.1

9

involved with the coupling/charging of small amino acids such as alanine, glycine and proline. It has been shown that the synthetase interacts with the acceptor and anticodon regions of *tRNA* (Saks *et al*. 1994). The central codon of *tRNA* (see Table 2.3) for the corresponding synthetase is coupled to the amino acid chemical properties, that is: A = charged/polar, U = hydr amino acids arginine, cysteine and tryptophan, are closely related to tyrosine, hydrophobic am acid, respectively (Buechter and Schimmel 1993

handwritten annotation:
$- log [H+]$
$+7$ neutral
<7 larger conc of H+
= acid, proton donor
>7 smaller conc H+
= base, acceptor

2.2 Amino Acids

Proteins and peptides are linear polymers of the twenty naturally-occurring amino acids. Amino acids in protein sequences are usually written in a 3- or 1-letter code (see Fig. 2.1). All amino acids have a backbone of an amino and a carboxylic acid function at a central Carbon (C_α) atom. Nineteen of the twenty amino acids have a sidechain branch at the C_α atom. The sidechain atoms are designated as β, γ, δ, ϵ, etc., in order from the C_α atom. Because of this sidechain branch at the C_α atom, these amino acids can exist as the enantiomers L (*levorotatory*) and D (*dextrorotatory*) (J. Biol. Chem. 169, 237, 1947) or S (*sinister*) and R (*richtig*), respectively, in the Cahn-Ingold-Prelog notation (Cahn *et al*. 1956). The amino acids in proteins are the L enantiomer. The conformations and atom designations of amino acids have been defined by the IUPAC-IUB Commission on Biochemical Nomenclature (Biochemistry 9, 3471-3479, 1970 and Eur. J. Biochem 53, 1-14, 1975) and their coordinates as well as the average bond distances and bond angles, observed experimentally, are available in the literature (Creighton 1983, Engh and Huber 1991, Laskowski *et al*. 1993; see Appendix 2 for values and references).

Amino acids can be grouped into five classes based on similar physical-chemical properties of solubility, *pKa*, size, shape, charge, and functional group type. The common properties of amino acids are listed in Fig. 2.1: group I is glycine, which is the only amino acid without a sidechain and is not chiral; group II is proline, which is the only amino acid where the sidechain forms a cyclic ring with the backbone, imposing a conformation restraint on backbone rotation and determining that it cannot form a H-bond with another residue in the backbone; group III amino acids A, S, T, C can be classified as small and polar; group IV amino acids D, N, E, Q, K, R, H are charged and polar or hydrophilic; and group V amino acids V, I, L, W, F, Y, M are nonpolar or hydrophobic. The *pK* of free C-terminal carboxyl groups is in the range of 1.8-2.3, the free N-terminus has a *pK* range of 8.8-10.8 depending on the residue, and the sidechains have the following *pK*: R (12.0-12.5), K (10.0-10.8), D (3.9-4.4), E (4.0-4.4), H (6.0-6.5), C (8.3-8.5), Y (10.0-10.1), S (13) and T (13). At physiological pH the important residues are the charged residues D, E, K, R and H.

Amino Acid Structure	Residue Name	Group	W_t	H	A
$^-OOC\!-\!C(H)(NH_3^+)\!-\!H$	Glycine Gly G	I	57.1	5.7	88.1
$^-OOC\!-\!C(H)(NH\!-\!CH_2)\!-\!CH_2\!-\!CH_2$ (proline ring)	Proline Pro P	II	97.1	2.1	146.8
$^-OOC\!-\!C(H)(NH_3^+)\!-\!CH_2\!-\!OH$	Serine Ser S	III	105.1	6.5	129.8
$^-OOC\!-\!C(H)(NH_3^+)\!-\!CH(CH_3)\!-\!OH$	Threonine Thr T	III	101.1	5.2	152.5
$^-OOC\!-\!C(H)(NH_3^+)\!-\!CH_3$	Alanine Ala A	III	71.1	2.1	118.1
$^-OOC\!-\!C(H)(NH_3^+)\!-\!CH_2\!-\!SH$	Cysteine Cys C	III	103.2	1.4	146.1

Structure	Amino acid				
$^-OOC\cdots C\cdots CH_2-COO^-$ with H and NH_3^+	Aspartic Asp D	IV	115.1	10.0	158.7
$^-OOC\cdots C\cdots CH_2-CH_2-COO^-$ with H and NH_3^+	Glutamic Glu E	IV	129.1	7.8	186.2
$^-OOC\cdots C\cdots CH_2-CONH_2$ with H and NH_3^+	Asparagine Asn N	IV	114.1	7.0	165.5
$^-OOC\cdots C\cdots CH_2-CH_2-CONH_2$ with H and NH_3^+	Glutamine Gln Q	IV	128.0	6.0	193.2
$^-OOC\cdots C\cdots CH_2-CH_2-CH_2-NH_3^+$ with H and NH_3^+	Lysine Lys K	IV	128.2	5.7	225.8
$^-OOC\cdots C\cdots CH_2-CH_2\cdot CH_2-NH-\overset{NH_2^+}{\underset{\|}{C}}-NH_2$ with H and NH_3^+	Arginine Arg R	IV	156.2	4.2	256.0
$^-OOC\cdots C\cdots C$ (imidazole ring: N=CH, NH, CH) with H and NH_3^+	Histidine His H	IV	137.1	2.1	202.5
$^-OOC\cdots C\cdots CH_2-$ (phenol ring)$-OH$ with H and NH_3^+	Tyrosine Tyr Y	V	163.2	-1.9	236.8

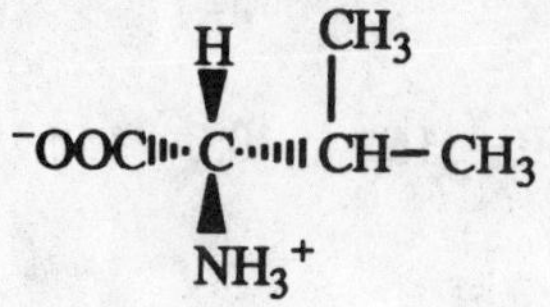 Valine Val V	V	117.1	-3.7	164.5
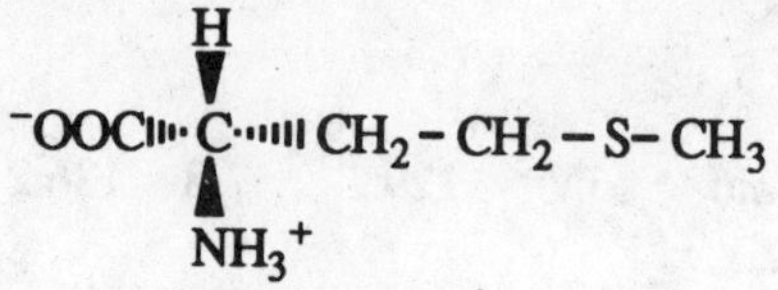 Methionine Met M	V	131.2	-4.2	203.4
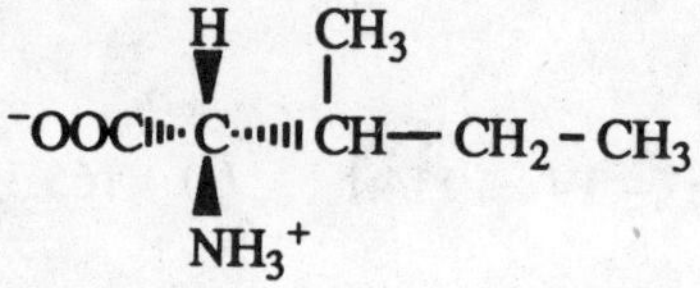 Isoleucine Ile I	V	131.2	-8.0	181.0
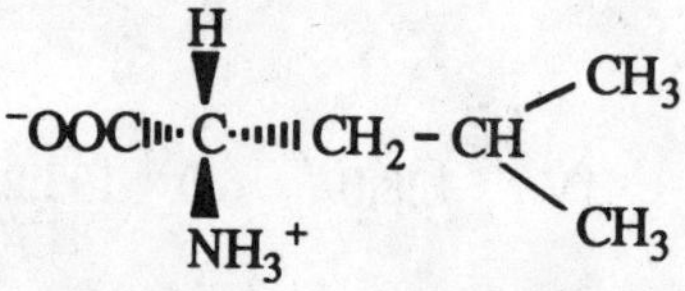 Leucine Leu L	V	113.2	-9.2	193.1
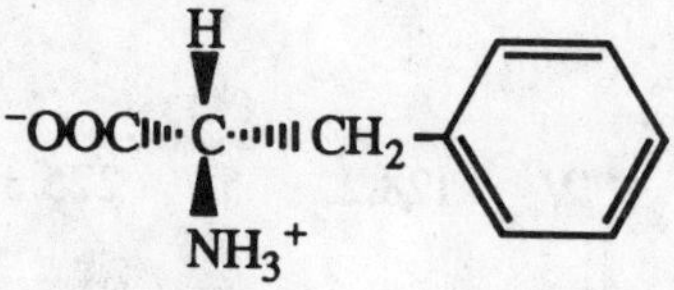 Phenyl–alanine Phe F	V	147.2	-9.2	222.8
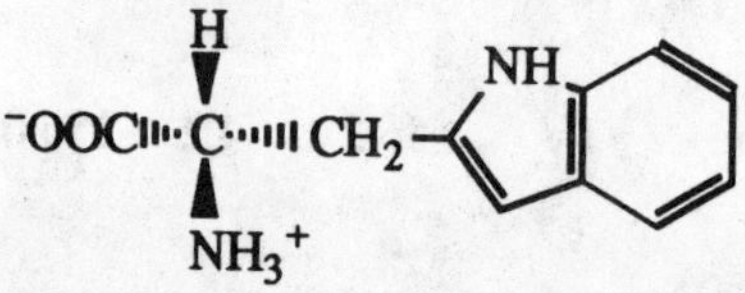 Tryptophan Trp W	V	186.2	-10.0	266.3

Figure 2.1. Properties of the naturally-occurring amino acids. The symbols W_t, H, and A stand for residue weight (molecular weight - 18, in Daltons), hydrophilicity (dimensionless), and accessible surface area (in $Å^2$). The structures are drawn in the ionized form at physiological pH. The peptide and protein weights are evaluated as $\Sigma W_{t(i)} + 18.0$, where the summation over i extends to all the residues in the sequence and $W_{t(i)}$ denotes the residue weights given in this table. The hydrophilicities and areas have been taken from the work of Parker *et al.* (1986) and Rose *et al.* (1985), respectively.

Bacteria can synthesize all twenty amino acids whereas humans cannot synthesize histidine, isoleucine, leucine, lysine, methionine, phenylalanine, threonine, tryptophan and valine. These amino acids are denoted as essential amino acids and must be derived from external sources. The amino acids, synthesized as intermediates in the citric acid cycle, may be grouped in six bio-synthetic families: the 3-phosphoglycerate family (which produces serine, glycine and cysteine), the pyruvate family (which produces alanine, valine and leucine), the alpha-keto-glutarate family (which produces glutamate, glutamine, proline and arginine), the oxaloacetate family (which produces aspartate, asparagine, methionine, threonine, isoleucine and lysine), the phosphoenolpyruvate and erythrose 4-phosphate families (which produce phenylalanine, tyrosine and tryptophan) and the ribose 5-phosphate family (which produces histidine). Amino acids also play an important role in the biosynthesis of other biomolecules, such as nucleosides, sphingosine, histamine, melanin, serotonin, nicotinamide and porphyrin.

2.3 Peptide and Protein Chains

2.3.1 Primary Structure

The primary structure refers to the linear arrangement of amino acids. Sequences are written from left to right with the N-terminus to the left and the C-terminus of the peptide to the right. Residues are located in the sequence by their residue number or by the letter i: residues following the residue i are designated $i+1$, etc. and residues before residue i are designated $i-1$, etc. Peptides are formed by a covalent link between the carboxyl Carbon C of one residue (i) and the backbone Nitrogen of the next residue ($i+1$). This chemical bond, formed by the loss of water, is called the *peptide bond*. The structure of the peptide backbone (N, C_α, C, O) is similar for all peptides and proteins. The backbone can be viewed as a series of planes (the peptide bond) connected at the C_α atom, with the peptide bond atoms (C, O, N, H) restricted to a plane because of the double-bond character of the amide bond (Fig. 2.2). The unique property of proteins, compared to other freely jointed polymers, is that the peptide bond is restricted to a *trans* conformation in most cases, although there are known examples of *cis* peptide bonds, particularly if the residue following the peptide bond is proline. Generally, 2 to 30 of these linear arrangements of amino acids or residues are called *peptides*, while those with more than 30 residues are called *proteins*. Most proteins are in the range of 200-600 amino acid residues, although there are some proteins with much higher number of residues.

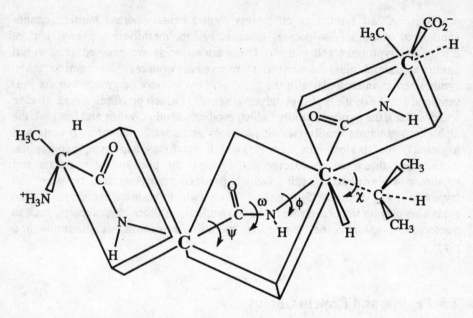

Figure 2.2. Schematic representation of the tetrapeptide *AGVA*, illustrating the peptide bond planes. The distance between consecutive C_α atoms (in bold face) is 3.8 Å.

The conformational flexibility of the peptide and protein backbone is due mainly to the rotation about the $N\text{-}C_\alpha$ and $C_\alpha\text{-}C$ bonds for each amino acid residue. These rotations describe torsion or dihedral angles, referred to as ϕ and ψ angles, and take place on a nanosecond timescale (Glushko *et al.* 1972). The overall conformation of a protein is conveniently viewed (Fig. 2.3) in a two dimensional plot of ϕ *versus* ψ, denoted as a Ramachandran plot (Ramachandran and Sasisekharan 1968), with the convention that a clockwise rotation of ϕ or ψ gives a positive angle when viewed from the *N*-terminus of a residue. The majority of protein conformations observed are clustered in the area of right-hand helical (-60°/-40°) and extended or sheet (-140°/140°) conformations. It is observed that the ϕ,ψ angles for most amino acid residues in proteins are restricted and can be grouped into three main residue conformational groups: group I (glycine) is populated in three main areas (-60°/-40°, 60°/40°, 180°/180°); group II (proline) is restricted to $\phi = -70°\pm20$ and $\psi = -40°$ and 140°, because of the backbone ring structure; the conformations of group III amino acids are generally found in three main areas (-60°/-40°, 60°/40°, -140°/140°). The angle about the C_α to C_β bond (χ_1) is usually found to cluster in three conformational groups, 180°, -60° and 60° (Janin *et al.* 1978, Ponder and Richards 1987), usually designated as *gauche plus* (-60°, *g*+), *gauche* (60°, *g*) and *trans* (180°, *t*) conformations. Torsion about the peptide bond angle (ω) is usually restricted to the trans conformation of 180°±5°.

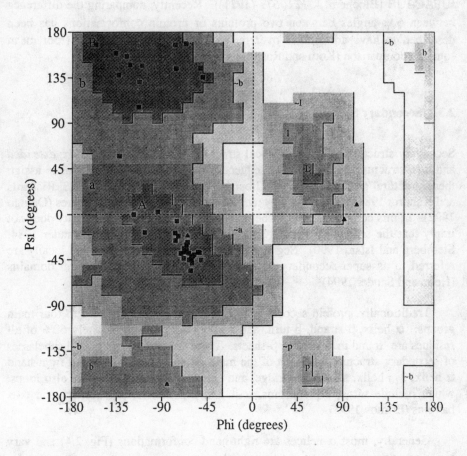

Figure 2.3. Ramachandran plot of φ/ψ values for the residues of the protein
Crambin. As for most proteins, 92% of these residues are confined to
the areas designated A, B, L. Glycines are shown as triangles. The
plot was produced using PROCHECK (Morris *et al*. 1992) and plotted
using XPSVIEW (Adobe Systems Display Postscript Document
Previewer, Version 1.1.1).

The convention to describe peptide conformation has been defined by IUPAC-IUB [Biochem. J. *121*, 577 (1971)]. Recently, comparing the difference between ϕ,ψ angles between two proteins or protein conformations has been described to have advantages over the more commonly used *rms* (root mean squared) comparison (Korn and Rose 1994).

2.3.2 Secondary Structure

Secondary structure refers to the local structure or the location of *helix, extended* and *turn* structures in the primary sequence. There are several methods to assign these structural groups (Levitt and Greer 1977, Kabsch and Sander 1983, Richards and Kundrot 1988) as well as many methods to compare protein structures (Orengo 1992). In this connection it must be noted that local sequence similarity does not imply that the secondary structures will be identical (Kabsch and Sander 1984, Sternberg and Islam 1990). Segments of secondary structure that fold locally are referred to as super-secondary structures, folding units, or structural domains (Holm and Sander 1994).

Traditionally, protein secondary structures have been assigned to four main groups: α-helix, β-strand, β-turn, and random coil. Approximately 60% of all residues are found in helices or β-sheets. There are several recognized subclasses of secondary structures for each of the main groups, such as left- and right-hand α-helix, 3_{10}-helix, π-helix, β-bridge, and extended strand. There are also loops, which include turns, β-β hairpins, α-β and β-β arches, β-β corners and α-α hairpins (Efimov 1993).

Generally, most α-helices are right-hand conformations (Fig. 2.4) and vary from 4 to 15 residues, with a length estimated as (number of residues $\times$ 1.5 Å). Most alpha helices are amphipathic, which means that one face of the helix is hydrophobic (nonpolar) and the other is hydrophilic (polar and charged), and this pattern is thought to determine its packing. Helices generally have i, $(i+3)$, $(i+4)$ hydrophobic repeats with negative-charged sidechains at the N-terminus and positive-charged sidechains at the C-terminus. Because all of the backbone carbonyl functional groups are aligned parallel to the direction of the helix, the resulting helix dipole is thought to be an important factor in protein folding and helix stabilization (Hol *et al.* 1981).

β-strands or β-sheets or extended structures (Fig. 2.4) usually vary from 3 to 9 residues, with a length calculated as (number of residues $\times$ 3.8 Å). Isolated strands are called β-*strands* while two or more parallel or antiparallel strands are denoted as β-*sheets*. In many cases one side of the strand is hydrophobic with a repeat of i, $i+2$ and the opposite side of the plane is hydrophilic. Individual extended strands

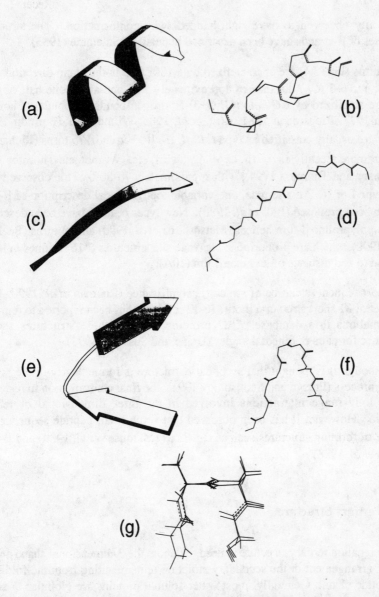

Figure 2.4. Representations of α-helix, β-strand, and β-turns structures. The ribbon diagrams were obtained with MOLSCRIPT (Kraulis 1990) and the stick representations were obtained using INSIGHTII (BioSym Technologies). Figures (*a*) and (b) represent an α-helix, (*c*) and (d) show a β-strand, (*e*) and (*f*) represent two β-strands separated by a β-turn, and (g) shows a four residue type I turn superimposed on a type II turn. The major difference between type I and type II turns is the 180° flip of the plane of the peptide bond between residues *i*+1 and *i*+2.

are usually observed to have right-handed twist conformation. The structural properties of β-strands have been described in detail by Salemme (1983).

β-turns (Fig. 2.4) are characterized by a 180° turn in the chain direction, have an i (C_α) to i+3 (C_α) distance of approximately 7 Å, and about one half of these turns are stabilized by $CO(i)$ to NH (i+3) Hydrogen-bonds (H-bonds) (Chou and Fasman 1977, Richardson 1981, Rose *et al.* 1985, Wilmot and Thornton 1988). β-turns are usually classified as type *I, I', II, II', IV, VIa* and *VIb* turns (Richardson 1981) or more recently as $\alpha_R\alpha_R$, $\alpha_L\gamma_L$, $\beta\gamma_L$, $\epsilon\alpha_R$, etc. (Wilmot and Thornton 1990, Hutchinson and Thornton 1994), with approximately 70-80% of the observed turns being type I or II. An alternate, one-variable topographical description of β-turns has also been proposed (Ball *et al.* 1990). New types of turns have been described, such as polyproline β-turn helices (Matsushima *et al.* 1990) and ω-turns (Beck and Berry 1990), which are prohormone cleavage site structures 6-16 residues in length with end to end distance of less than 1 nm (10Å).

Short sequences can be of structural significance (Rooman *et al.* 1990, 1992) and a set of 81 short structural motifs (6-12 residues) has been reported to represent conformations in a database of 82 proteins and that these structures show a preference for pairs of dihedral angles (Unger and Sussman 1993).

Although it is thought that secondary structure is formed early in the protein folding process (Ptitsyn and Semisotnov 1991), the final conformation may depend on the long range interactions involved in the three dimensional or tertiary structure. However, it has been observed that some small peptide sequences are capable of forming structures such as α-helices (Marqusee *et al.* 1989) and β-turns (Dyson *et al.* 1988).

2.3.3 Tertiary Structure

The designation tertiary structure is used to denote the 3-dimensional shape and the relative arrangement of the secondary structure units resulting from the folding of the peptide chain. Generally, most water soluble proteins are globular in shape. Exceptions, for example, are the *Y*-shaped immunoglobulins, the twisted, double-spring-like shape of tropomyosin molecules (Sodek *et al.* 1972, Cohen and Parry 1986), and cylindrical, helical bundles in membrane proteins. A graphic description of the relative sizes (Fig. 2.5) of small molecules to large proteins (like glutamine synthetase) has recently been presented (Goodsell and Olson 1993).

It is generally believed that there are approximately 1,000 distinct folding domains (Dorit *et al.* 1990, Chothia 1992), because of the limited number of ways in which helices and strands can fold (Finkelstein and Ptitsyn 1987). Recently,

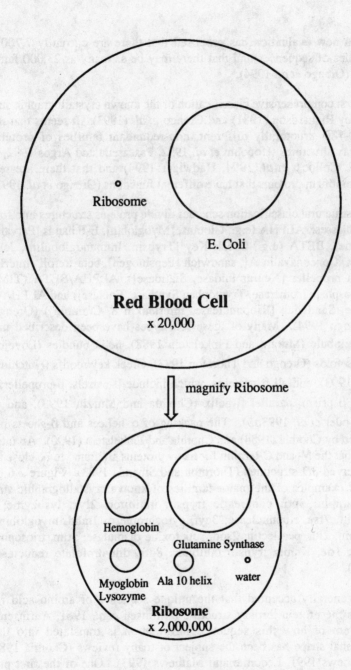

Figure 2.5. Comparison of protein dimensions with a red blood cell (magnified 20,000 times) and a ribosome (magnified 2,000,000 times).

however, a new evaluation has proposed that there are currently 7,700 known superfamilies of sequences and that there may be as many as 23,000 families of sequences (Orengo *et al.* 1994).

The first comprehensive classification of the known crystallographic structures was done by Richardson (1981) and Orengo *et al.* (1993). It seems that there are about 100-180 structurally different, non-redundant families of proteins with known *X*-ray structures (Hobohm *et al.* 1992, Pascarella and Argos 1992, Boberg *et al.* 1992, Colloc'h *et al.* 1994, Hadwiger 1994) and that there are recurring structural motifs in proteins that have different functions (Orengo *et al.* 1993).

Most structural classification schemes divide proteins structures into four main structural classes: ALPHA (e.g., Globins, [Myoglobin], E-F hands [Parvalbumin], Cytochrome), BETA (e.g., Greek Key [Trypsin, Immunoglobulin], Jelly Roll [Poliovirus, Conconavalin A], sandwich [Pepsinogen], beta trefoil [Interleukin 1 beta], beta propeller [Neuraminidase, Sialidase]), ALPHA/BETA (TIM barrel [Triose Phosphate Isomerase, Tryptophan Synthase, Enolase]) and ALPHA+BETA (Lysozyme, Sandwich [Ribonuclease, Interleukin 8, Crambin]) (Orengo *et al.* 1993, Orengo 1994). Many of these fold types have been described in detail: α-helical globule (Murzin and Finkelstein 1988), helix bundles (Lovejoy *et al.* 1993), α+β folds (Orengo and Thornton 1993), Greek key motifs (Hutchinson and Thornton 1993), and all β proteins, which include β-barrels, β-propellers, barrel sandwich, β-prism, parallel β-helix (Chothia and Murzin 1993), and parallel β-helix (Yoder *et al.* 1993*ab*). The packing of α-helices and β-sheets has been summarized by Chothia (1984) and Chothia and Finkelstein (1990). An interesting feature is that the *N*- and *C*-termini for many proteins are found to be close together in the observed 3*D*-structures (Thornton and Sibanda 1983). Figure 2.6 presents illustrative examples of the major families of known crystallographic structures (1cbn, crambin; 5pti, pancreatic trypsin inhibitor; 2lyz, lysozyme; 2hmq, hemerythrin; 7rsa, ribonuclease; 3cyt, cytochrome c; 1mbd, myoglobin; 1cpc, phycocyanin; 2ltn, pea lectin; 2sod, superoxide dismutase; 5tim, triosephosphate isomerase; 1tie, Kunitz trypsin inhibitor; 8dfr, dihydrofolate reductase; 1fxi, ferredoxin).

It is generally accepted that the unique sequence of amino acid residues determines the protein tertiary structure (Anfinsen *et al.* 1961, Anfinsen 1973). The problem of how this sequence information is translated into the final 3-dimensional shape has been the subject of many reviews (Chothia 1984, Kim 1990, Mathews 1991, Lecomte and Mathews 1993). One of the first problems addressed was the so-called *Levinthal paradox* (Levinthal 1968, 1969): How can the total number of possible conformations be sampled for proteins to fold in the experimentally observed millisecond to minutes folding times? (Jaenicke 1987).

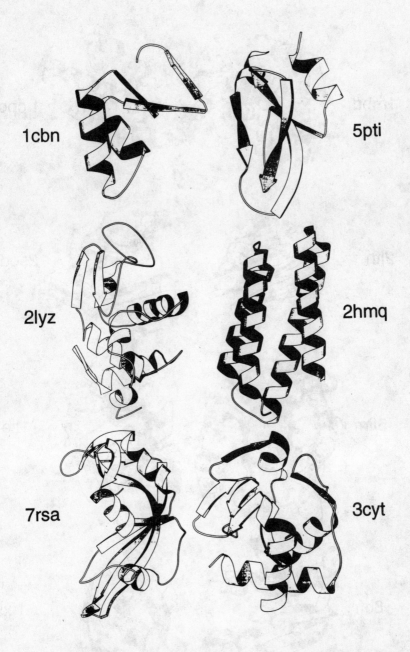

1cbn

5pti

2lyz

2hmq

7rsa

3cyt

Figure 2.6.   Representative ribbon diagrams of protein structures. Each protein is identified by its code in the Brookhaven Protein Data Bank. The diagrams have been obtained with MOLSCRIPT (Kraulis 1990).

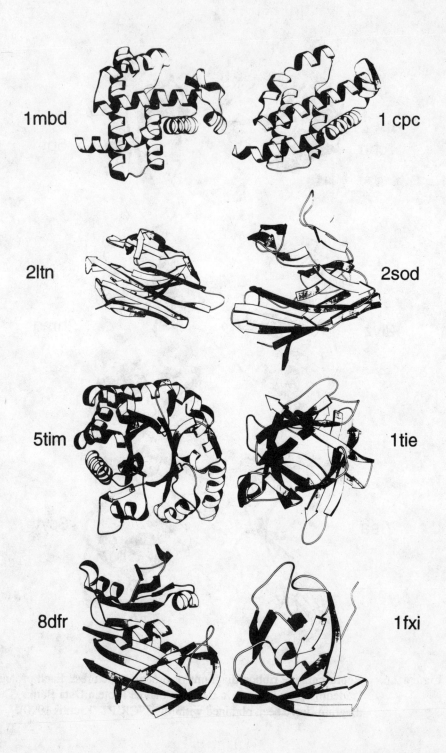

1mbd

1 cpc

2ltn

2sod

5tim

1tie

8dfr

1fxi

Several models have been proposed for protein folding. Much of the early work used theory developed for polymers (Flory 1953, 1989) and for the hard-sphere potentials for peptides (Ramachandran *et al*. 1963). More recently other models have been proposed: the diffusion-collision model (Karplus and Weaver 1979), which considers proteins as divided into microdomains each of which can search many conformations, the framework model (Kim and Baldwin 1982, Baldwin 1989), several types of 2- and 3-dimensional lattice models, which have been used as simplified tools to study the folding (Skolnick and Kolinski 1989, Chan and Dill 1990ab, 1991, Covell and Jernigan 1990, Shakhnovich and Gutin 1990, 1991), and the jigsaw puzzle model which states that protein folding follows multiple pathways (Harrison and Durbin 1985).

It has been proposed that the first step in protein folding is a hydrophobic collapse followed by secondary structure formation and build-up of the final tertiary structure (Dill 1990, Honig *et al*. 1993). This hydrophobic collapse, involving hydrophobic residues is thought to form a compact partially folded structure or molten globule (Dobson 1994). The fact that hydrophobic residues are conserved during evolution supports the idea that they may be important in the folding process (Plochocka *et al*. 1988). Folding pathways are thought to be ordered with rapid formation of secondary structure units, which rapidly diffuse towards each other (Ptitsyn 1973, 1987). Recently it has been proposed that the Levinthal paradox can be explained if restrictions are placed on the number of allowed conformations during the folding process (Zwanzig *et al*. 1992).

It is debated whether the protein folding is a thermodynamic (towards a global minimum of lowest energy) or a kinetic (towards the state formed more rapidly) process, with the possibility that a kinetic pathway, yielding the thermodynamically most stable state, could be followed. In some cases, the problem is complicated by the formation of complexes with other proteins (*chaperonins*) (Gething and Sambrook 1992) in the *in vivo* folding process, in order to avoid unfavorable pathways. [See also the work of Murphy and Gill (1991) on the thermodynamics of protein folding, the reviews of King (1989) and Moult and Unger (1991), the experimental studies of Weismann and Kim (1991) and Dobson *et al*. (1994), and the studies of Daggett and Levitt (1993) on protein unfolding. The studies of Sosnick *et al*. (1994) on cytochrome *C* have led to the interesting proposal that intermediates and molten globules are kinetic products and are not related to the protein folding process.]

Whereas protein folding may involve intermediates, the transition between the native and denatured state for simple cases is thought to be first order. This means that, under physiological conditions, only the native and denatured state are present in significant amounts during the folding process (Privalov 1989). While the native state or the structure observed experimentally under physiological conditions is a collection of structurally and energetically similar conformational states (Fauenfelder *et al*. 1991), there is little consensus as to what defines the denatured state. However, there is agreement that the denatured state is not necessarily a completely random structure because volume effects and local interactions reduce the total possible number of random conformations for a denatured protein.

2.3.4 Quaternary Structure

Quaternary structure, the highest level of protein structure, defines how the tertiary structure of monomer proteins assemble or aggregate to form a complex (Jaenicke 1987, Horton and Lewis 1992). The monomers can be single-chain proteins or multi-chain proteins covalently linked by disulfide bridges. Oligomers are non-covalent complexes of similar or different protein chains. In many cases, these complexes are required for the proper biological activity of a protein, with enzymatic active sites formed at the interface.

Examples of higher-order quaternary structures in proteins are the dimers of superoxide dismutase and of triose phosphate isomerase, the tetramers of hemoglobin and immunoglobulin, the hexamers of insulin, the twelve subunits of aspartate transcarbamylase, the 17 subunits of TMV protein disks and the 180 subunits of Tomato Bushy Stunt Virus coat protein.

Factors that have been proposed to influence protein association are temperature-induced complementarity (Leikin and Parsegian 1994) and hydrodynamic steering (Brune and Kim 1994).

2.4 Post-Translation Modifications

Most proteins are synthesized *in vivo* with free *N*- and *C*-termini and no backbone or sidechain modifications. However, many proteins undergo post-translational modifications that change their structure or functional properties. Some proteins are cleaved enzymatically to produce one or more proteins with new activities. Those precursors are called pro-proteins or prepro-proteins if they are synthesized with *N*-terminal signal peptide extensions on the pro-protein. For example, serum albumin and insulin are synthesized as prepro-proteins. After the signal peptide is removed enzymatically, the pro-protein is further cleaved to produce the active protein. Other examples of prepro and pro-proteins are trypsinogen, chymotrypsinogen, and pepsinogen, which are the inactive precursors of the enzymes trypsin, chymotrypsin and pepsin and are denoted as zymogens.

There are post-translational modifications of proteins to give over 200 different amino acid derivatives. More than 50% of eukaryotic proteins are produced as *N*-acetylated proteins, and other *N*-terminal residues are blocked by formyl, pyruvoyl, palmitoyl, fatty acids, alpha-keto acyl, glucuronyl, and methyl groups. Many peptide hormones are modified as *C*-terminal amide peptides and some proteins, located at the cell surface, form a *C*-terminal complex with glycosyl-phosphatidyl inositol. Attachment of carbohydrate groups is one of the most frequent modifications in eukaryotic proteins, usually at the sidechain carbonyl amide of asparagine in the tripeptide sequences *NXS, NXT* or *NXC*, and at the side-chain hydroxyl function of threonine and serine. Sulfonation of tyrosine and phosphorylation of tyrosine, serine and threonine are common and in some proteins proline is converted to hydroxy-proline. Post-translational modification of proteins to *D*-amino acids increases the biological diversity of proteins (Mor *et al*. 1992).

Some of the most common non-enzymatic modifications of amino acids are oxidation of cysteine to cystine (which occurs by disulfide exchange with glutathione), oxidation of cysteine to cysteic acid, and oxidation of methionine to methionine sulfone. Other non-enzymatic modifications include deamidation of asparigine and glutamine to aspartic and glutamic acid.

Theoretical Formulation

The chapters in this section present the theoretical foundations for the various techniques used in the computer simulations of protein structures and of their interactions, described in subsequent sections.

A perspective of the basic concepts and formulations needed for a quantum-chemical study of molecules, whether isolated or involved in interactions, is given in Chapter 3. It is not intended that this chapter should constitute a complete exposition of Quantum Mechanics. Rather, its purpose is to outline the conceptual flow inherent in a self-consistent field/configuration interaction calculation: Schrödinger equation $\rightarrow$ independent-particle approximation $\rightarrow$ (perturbation theory/variational principle) $\rightarrow$ self-consistent field theory $\rightarrow$ expansion approximation $\rightarrow$ configuration interaction treatment. The reader interested in a complete review of the present state of research in Computational Chemistry, in general, and in self-consistent field theory, in particular, is referred to the work edited by Fraga (1992) and Carbo and Klobukowski (1991).

Quantum-mechanical calculations yield results which are temperature-independent. This dependence may be introduced through consideration of Statistical Mechanics, where use can be made of the quantum-mechanical results, and Chapter 4 presents a review of the concepts in Quantum Statistical Mechanics, with emphasis on the Monte Carlo method.

These two chapters would suffice if the corresponding calculations were practically feasible. Such is not the case for large peptides and proteins, as emphasized again and again throughout the text and, consequently, recourse must be made to very drastic approximations, embodied in the Molecular Mechanics formulation, discussed in Chapters 5 and 6. The purpose of the brief Chapter 7 is to focus the attention on two important questions: Are all the necessary tools for the simulations already available? Is it possible to discriminate among the many approaches and software packages?

Quantum Mechanics was supposed to replace Newtonian Mechanics in the study of the microscopic properties of molecules. And it is somewhat amusing that, after so many years since the birth of Quantum Mechanics, the study of proteins through Molecular Mechanics is performed in most cases at the level of Newtonian Mechanics, with disregard for electrons and bonds and where the simulation is applied to a set of points subjected to a certain force field.

3 Quantum Mechanics

The quantum-chemical study of the ground state of a peptide should be performed, as for any other molecule, through the solution of the corresponding Schrödinger equation. The above statement implies that the relativistic corrections may be ignored as only light atoms (H, C, N, O and S) are involved as a rule. This assumption, however, is not acceptable if the energy levels of the system are to be determined.

3.1 The Schrödinger Equation

The time-independent Schrödinger equation (Schrödinger 1926abcde) may be written, in the usual notation, as

$$\mathcal{H}\Psi_I = \Psi_I E_I \tag{3.1}$$

where $\mathcal{H}$ denotes the Hamiltonian (or energy) operator of the system under study. Ψ_I is the eigenfunction (or state function) of the I-th state, with corresponding eigenvalue (or energy) E_I.

The meaning of the above equation, as for any other eigenvalue equation, is that operating with the Hamiltonian operator on any of its eigenfunctions will reproduce that same eigenfunction multiplied by a constant, which is the corresponding eigenvalue. This statement, naturally, does not hold for functions other than the eigenfunctions of the Hamiltonian operator. Operation with the Hamiltonian operator on the function simply implies performing the mathematical operations embodied in the operator, as described below.

3.1.1 The Hamiltonian Operator

The Hamiltonian operator includes all the energy terms arising from the motion of the electrons and the nuclei and the electrostatic interactions between them.

After separating the motion of the centre of mass and assuming that it is at rest (at the centre of coordinates) the terms to be included are the electron kinetic, electron-nucleus attraction, electron-electron repulsion, and nucleus-nucleus repulsion energy terms. An additional term, denoted as specific mass effect and which appears when separating the motion of the centre of mass, is usually omitted. Operating within the atomic unit system (see Appendix 1) one can then write

$$\mathcal{H} = -\frac{1}{2} \sum_{\mu} \nabla_{\mu}^2 - \sum_{a} \sum_{\mu} (Z_a / r_{\mu a}) + \sum_{\mu} \sum_{\nu > \mu} (1 / r_{\mu \nu}) + \sum_{a} \sum_{b > a} (Z_a Z_b / R_{ab}) \qquad (3.2)$$

where the summations over μ, ν extend to all the electrons of the system and the summations over a, b extend to all the nuclei of the system; $r_{\mu a}$ denotes the separation of electron μ from nucleus a, $r_{\mu \nu}$ denotes the separation of electron μ from electron ν, and R_{ab} denotes the separation of nucleus a from nucleus b. Z_a is the nuclear charge of atom a.

The terms of this equation correspond, in that order, to the kinetic energy of the electrons, the electrostatic attraction energy of the electrons by the nuclei, the electrostatic repulsion energy between the electrons, and the electrostatic repulsion energy between the nuclei. The operator ∇_{μ} involved in the electron kinetic energy term is related

$$-i\nabla_{\mu} = \mathbf{p}_{\mu}$$

to the linear momentum vector operator $\mathbf{p}$, so that in Cartesian coordinates one can write

$$-\nabla_{\mu}^2 = \mathbf{p}_{\mu} \cdot \mathbf{p}_{\mu} = \partial^2 / \partial x_{\mu}^2 + \partial^2 / \partial y_{\mu}^2 + \partial^2 / \partial z_{\mu}^2$$

Within the present approximation, given a molecular geometry, the solution of the Schrödinger equation may be performed for the Hamiltonian operator from which the nucleus-nucleus repulsion energy terms have been omitted. [These terms must be included, however, whenever a geometry optimization is to be performed, that is, when the most stable geometry is to be determined.] The corresponding solutions, Ψ_I and $E_{I(e)}$, depend on the molecular geometry. $E_{I(e)}$ denotes the electronic energy and the corresponding molecular energy, $E_{I(m)}$, is obtained by adding the nuclear repulsion energy. The equilibrium conformation is the one corresponding to the lowest molecular energy.

3.1.2. Eigenfunctions and Eigenvalues

The functions Ψ_I must be well-behaved: that is, they must be monotonic and single-valued at each point of space, tend to zero at infinity, and be quadratically integrable. Because the Schrödinger equation is not affected if the function Ψ_I is multiplied by a constant, it is customary (but not necessary) to assume that the eigenfunctions are normalized (to unity); that is,

$$<\Psi_I|\Psi_I> = \int \Psi_I^*\Psi_I d\tau = 1$$

with integration, over all the electron coordinates, extending to all space; Ψ_I^* denotes the complex conjugate of Ψ_I and $d\tau$ denotes the volume element. The bra-ket notation, $<\Psi_I|\Psi_I>$, will be used hereafter for simplicity.

A function Ψ_I is said to be non-degenerate when it is the only function associated with the eigenvalue E_I. On the other hand, the n functions Ψ_{I1}, Ψ_{I2}, ..., Ψ_{In} are said to be degenerate if they all share the common eigenvalue E_I, which is then said to have an n-fold degeneracy.

Functions corresponding to different eigenvalues are orthogonal; that is, they satisfy the condition

$$<\Psi_I|\Psi_J> = 0$$

Degenerate functions are either orthogonal or may be orthogonalized.

Consequently, hereafter, it will be assumed that the eigenfunctions for the system under consideration constitute an orthonormal set, i.e., they satisfy the condition

$$<\Psi_I|\Psi_J> = \delta_{IJ} \tag{3.3}$$

where δ_{IJ} is a Kronecker delta ($\delta_{IJ} = 1$ for $I=J$; $\delta_{IJ} = 0$ for $I \neq J$).

Multiplying Eq. (3.1) on the left with Ψ_I^* and integrating yields

$$<\Psi_I|\mathcal{H}|\Psi_I> = <\Psi_I|\Psi_I>E_I \equiv E_I \tag{3.4}$$

taking into account that Ψ_I is normalized. The eigenvalue E_I is said, according to this result, to represent the energy expectation value of $\mathcal{H}$ for Ψ_I. Similarly, multiplying Eq. (3.1) on the left with Ψ_J^* and integrating over all space yields

$$<\Psi_J|\mathcal{H}|\Psi_I> = <\Psi_J|\Psi_I>E_I = 0 \tag{3.5}$$

because of the orthogonality condition. This result is described by saying that eigenfunctions corresponding to different eigenvalues do not interact. Similarly, it can be seen that orthogonal degenerate eigenfunctions will not interact either.

Using the same terminology as above, it can be said that the normalization condition defines the expectation value of the identity operator for Ψ_I. Both the identity and the Hamiltonian operators are said to be Hermitian, a property which in the present case (for real functions) translates into

$$<\Psi_I|\Psi_J> \equiv <\Psi_J|\Psi_I> \tag{3.6a}$$

$$<\Psi_I|\mathcal{H}|\Psi_J> \equiv <\Psi_J|\mathcal{H}|\Psi_I> \tag{3.6b}$$

3.2. Orbitals and Spin-orbitals

Taking into account the comments made above regarding the contribution from the nuclear repulsion terms, it will be assumed hereafter (unless stated otherwise) that the electronic Hamiltonian operator is being considered. After removing the term corresponding to the nuclear repulsion from Eq. (3.2), the electronic Hamiltonian may be written as

$$\mathcal{H} = \sum_{\mu} \mathcal{H}_{\mu} + \sum_{\mu} \sum_{\nu \neq \mu} (1/r_{\mu\nu}) \tag{3.7}$$

where

$$\mathcal{H}_{\mu} = -\frac{1}{2} \nabla_{\mu}^2 - \sum_{a} (Z_a / r_{\mu a}) \tag{3.8}$$

represents the one-particle Hamiltonian operator corresponding to electron μ. The one-electron operators $\mathcal{H}_1, \mathcal{H}_2, \dots \mathcal{H}_{\mu}, \dots$, are all formally identical, differing only in the label for the electron being considered.

Let us now consider these one-particle Hamiltonian operators in detail. First, it is evident that they also represent energy operators, as each one includes the kinetic energy of one electron and the attraction energy of that electron by the nuclei. Second, for each one of these Hamiltonian operators one can write the corresponding eigenvalue equation, formally analogous to Eq. (3.1). For example, the one-electron Schrödinger equation corresponding to the one-electron Hamiltonian operator $\mathcal{H}_{\mu}$ (for electron μ) is

$$\mathcal{H}_{\mu}\phi_i(\mu) = \phi_i(\mu)\varepsilon_i \qquad (3.9)$$

where the one-electron eigenfunctions $\phi_i(\mu)$ and associated eigenvalues ε_i are denoted, respectively, as orbitals and orbital energies. At this point it must be emphasized that, always within the context of the Schrödinger equation, orbitals have only meaning insofar as they represent solutions of the one-electron problem. It follows, from the formal identity of all the one-electron Hamiltonian operators associated with the system under study, that all such operators have the same set of solutions (differing only in the electron labels).

If the last term (i.e., the electron repulsion) is omitted from Eq. (3.7), it can be seen immediately that the Hamiltonian operator

$$\mathcal{H} = \sum_{\mu} \mathcal{H}_{\mu}$$

has as solutions the corresponding products of orbitals:

$$\mathcal{H}\{\phi_i(1)\phi_j(2)...\phi_k(\mu)...\} = \{\mathcal{H}_1 + \mathcal{H}_2 + ... + \mathcal{H}_{\mu} + ...\}\{\phi_i(1)\phi_j(2)...\phi_k(\mu)...\}$$

$$= \{\mathcal{H}_1\phi_i(1)\}\phi_j(2)...\phi_k(\mu)... + \phi_i(1)\{\mathcal{H}_2\phi_j(2)\}...\phi_k(\mu)... + ...$$

$$+ \phi_i(1)\phi_j(2)...\{\mathcal{H}_{\mu}\phi_k(\mu)\}... + ...$$

$$= \{\phi_i(1)\varepsilon_i\}\phi_j(2)...\phi_k(\mu)... + \phi_i(1)\{\phi_j(2)\varepsilon_j\}...\phi_k(\mu)... + ...$$

$$+ \phi_i(1)\phi_j(2)...\{\phi_k(\mu)\varepsilon_k\}... + ...$$

$$= \{\phi_i(1)\phi_j(2)...\phi_k(\mu)...\}\{\varepsilon_i + \varepsilon_j + ... + \varepsilon_k + ...\} \qquad (3.10)$$

with an eigenvalue equal to the sum of the corresponding orbital energies. In the development of the above expression, use has been made of the fact that the operator $\mathcal{H}_1$ will act only on the orbital occupied by electron 1, the operator $\mathcal{H}_2$ will act only on the orbital occupied by electron 2, etc.

The above development represents the mathematical formulation for the so-called independent-particle approximation. This approximation corresponds to a hypothetical case, which will not occur in reality, where there are no interactions between the electrons. Its significance is to be found in the fact that it will help in developing the formulation for the ultimate determination of the eigenfunction Ψ.

Even if this assumption were valid, the description of the electrons in terms of orbitals would not be complete. The correct description requires the consideration of an additional function, the so-called spin-function. There are two spin functions, α and β, which are assumed to be orthonormal; that is, $\langle\alpha|\alpha\rangle = \langle\beta|\beta\rangle = 1$ and

$\langle\alpha|\beta\rangle = \langle\beta|\alpha\rangle = 0$, where integration is performed over the spin space. One can generate two spin-orbitals from each orbital; for example, from the orbital ϕ_i one would obtain the two spin-orbitals $\phi_i\alpha$ and $\phi_i\beta$.

The result obtained in Eq. (3.10) still holds for spin-orbitals because the one-electron Hamiltonian operators are spin-independent (that is, they do not include terms depending on the spin coordinates) but it is not acceptable because of the Pauli exclusion principle and the indistinguishability of the electrons. The Pauli principle states that no two electrons may be described, under any circumstances, by the same spin-orbital. Consequently, the total function of the system must be antisymmetric in every pair of electrons.

The construction of an antisymmetric function is achieved very simply if it is written as a determinant, as illustrated below for the case of the ground state (still within the independent particle approximation). This development requires consideration of the *Aufbau* principle, as follows. Let us assume that all the solutions, ϕ_i and ε_i, of Eq. (3.9) have been obtained. These solutions may be numbered according to the increasing values of ε_i; that is, $\varepsilon_1 < \varepsilon_2 < ... < \varepsilon_n < ....$ If the system consists of 2N electrons and one takes into account that each orbital may accommodate two electrons, one with α and the other with β spin, the ground state of the system may be written, disregarding the Pauli principle, as

$$\Psi = \{\phi_1(1)\alpha(1)\}\{\phi_1(2)\beta(2)\}\{\phi_2(3)\alpha(3)\}\{\phi_2(4)\beta(4)\}$$

$$...\{\phi_N(2N-1)\alpha(2N-1)\}\{\phi_N(2N)\beta(2N)]$$

In short, one can say that the electron configuration of the ground state is $\phi_1^2 \phi_2^2 ... \phi_N^2$, where the superscript gives the occupancy.

Any other configuration, obtained by replacing any of the first N orbitals with a higher one, will result in a function with a higher energy, i.e., a function corresponding to an excited state. This procedure of construction of configurations and corresponding functions is the expression of the *Aufbau* principle.

The above function, however, is not correct because it is not antisymmetric. A proper antisymmetric function may be obtained in the form of the following determinant:

$$\Psi = \begin{vmatrix} \phi_1(1)\alpha(1) & \phi_1(2)\alpha(2) & \cdots & \phi_1(2N)\alpha(2N) \\ \phi_1(1)\beta(1) & \phi_1(2)\beta(2) & \cdots & \phi_1(2N)\beta(2N) \\ & & & \\ \cdots & \cdots & \cdots & \cdots \\ \phi_N(1)\beta(1) & \phi_N(2)\beta(2) & \cdots & \phi_N(2N)\beta(2N) \end{vmatrix} \qquad (3.11)$$

The function will vanish identically if two spin-orbitals are identical; this is a consequence of the property that exchange of two rows (or two columns) of a determinant will change the sign of the value of the determinant.

This function satisfies the Pauli principle and is still an eigenfunction of the Hamiltonian operator $\mathcal{H} = \Sigma \mathcal{H}\mu$, with the same eigenvalue, $(\varepsilon_1 + \varepsilon_2 + ... + \varepsilon_k + ...)$, as the original monomial product of orbitals. [This function and eigenvalue will be denoted as $\Psi^{(0)}$ and $E^{(0)}$, within the context of perturbation theory, to be discussed below.] It must be emphasized at this point that the function Ψ above is not eigenfunction of the complete Hamiltonian operator.

3.3 Perturbation Theory for Non-degenerate States

The independent-particle approximation cannot be applied to real systems, since there are electron-electron repulsions. This approximation suggests, however, a possible way of proceeding towards the solution of the Schrödinger equation. The corresponding mathematical formalism is the one embodied in perturbation theory.

Perturbation theory may be applied in those cases where the Hamiltonian operator under consideration may be written as

$$\mathcal{H} = \mathcal{H}^{(0)} + \lambda \mathcal{H}^{(1)} \tag{3.12}$$

where $\mathcal{H}^{(0)}$ is the so-called zero-order Hamiltonian operator and $\mathcal{H}^{(1)}$ is the perturbation term that makes it impossible to obtain the exact solution of the corresponding eigenvalue equation; λ is simply an ordering parameter. It is assumed (as seen in Section 3.2) that the eigenvalue equation corresponding to the zero-order Hamiltonian operator

$$\mathcal{H}^{(0)} \Psi_I^{(0)} = \Psi_I^{(0)} E_I^{(0)}$$

is solvable, yielding the zero-order solutions, $\Psi_I^{(0)}$ and $E_I^{(0)}$. In principle, perturbation theory should only be applied to those cases where the energy contribution arising from the perturbation term is small. This consideration is of no concern here because it is not intended to proceed with the practical application of perturbation theory. Rather, the development of perturbation theory will be used, in conjunction with the variational principle (see below), in order to establish the foundation of the self-consistent field method, described in Section 3.4.2.

Formally, perturbation theory may be applied to the Hamiltonian under consideration, taking into account the developments of the preceding section. The Hamiltonian operator, Eqs. (3.2) and (3.7), may be written, towards a perturbation theory treatment as in Eq. (3.12), with

$$\mathcal{H}^{(0)} = \sum_{\mu} \mathcal{H}_{\mu}$$

$$\lambda \mathcal{H}^{(1)} = \sum_{\mu} \sum_{\nu \neq \mu} (1/r_{\mu\nu})$$

That is, we are reinstating the terms, corresponding to the electrostatic repulsion energy between the electrons, which were omitted in order to develop the independent-particle approximation. The zero-order solutions $\Psi_I^{(0)}$ are known, being expressed in terms of the spin-orbitals, and the zero-order energies $E_I^{(0)}$ are given in terms of the corresponding orbital energies. It is understood, of course, that the functions $\Psi_I^{(0)}$ must be antisymmetric (see Section 3.2).

3.3.1. Zero- and First-order Approximations

The perturbation theory is based on the expansion of the function to be determined, Ψ_I, as well as of its corresponding eigenvalue, E_I, in terms of λ. One can write

$$\Psi_I = \sum_n \Psi_I^{(n)} \lambda^n \tag{3.13a}$$

$$E_I = \sum_n E_I^{(n)} \lambda^n \tag{3.13b}$$

where n takes the values 0, 1, 2, ... and $\Psi_I^{(n)}$ and $E_I^{(n)}$ denote, correspondingly, the n-th order corrections towards the exact Ψ_I and E_I. As mentioned above, the zero-order corrections $\Psi_I^{(0)}$ and $E_I^{(0)}$, are known.

Substitution of the above expansions into the eigenvalue equation yields

$$(\mathcal{H}^{(0)} + \lambda \mathcal{H}^{(1)}) \sum_n \Psi_I^{(n)} \lambda^n = \{\sum_m \Psi^{(m)} \lambda^m\} \{\sum_n E_I^{(n)} \lambda^n\}$$

Collecting together all the terms associated with a given power of λ, the above expression may be rewritten as

$$\sum_n \lambda^n \{\mathcal{H}^{(0)} \Psi_I^{(n)} + \mathcal{H}^{(1)} \Psi_I^{(n-1)} - \sum_m \Psi_I^{(n-m)} E_I^{(m)}\} = 0$$

In order for this expression to be satisfied for all values of λ, it is necessary that the coefficients of each power of λ will vanish. Therefore, in general, for the n-th correction one obtains the condition

$$\mathcal{H}^{(0)}\Psi_I^{(n)} + \mathcal{H}^{(1)}\Psi_I^{(n-1)} = \sum_m \Psi_I^{(n-m)}E_I^{(m)}$$

In particular, for the first- and second-order corrections one would have

$$\mathcal{H}^{(0)}\Psi_I^{(1)} + \mathcal{H}^{(1)}\Psi_I^{(0)} = \Psi_I^{(1)}E_I^{(0)} + \Psi_I^{(0)}E_I^{(1)} \tag{3.14a}$$

$$\mathcal{H}^{(0)}\Psi_I^{(2)} + \mathcal{H}^{(1)}\Psi_I^{(1)} = \Psi_I^{(2)}E_I^{(0)} + \Psi_I^{(1)}E_I^{(1)} + \Psi_I^{(0)}E_I^{(2)} \tag{3.14b}$$

The problem is now reduced to determining the various corrections. Once this is done, the ordering parameter λ is absorbed into the perturbation term.

The determination of the corrections is performed through expansion in terms of the complete set of unperturbed functions $\Psi_J^{(0)}$. For example, for the first-order correction one can write

$$\Psi_I^{(1)} = \sum_J \Psi_J^{(0)}a_{JI} \tag{3.15a}$$

$$\mathcal{H}^{(1)}\Psi_I^{(0)} = \sum_J \Psi_J^{(0)}d_{JI} \tag{3.15b}$$

where a_{JI} and d_{JI} are appropriate expansion coefficients. From Eq. (3.15b) one can see that, multiplying it on the left with $\Psi_K^{(0)*}$ and integrating, one obtains

$$<\Psi_K^{(0)} \mid \mathcal{H}^{(1)} \mid \Psi_I^{(0)}> = \sum_J <\Psi_K^{(0)} \mid \Psi_J^{(0)}>d_{JI} = d_{KI}$$

where use has been made of the orthonormality condition, $<\Psi_K^{(0)} \mid \Psi_J^{(0)}> = \delta_{KJ}$, of the zero-order functions. The above result, which may be rewritten in short as

$$d_{KI} = <\Psi_K^{(0)} \mid \mathcal{H}^{(1)} \mid \Psi_I^{(0)}> \equiv H_{KI}^{(1)} \tag{3.16}$$

shows that the expansion coefficients d_{KI} are the matrix elements of the perturbation term $\mathcal{H}^{(1)}$ over the zero-order functions $\Psi_K^{(0)}$ and $\Psi_I^{(0)}$.

Substitution of Eqs. (3.15) and (3.16) into Eq. (3.14a), and taking into account Eq. (3.16), leads to

$$\sum_J \mathcal{H}^{(0)}\Psi_J^{(0)}a_{JI} + \sum_J \Psi_J^{(0)}H_{JI}^{(1)} = E_I^{(0)}\sum_J \Psi_J^{(0)}a_{JI} + \Psi_I^{(0)}E_I^{(1)}$$

which can be rewritten as

$$\sum_J \mathcal{H}^{(0)}\Psi_J^{(0)}a_{JI} = \Psi_I^{(0)}E_I^{(1)} + \sum_J \Psi_J^{(0)}(a_{JI}E_I^{(0)} - H_{JI}^{(1)})$$

or

$$\sum_J \Psi_J^{(0)}E_J^{(0)}a_{JI} = \Psi_I^{(0)}E_I^{(1)} + \sum_J \Psi_J^{(0)}(a_{JI}E_I^{(0)} - H_{JI}^{(1)})$$

The functions $\Psi_J^{(0)}$ are linearly independent and therefore the coefficients of each $\Psi_J^{(0)}$ on both sides of the equation must be equal. For $J \equiv I$ one obtains

$$E_I^{(1)} = H_{II}^{(1)}$$

while for $J \neq I$ it is

$$a_{JI} = H_{IJ}^{(1)} / (E_I^{(0)} - E_J^{(0)})$$

That is, to first-order one can write then

$$\Psi_I = \Psi_I^{(0)} + \Psi_I^{(1)} = \Psi_I^{(0)} + \sum_{J \neq I} \Psi_J^{(0)}H_{JI}^{(1)} / (E_I^{(0)} - E_J^{(0)}) \tag{3.17a}$$

$$E_I = E_I^{(0)} + E_I^{(1)} = E_I^{(0)} + H_{II}^{(1)} \tag{3.17b}$$

which shows that *the energy may be evaluated, to first-order, if* $\Psi_I^{(o)}$ *is known but that the evaluation of the function, also to first order, requires a knowledge of all the unperturbed functions.*

Proceeding in a similar fashion one can obtain the expressions for the second- and higher-order corrections, which become more and more complicated. For example, the second-order correction to the energy is

$$E_I^{(2)} = \sum_{J \neq I} H_{IJ}^{(1)}H_{JI}^{(1)} / (E_I^{(0)} - E_J^{(0)}) \tag{3.17c}$$

3.4 Variational Method

Perturbation theory constitutes a very powerful tool but its practical application requires a considerable effort. A solution to this difficulty may be found through the variational principle.

3.4.1. The Variational Principle

Let us consider Eq. (3.4), which represents the integrated form of the eigenvalue equation. Giving an arbitrary infinitesimal, non-vanishing variation to the function Ψ_I one obtains

$$(E_I + \delta E_I) < \Psi_I + \delta\Psi_I \mid \Psi_I + \delta\Psi_I > = < \Psi_I + \delta\Psi_I \mid \mathcal{H} \mid \Psi_I + \delta\Psi_I >$$

where δE_I is the energy change caused by the change $\delta\Psi_I$ in the function. Taking into account that Ψ_I is normalized, the above equation may be rewritten, assuming without any loss of generality that Ψ_I and $\delta\Psi_I$ are real, as

$$\delta E_I = <\delta\Psi_I \mid \mathcal{H} - E_I \mid \delta\Psi_I > / <\Psi_I + \delta\Psi_I \mid \Psi_I + \delta\Psi_I >$$

which shows that a first-order change in the function produces a change in the energy which is second-order in $\delta\Psi_I$. That is: *one can approximate more accurately the energy than the function.*

If one expands $\delta\Psi_I$ in terms of the complete set of functions Ψ_J, with appropriate expansion coefficients c_{JI}, one obtains

$$\delta E_I <\Psi_I + \delta\Psi_I \mid \Psi_I + \delta\Psi_I > = < \sum_J \Psi_J c_{JI} \mid \mathcal{H} - E_I \mid \sum_K \Psi_k c_{KI} > = \sum_{J \neq I} (E_J - E_I) c_{JI}^2$$

The expectation value on the *lhs* of this equation is always positive. When the I-th state is the ground state, then $E_J - E_I > 0$, for every $J \neq I$, and therefore the *rhs* is positive. Therefore, when considering the ground state, a change in the function will always raise the energy. That is, *the energy corresponding to a function which represents an approximation to the exact function of the ground state constitutes an upper bound of its exact energy.*

This result is the basis for the variational treatments for ground states, whereby approximate functions are varied under a minimum energy criterion, knowing that it will not be possible to obtain an energy lower than the exact value.

3.4.2. Self-consistent Field Theory

One can then, in principle, propose an arbitrary, approximate function, containing some variational parameters. Variations of those parameters, which result in a lowering of the energy expectation value, will produce improved functions. Therefore, if the original function has the necessary flexibility so that it could ultimately transform into the exact function, the above procedure would lead in the end to the exact function and energy. Without any additional guidelines, the procedure would be rather cumbersome. Fortunately, the results obtained above in the development of perturbation theory will allow us to formulate a rational procedure, easily implemented for numerical calculations.

Let us inspect Eq. (3.17b). Taking into account Eq. (3.16) it can be written as

$$E_I = E_I^{(0)} + H_{II}^{(1)} = <\Psi_I^{(0)}|\mathcal{H}^{(0)}|\Psi_I^{(0)}> + <\Psi_I^{(0)}|\mathcal{H}^{(1)}|\Psi_I^{(0)}>$$

$$= <\Psi_I^{(0)}|\mathcal{H}^{(0)} + \mathcal{H}^{(1)}|\Psi_I^{(0)}> = <\Psi_I^{(0)}|\mathcal{H}|\Psi_I^{(0)}>$$

That is: The energy may be approximated, to first order, as the expectation value of the complete Hamiltonian operator over the unperturbed (zero-order) function. The corresponding value may be rather poor if the contribution from the perturbation term is appreciable. But, taking into account the variational principle, we may now decide to take $\Psi_I^{(0)}$ as a starting point for a variational treatment, as described below. The discussion here will be restricted to the case of closed-shell systems, that is, systems in which all the orbitals are fully occupied.

The trial function, denoted as Φ for simplicity, is of the form given by Eq. (3.11). It is now expressed in terms of orbitals to be determined under a minimum energy criterion, with the added condition that they will be orthonormal (Roothaan 1951), that is, normalized and linearly independent. Expansion of the energy expectation value yields

$$E = <\Phi|\mathcal{H}|\Phi> = 2\sum_i H_{ii} + \sum_{ij}(2J_{ij} - K_{ij})$$

with

$$H_{ii} = <\phi_i|H|\phi_i> = \int \phi_i^*(\mu)H_\mu\phi_i(\mu)d\tau_\mu$$

$$J_{ij} = <\phi_i\phi_i|1/r_{\mu\nu}|\phi_j\phi_j> = \iint \phi_i^*(\mu)\phi_i(\mu)(1/r_{\mu\nu})\phi_j^*(\nu)\phi_j(\nu)d\tau_\mu d\tau_\nu$$

$$K_{ij} = <\phi_i\phi_j|1/r_{\mu\nu}|\phi_j\phi_i> = \iint \phi_i^*(\mu)\phi_j(\mu)(1/r_{\mu\nu})\phi_j^*(\nu)\phi_i(\nu)d\tau_\mu d\tau_\nu$$

where $d\tau$ denotes the volume element. The integrals J_{ij} and K_{ij}, denoted respectively as Coulomb and exchange integrals, may be expressed as

$$J_{ij} = <\phi_i | J_j | \phi_i> = <\phi_j | J_i | \phi_i>$$

$$K_{ij} = <\phi_i | K_j | \phi_i> = <\phi_j | K_i | \phi_j>$$

where J_j and K_j (and J_i and K_i) are the corresponding Coulomb and exchange operators, for which one can write

$$J_j\phi_i = \left\{ \int (1/r_{\mu\nu})\phi_j^*(\nu)\phi_j(\nu)d\tau_\nu \right\}\phi_i(\mu)$$

$$K_j\phi_i = \left\{ \int (1/r_{\mu\nu})\phi_j^*(\nu)\phi_i(\nu)d\tau_\nu \right\}\phi_j(\mu)$$

The variation in the energy expectation value corresponding to an infinitesimal variation $\delta\phi_i$ of the orbitals is, after some manipulation,

$$\delta E = 2 \sum_i <\delta\phi_i | H + \sum_j (2J_j - K_j) | \phi_i >$$
$$+ 2 \sum_i <\delta\phi_i^* | H^* + \sum_j (2J_j^* - K_j^*) | \phi_i^* > \tag{3.18}$$

Similarly, the orthonormality conditions $<\phi_i|\phi_j> = \delta_{ij}$ lead to the restrictions

$$< \delta\phi_i | \phi_j > + < \delta\phi_j^* | \phi_i^* > = 0 \tag{3.19}$$

to be satisfied by the orbital variations $\delta\phi_i$.

The necessary, though not sufficient, condition so that E may reach its absolute minimum is that $\delta E = 0$ for any values of the orbital variations $\delta\phi_i$ which are compatible with the restrictions represented by Eq. (3.19). The two conditions are combined together by means of the Lagrangian multipliers technique. One obtains, again after some manipulation,

$$2 \sum_i <\delta\phi_i | [H + \sum_j (2J_j - K_j)] \phi_i - \sum_j \phi_j\varepsilon_{ji} >$$
$$+ 2 \sum_i <\delta\phi_i^* | [H^* + \sum_j (2J_j^* - K_j^*)] \phi_i^* - \sum_j \phi_j^*\varepsilon_{ij} > = 0$$

which leads to the two conditions

$$[H + \sum_j (2J_j - K_j)] \, \phi_i = \sum_j \phi_j \varepsilon_{ji} \qquad (3.20a)$$

$$[H^* + \sum_j (2J_j^* - K_j^*)] \, \phi_i^* = \sum_j \phi_j^* \varepsilon_{ij} \qquad (3.20b)$$

It can be shown that the Lagrangian multipliers are the elements of a Hermitian matrix, i.e., $\varepsilon_{ij} = \varepsilon_{ji}^*$, which implies that the two Eqs. (3.20) are equivalent; that is, it is only necessary to solve one of them, say, Eq. (3.20a).

That equation may be written, in short, as

$$F\phi_i = \sum_j \phi_j \varepsilon_{ji} \qquad (3.21)$$

where

$$F = H + \sum_j (2J_j - K_j)$$

represents the Hartree-Fock operator. There is one equation of the type of Eq. (3.21) for each occupied orbital, i.e.,

$$F\phi_1 = \sum_j \phi_j \varepsilon_{j1}$$

$$F\phi_2 = \sum_j \phi_j \varepsilon_{j2}$$

--- --------------

$$F\phi_n = \sum_j \phi_j \varepsilon_{jn}$$

and therefore one can represent them by a single equation

$$F\phi = \phi \varepsilon \qquad (3.22)$$

in matrix form; ϕ is a row vector, with elements $\phi_1, \phi_2, ..., \phi_n$ and ε is a square, symmetric matrix, with elements ε_{ji}. That is,

$$F(\phi_1\,\phi_2...\phi_n) = (\phi_1\,\phi_2...\phi_n) \begin{pmatrix} \varepsilon_{11} & \varepsilon_{12} & \cdots & \varepsilon_{1n} \\ \varepsilon_{21} & \varepsilon_{22} & \cdots & \varepsilon_{2n} \\ \cdots & \cdots & \cdots & \cdots \\ \varepsilon_{n1} & \varepsilon_{n2} & \cdots & \varepsilon_{nn} \end{pmatrix}$$

Using a unitary matrix U, i.e.,

$$\bar{U}\,U = 1 \tag{3.23}$$

(where $\bar{U}$ denotes the Hermitian conjugate of the matrix U and 1 is the identity matrix) one can transform the set of orbitals and Lagrangian multipliers to yield

$$\phi' = \phi\,U$$

$$\varepsilon' = \bar{U}\,\varepsilon\,U$$

Multiplying Eq. (3.22) on the right with U

$$F\,\phi\,U = \phi\,\varepsilon\,U$$

and using the condition given by Eq. (3.23) leads to

$$F\,\phi\,U = \phi\,1\,\varepsilon\,U = \phi\,U\,\bar{U}\,\varepsilon\,U$$

or

$$F\,\phi' = \phi'\varepsilon' \tag{3.24}$$

which is of the same form as Eq. (3.22). The difference, however, is that in Eq. (3.22) the Hartree-Fock operator, which acts on the orbitals ϕ_j, is defined in terms of those same orbitals while in Eq. (3.24) that same operator is acting on the transformed orbitals ϕ_j'. It can be shown, however, that the operators F, defined in terms of the orbitals ϕ_j, and F', defined in terms of the orbitals ϕ_j', are identical.

Taking into account that the matrix ε is Hermitian, it is possible to find a unitary matrix U such that ε is a diagonal matrix, with real diagonal elements. Therefore, dropping the prime for simplicity, one can finally write Eq. (3.24) as

$$F\,\phi = \phi\,\varepsilon_d \tag{3.25}$$

where ε_d denotes the diagonal matrix. The corresponding Hartree-Fock equation for any of the orbitals is then

$$F\phi_i = \phi_i\varepsilon_{ii} \tag{3.26}$$

and the Lagrangian multiplier ε_{ii} (or, in short, ε_i) associated with each orbital ϕ_i is customarily denoted as its orbital energy.

The designation of self-consistent field (SCF) theory, given to this formulation, arises from the characteristics of Eq. (3.25): the Hartree-Fock operator F is defined in terms of its eigenfunctions. That is, one needs to know those solutions in order to construct the operator. The way out of this dilemma is a procedure which will lead to self consistency: A set of trial orbitals is proposed, the operator F is constructed, and the eigenvalue equations are solved, yielding a new set of orbitals; these new orbitals are then used to construct a new operator F, etc., and the process is continued until the eigenfunctions (orbitals) obtained at a certain point are identical to the orbitals used in the construction of the operator F.

The exact solutions of the set of Eqs. (3.26) are denoted as Hartree-Fock orbitals and energies, respectively. The Hartree-Fock solutions are easily obtainable in the case of atoms through numerical integration (see, e.g., the work of Fisher 1977) but in the case of molecules recourse must be made of additional approximations, as discussed below.

3.5 The Expansion Approximation

The SCF method has become the workhorse for the mass production of results in Computational Chemistry due to the fact that, in its expansion approximation, it is very easily implemented in computer programs. There are, of course, practical difficulties (see below) and the study of large systems (such as peptides) is dependent on the availability of fast, large-capacity computers.

[Initially, this approximation was denoted as the LCAO approximation, where LCAO stands for linear combination of atomic orbitals. This designation has been abandoned as a rule and in the terminology currently in use one refers to the expansion of orbitals in terms of basis functions.]

3.5.1 Matrix Representation of the Hartree-Fock Equations

The expansion approximation consists of expressing the orbitals in terms of known functions, denoted as basis functions. That is, any orbital ϕ_i may be expressed as

$$\phi_i = \sum_p \chi_p c_{pi} \qquad (3.27)$$

where the summation over p extends to the set of chosen basis functions χ_p and c_{pi} denote the expansion coefficients, which are to be determined. That is, one has transformed the problem of determining the orbitals ϕ_i into a problem of determining the corresponding expansion coefficients. The above equation may be rewritten in matrix form as

$$\phi_i = \chi \, c_i \qquad (3.28)$$

where χ denotes a row vector, with elements χ_p, and c_i is a column vector, with elements c_{pi}. That is,

$$\phi_i = (\chi_1 \chi_2 \cdots \chi_t) \begin{pmatrix} c_{1i} \\ c_{2i} \\ \cdots \\ c_{ti} \end{pmatrix}$$

The complete set of orbitals may be represented correspondingly as

$$\phi = (\phi_1\, \phi_2 \cdots \phi_n) = (\chi_1 \chi_2 \cdots \chi_t) \begin{pmatrix} c_{11} & c_{12} & \cdots & c_{1n} \\ c_{21} & c_{22} & \cdots & c_{2n} \\ \cdots & \cdots & \cdots & \cdots \\ c_{t1} & c_{t2} & \cdots & c_{tn} \end{pmatrix} = \chi C$$

where C denotes the complete matrix of expansion coefficients, whose columns are the column vectors c_i. This matrix is not necessarily square: the number of basis functions to be used, t, must be at least equal to the number, N, of orbitals to be determined but preferably it should be larger (see below).

The reformulation of the *SCF* problem within the expansion approximation consists of obtaining the equation equivalent to Eq. (3.25) but in terms of the expansion coefficients. The procedure is analogous to the one presented above, with the difference that the variation of the orbitals is expressed in terms of a variation of the expansion coefficients (Roothaan 1951). One obtains in such a case

$$F C = S C \, \varepsilon_d \qquad (3.29)$$

where $\mathbf{F}$ and $\mathbf{S}$ represent the matrix representations, for the basis set used, of the Hartree-Fock and overlap operators, respectively. Their matrix elements are

$$F_{pq} = <\chi_p \,|\, F \,|\, \chi_q> = <\chi_p \,|\, H + \sum_j (2J_j - K_j) \,|\, \chi_q>$$

$$= <\chi_p \,|\, H \,|\, \chi_q> + 2\sum_j <\chi_p \,|\, J_j \,|\, \chi_q> - \sum_j <\chi_p \,|\, K_j \,|\, \chi_q>$$

$$S_{pq} = <\chi_p \,|\, \chi_q>$$

The term $<\chi_p|H|\chi_q>$ represents the element H_{pq} of the matrix representation $\mathbf{H}$ of the one-electron part, H, of the Hartree-Fock operator F. The terms corresponding to the Coulombic and exchange operators (see Section 3.4.2) may be expanded, taking into account their definitions. For example, for the Coulombic operator one has

$$<\chi_p \,|\, J_j \,|\, \chi_q> = \int \chi_q^*(\mu) \left\{ \int (1/r_{\mu\nu})(\chi(\nu)c_j)^* (\chi(\nu)c_j) d\tau_\nu \right\} \chi_q(\mu) d\tau_\mu$$

$$= \int \chi_p^*(\mu) \left\{ \bar{c}_j \left(\int (1/r_{\mu\nu}) \bar{\chi}(\nu)\chi(\nu) d\tau_\nu \right) c_j \right\} \chi_q(\mu) d\tau_\mu$$

$$= \bar{c}_j \left\{ \int \chi_p^*(\mu) \left(\int (1/r_{\mu\nu}) \bar{\chi}(\nu)\chi(\nu) d\tau_\nu \right) \chi_q(\mu) d\tau_\mu \right\} c_j$$

$$= \bar{c}_j J_{pq} c_j = J_{j,pq}$$

and similarly

$$<\chi_p \,|\, K_j \,|\, \chi_q> = \bar{c}_j K_{pq} c_j = K_{j,pq}$$

for the exchange operator. The integrals involved in these operators, $J_{rs,pq}$ and $K_{rs,pq}$, are stored as supermatrices, whose elements are the matrices $J_{j,pq}$ and $K_{j,pq}$, respectively. In this connection it is convenient to mention at this point how large the number of integrals over the basis functions can be for large sets of basis functions.

Equation (3.29) may be considered as a pseudo-eigenvalue equation, due to the presence of the overlap matrix $\mathbf{S}$. Its transformation into a proper eigenvalue equation may be achieved, say, by means of a Schmidt orthonormalization procedure. Defining a matrix $\mathbf{T}$ such that

$$\mathbf{S} = \mathbf{T}^\dagger \mathbf{T}$$

(where $\mathbf{T}^\dagger$ represents the transpose matrix) one can write

$$F(T^{-1}T)C = SC\varepsilon_d = (T^\dagger T)C\varepsilon_d$$

so that

$$(T^\dagger)^{-1}FT^{-1}TC = (T^\dagger)^{-1}T^\dagger TC\varepsilon_d = TC\varepsilon_d$$

This equation may be rewritten as

$$F'C' = C'\varepsilon_d \qquad\qquad (3.30)$$

with

$$F' = (T^{-1})^\dagger FT^{-1} \qquad C' = TC$$

The problem has now been reduced to performing a unitary transformation (diagonalization) on F'. That is, one has to find the matrix C', such that $(C')^\dagger C' = 1$, which will diagonalize F':

$$(C')^\dagger F'C' \equiv (C')^{-1}F'C' = (C')^\dagger C'\varepsilon_d = \varepsilon_d$$

(where use has been made of the fact that $(C')^\dagger \equiv (C')^{-1}$).

3.5.2 Eigenvectors and Orbital Energies

The practical solution of the problem is achieved in an iterative way, analogous to the one described above. Once the basis set has been chosen and all the required integrals have been evaluated, iterations are carried out until self-consistency for the expansion coefficients (eigenvectors): A set of trial vectors is chosen, all the matrices are constructed, and the Hartree-Fock matrix is diagonalized, yielding a new set of eigenvectors; these new expansion coefficients are then used to construct all the matrices, a new diagonalization is performed, and so on, until the diagonalization of F' yields the same vectors which were used in its construction. The correct set of eigenvectors, C, is then obtained,

$$C = T^{-1} C'$$

from the ones, C', evaluated in the final diagonalization. The elements of the diagonal matrix, ε_d, represent the corresponding orbital energies.

In order to justify the radical approximations, to be introduced in following chapters, it is convenient at this point to comment on the computing cost associated with an *SCF* calculation. The cost of the orthogonalization and diagonalization procedures depends only on the size of the matrices involved, i.e., the size of the set of basis functions. On the other hand, the cost of the construction of the matrices depends not only on the size of the basis set but also on the size of the system under study, i.e., its number of electrons.

The quality of the results depends on the size on the basis set and it is only in the limit, when using a complete set, that the Hartree-Fock solutions will be obtained (Fig. 3.1). [In actual calculations, the Hartree-Fock limit may be reached, say, to five decimals, with a limited basis set, properly chosen.] In all other cases, the solutions should be labelled as being *SCF* results, whose quality will depend on the basis set used. This point is emphasized here because of the improper use of the designation Hartree-Fock in the literature.

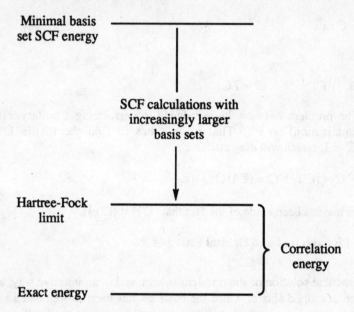

Figure 3.1 Schematic representation of the dependence of the *SCF* energies on the basis set

3.5.3 Practical Details

SCF calculations are performed almost in a routine way, with a variety of programs available in the literature. Small and intermediate-size systems may be easily studied within the framework of *SCF* theory but it is doubtful that it will be possible to tackle large peptides or proteins. Consequently there is no need here to present a detailed examination of the existing programs, the difficulties in the evaluation of the integrals, the selection of basis sets, etc. [The reader is referred, regarding existing programs, to the existence of the Quantum Chemistry Program Exchange, established at the Department of Chemistry, University of Indiana, Bloomington, Indiana, U.S.A. See also Section 12.3.5.]

For the purpose of this work it suffices to mention that *SCF* calculations are possible for individual amino acids and that the corresponding results, as discussed below, may be of use in the approximations developed for the study of peptides and proteins.

3.6 Beyond Hartree-Fock

The development of the preceding formulation has proceeded from the independent-particle approximation, through consideration of perturbation theory and application of the variational principle: the *SCF* formalism embodies the variational formulation of the energy to first-order and, therefore, it constitutes only an approximation. The possibility exists, however, of proceeding beyond, with an improvement of the results.

As already mentioned above, N orbitals are required for the description of the closed-shell ground state of a system with 2N electrons. Therefore, the minimal basis set to be used in an *SCF* calculation for that state should also consist of N basis functions. The corresponding results, however, will not be very satisfactory, as illustrated in Fig. 3.1, and it is necessary to use larger basis sets.

Let us assume that the *SCF* calculation has been performed using N' (>N) basis functions. As N' becomes larger and larger, the quality of the results increases, until the Hartree-Fock limit is reached. Within the framework of the independent-particle approximation, it is not possible to go beyond the Hartree-Fock limit. The difference between the Hartree-Fock energy and the exact, non-relativistic energy is denoted as correlation energy (see Fig. 3.1). This correlation energy, which may be appreciable, can only be accounted for by methods which take into account the correlation between the motion of the electrons. The use of correlated functions constitutes a very powerful technique but the discussion here will be restricted to the method of configuration interaction, which may be formulated as a logical extension of the *SCF* theory.

Let us assume that the orbitals obtained in the *SCF* calculation, with N' basis functions, are ordered according to the increasing values of their corresponding orbital energies. Then, the set of orbitals may be separated into two subsets: the lowest N orbitals, which will be occupied in the representation of the ground state of the system are denoted as occupied orbitals; the remaining (N'-N) orbitals, which are denoted as virtual orbitals, are simply an artifact resulting from the expansion approximation. As such, they do not have an actual physical meaning but they are of use in configuration interaction treatments.

[Particular attention has always been attached to the highest-occupied molecular orbital (HOMO) and the lowest-unoccupied molecular orbital (LUMO, also denoted as LEMO, which stands for lowest-empty molecular orbital). Their orbital energies, with sign changed, may be seen to represent approximations to the ionization potential and the electron affinity of the system.]

The virtual (unoccupied) orbitals may be used, together with the occupied orbitals, in the generation of approximate functions for the excited states of the system. For example, while the configuration

$$(\phi_1)^2(\phi_2)^2....(\phi_N)^2$$

corresponds to the ground state, the configurations

$$(\phi_1)^2(\phi_2)^2....(\phi_N)^1(\phi_{N+1})^1$$

$$(\phi_1)^2(\phi_2)^2....(\phi_N)^1(\phi_{N+2})^1$$

$$(\phi_1)^2(\phi_2)^2....(\phi_{N+1})^2$$

$$(\phi_1)^2(\phi_2)^2....(\phi_{N+1})^1(\phi_{N+2})^1$$

----

may be considered as representing excited states.

The corresponding functions and energies may be easily obtained from the *SCF* results for the ground state. That is, an approximate description of the energy levels of the system may be obtained, with very little additional effort, from the *SCF* results for the ground state.

This description is approximate because Eq. (3.5) will not, as a rule, be satisfied. That is, the matrix elements $<\Phi_I|\mathcal{H}|\Phi_J>$ will not be necessarily zero. In other words, the configurations described in terms of occupied/virtual orbitals still interact. This fact is the basis for the configuration interaction (*CI*) treatments developed within the framework of the variational principle.

In a *CI* calculation, the function of the ground state is approximated as

$$\Psi = \sum_J \Phi_J c_J$$

where Φ_1 will correspond to the *SCF* function previously obtained for the system and the remaining functions correspond to excited states (of the same symmetry and multiplicity designations as the ground state); the c_J are the expansion coefficients to be determined in the *CI* calculation. The corresponding variational treatment results in a formulation capable of yielding an improved description of the energy spectrum of the system, with particular improvement of the description for the ground state. A detailed examination of the corresponding formulation is beyond the scope of this work, because it is very doubtful that *CI* calculations will be possible, at least in the near future, for large peptides and proteins. Here it will suffice to mention that the quality of the results will increase with the number of excited configurations used. The number of such configurations which will be available for a *CI* calculation depends, as mentioned above, on the number of virtual orbitals obtained in the *SCF* calculation, which in turn depends on the number of basis functions used. It is immediately seen that a high-quality *CI* calculation will be extraordinarily expensive in terms of computing time.

Another warning must be put forward, in the sense that caution must be exercised when dealing with *SCF CI* results. The *CI* treatment will certainly introduce a correction for (part of) the missing correlation energy, but that does not mean that the corresponding total energy will be lower than the Hartree-Fock energy. That is, a *CI* calculation, performed on top of a rather poor *SCF* calculation (say, a minimal basis set *SCF* calculation) may still yield a very poor total energy.

For completeness, and without any intention to go into further details, it should be mentioned that a very powerful treatment may be achieved by a combination, in a single step, of the *SCF* and *CI* procedures, in what is denoted as a multi-configuration *SCF* (*MC SCF*) method. [See the work of Carbo and Riera (1978) and Carbo and Klobukowski (1991) for formulations, results, and references.]

3.7 Analysis of the Results

Taking into account all the comments made above regarding the calculations at a more sophisticated level, we will limit our inspection of the results to the simple case of *SCF* calculations for closed-shell systems.

The *SCF* calculation yields directly the eigenvectors (i.e., the expansion coefficients) and the orbital energies. In sophisticated *SCF* calculations, which include geometry-optimization, the results will correspond to the conformation predicted theoretically to be most stable (at the level afforded by the basis set used).

A calculation with geometry optimization may be of interest for simple peptides (say, dipeptides and tripeptides) but it will not be readily feasible for larger peptides and proteins. One must emphasize, however, that if such calculations were computationally possible, they would provide the best approach to the problem of predicting protein structures, within the normal limitations of a quantum chemical approach (as discussed in the following chapter). In its crudest form, a geometry optimization may be performed by repeating the *SCF* calculations for a number of conformations, until the one with the lowest energy is found. Efficient programs exist, however, that perform the search for the most stable geometry in an automatic way. [See Section 3.5.3].

Because an *SCF* calculation, with geometry optimization, is not possible at this stage for a protein and the orbital energies are of no use in the simulation of protein structures, we will restrict the discussion here to the information that may be obtained from the eigenvectors [see Eqs. (3.27) and (3.28)].

3.7.1 Population Analysis

Let us consider Eq. (3.27) in more detail in the case of a molecule consisting of various nuclei. In such a case the basis set will consist of basis functions centered on the various atoms and one can write, for the i-th orbital,

$$\phi_i = \sum_{r_k} \chi_{r_k} c_{r_k i}$$

where the summation extends to all the appropriate basis functions (as determined by symmetry considerations) and k labels the atoms in the system. Assuming that the orbitals are normalized and denoting by n_i the number of electrons occupying orbital ϕ_i (1 or 2, for non-degenerate orbitals), the electronic population of the orbital may be written as

$$n_i <\phi_i | \phi_i> \equiv n_i < \sum_{r_k} \chi_{r_k} c_{r_k i} | \sum_{s_\ell} \chi_{s_\ell} c_{s_\ell i} >$$

$$= n_i \sum_{r_k} c_{r_k i}^2 + 2n_i \sum_{r_k} \sum_{s_\ell} S_{r_k s_\ell} c_{r_{ki}} c_{s_{\ell i}} \tag{3.31}$$

(with $\ell > k$), where use has been made of the assumption that the basis functions are normalized. $S_{r_k s_\ell}$ denotes the overlap integral

$$S_{r_k s_\ell} = <\chi_{r_k} | \chi_{s_\ell}>$$

between the two basis functions χ_{r_k} and χ_{s_ℓ}. Equation (3.31) may be used (Mulliken 1955abcd) for an analysis of the electronic population in a molecule.

The distinction can be made between gross and overlap populations. The latter may be used in a discussion of relative bond strengths but they are of no interest for the purpose of this work and our attention will concentrate on the gross populations.

Various types of gross populations may be postulated:

partial gross population in ϕ_i and χ_{r_k}

$$N(i;r_k) = n_i c_{r_k i}(c_{r_k i} + \sum_{\ell \neq k} c_{s_\ell i} S_{r_k s_\ell})$$

subtotal gross population in ϕ_i on atom k

$$N(i;k) = \sum_r N(i;r_k)$$

total gross population on atom k

$$N(k) = \sum_i N(i;k) = \sum_r N(r_k)$$

total gross population in ϕ_i

$$N(i) = \sum_{r_k} N(i;r_k) = \sum_k N(i;k) \equiv n_i$$

total gross population in χ_{r_k}

$$N(r_k) = \sum_i N(i;r_k)$$

total population

$$N_t = \sum_i n_i = \sum_i N(i) = \sum_{r_k} N(r_k)$$

The total gross population on atom k is of particular interest. Denoting by Z_k the nuclear charge of that atom, one can define its effective charge, q_k, as

$$q_k = Z - N(k)$$

Effective charges have become extremely popular over the years, with a variety of applications. They are also widely used in many of the approaches for the simulation of protein structures and for that reason it will be appropriate to examine them in detail.

Before proceeding to do so it is necessary to point out that the Mulliken population analysis has been chosen here because of its simplicity and wide use. There are, at present, more powerful techniques for the analysis of the electronic densities in molecules (Reed *et al.* 1985, 1988, Weinhold and Carpenter 1988, Breneman and Wiberg 1990, Boyd and Ugalde 1992).

3.7.2 Effective Charges and Atom Classes

Effective charges play a vital role in most of the potential energy functions used in simulations of protein structures. We want to emphasize the importance of effective charges before proceeding to point out their deficiencies as well as the deficiencies in their use. Without an alternative, effective charges will continue to be used and therefore the user of any given potential energy function (and consequently of a software package based on the latter) should be aware of the potential risks.

Let us start by saying that the effective charges, derived from a Mulliken population analysis, arise from a given partition, as adopted in Eq. (3.31) and the successive expressions. Other partitions may be adopted in order to define effective charges. However, the effective charges obtained within the context of any partition scheme will not represent real physical properties of the system.

In addition, effective charges obtained from *SCF* results will show a dependence on the basis set used in the *SCF* calculation. This dependence could be removed through a semiempirical scaling (Smith 1977, Wiberg 1981) but it represents additional work which, in any case, will not correct for other deficiencies.

These other deficiencies arise when trying to use the effective charges. One is clearly faced with a vicious circle: One is trying to study a given system and for that purpose a potential energy function, which involves the use of the appropriate effective charges, is needed. If those charges are available, the implication is that an *SCF* calculation has been performed for the system under consideration. But, if such is the case, there is then no need for an additional simulation. On the other hand, if the *SCF* results are not available, as it is the case when trying to simulate the structure of a protein, then one can neither construct the corresponding potential energy function nor perform the simulation.

The solution to this dilemma is to be found in an additional approximation: the use of average effective charges. It is in this connection that the definition of atom classes by Clementi (1980) represents a very significant contribution, as seen in the description below.

The molecular *SCF* energy of a system may be expressed as

$$E = \sum_k E_k + \sum_k \sum_\ell{}' E_{k\ell} + \sum_k \sum_\ell \sum_m{}' E_{k\ell m} + \sum_k \sum_\ell \sum_m \sum_n{}' E_{k\ell mn}$$

where k, ℓ, m, n denote the atoms in the molecule and the prime in the summations over ℓ, m, and n indicates that $\ell \neq k$, $m \neq \ell \neq k$, $n \neq m \neq \ell \neq k$. The four terms in the above expression represent then the one-, two-, three-, and four-centre contributions to the molecular energy. $E_{k\ell m}$ and $E_{k\ell mn}$ are smaller than E_k and $E_{k\ell}$ (which includes the nuclear repulsion energy terms).

The one-centre terms, E_k, represent the energies of the atoms when forming part of the molecule and they may be evaluated from the corresponding contributions (i.e., for each atom) of the orbital energies and the one-electron term in the Hartree-Fock operator.

These one-centre energies, E_k, denoted by Clementi as molecular orbital valency states (MOVS) may be used as a criterion in order to characterize the atoms in a molecule. That is, similar values of the MOVS for a given atom in different molecules may be taken as to indicate that the molecular environments of that atom in the various molecules are similar. It is on the basis of this criterion that Clementi was able to define atom classes, in close parallel with what chemical intuition would suggest.

Once the atom classes have been defined it is possible to define more meaningful effective charges. Clementi and co-workers (Clementi *et al.* 1977, Scordarmaglia *et al.* 1977, Bolis and Clementi 1977) performed accurate *SCF* calculations for the twenty naturally-occurring amino acids (as well as hydroxyproline) and the four bases of *DNA* (adenine, cytosine, guanine, and thymine), which yielded the effective charges for the atoms in the various molecules. Those results were then used to define the average effective charges for the different classes of atoms. Table 3.1 collects the corresponding values together with the specifications of the atom classes.

Table 3.1 Atom classes and effective charges[a]

Atom	Class[b]	q	Class specifications
H	23	.060	SH
	2	.204	CH, CH_2
	3	.205	CH_3
	16	.253	CH (ring C)
	1	.266	NH_2
	29	.311	in H_2O
	4	.404	COOH
C	6	-.608	CH_3

(continued...)

Table 3.1 (continued)

	21	-.503	CH_2 (α to S)
	7	-.383	CH_2
	17	-.218	ring
	8	-.135	CH
	28	-.090	ring (α to CH_3)
	14	-.078	ring (fully substituted or next to N)
	19	-.032	ring (junction, geminal to C)
	18	.008	COH (alcoholic or phenolic)
	25	.012	ring (geminal to N)
	20	.187	ring (junction, geminal to N)
	24	.310	ring (α to NH_2)
	26	.423	CO (ring)
	5	.511	COOH
N	13	-.630	$CONH_2$
	11	-.554	NH_2
	15	-.473	ring (with H)
	12	-.317	ring (without H)
O	30	-.622	in H_2O
	9	-.539	OH (in COOH)
	10	-.409	CO (in COOH)
	27	-.380	CO (attached to ring)
S	22	.123	SH

[a]Clementi *et al.* (1977), Scordamaglia *et al.* (1977).

[b]Order labels given by Clementi and co-workers. H and O in H_2O have been given the labels 29 and 30, respectively.

The use of average effective charges introduces, however, a new problem. When using average effective charges for a given molecule, its total charge will not be correctly reproduced, except by coincidence. This is a serious deficiency which must be corrected. The correction may be performed as follows (Fraga 1982). Let us denote by Q_t, Q_p, and Q_n the values predicted, using the average effective charges, for the total charge of the molecule and for the total positive and negative contributions, respectively (with $Q_t = Q_p + Q_n$). If Q is the true total charge of the molecule, the effective charge q_k (as given in Table 3.1) of the k-th atom must be transformed into

$$q_k^{\cdot} = q_k \left\{ 1 + c \, \frac{q_k}{|q_k|} \right\}$$

with

$$c = \frac{Q - Q_t}{Q_p - Q_n}$$

[See also the work of Bidacovich *et al.* (1990) for an additional discussion on other possible renormalization procedures.]

3.8 Molecular Interactions and Associations

Proteins, whether in biological systems or in industrial processes, are not isolated, but rather in interaction with their environment. Particular mention must be made in this connection of their solvation as well as of their involvement in recognition problems, wherein molecular associations, even if transitory, are formed.

The designations *system* and *molecule*, used throughout this chapter may be taken to denote either a single, isolated molecule or a collection of molecules, such as a solvated protein or two or more proteins interacting with each other. That is, the formulation presented is equally valid for all cases, independent of the size and composition of the system being considered. There are, however, some practical considerations that merit examination.

Particular emphasis has been placed, throughout this chapter, on the practical difficulties to be encountered in a theoretical calculation for a large system, such as a protein. It follows, naturally, that even greater difficulties (need of extremely large computers, considerable computing time, etc.) would be faced when dealing, say, with a solvated protein or an antigen-antibody association.

For the sake of the discussion, however, we will assume that such calculations could be performed, in order to point out the possible risks in the evaluation of interaction energies.

3.8.1 Binding Energy

Let us assume that we are interested in studying the molecular association of two molecules (say, two proteins or a protein in interaction with a non-peptide molecule), which we will denote as A and B.

The binding energy, ΔE, which measures the stability of the molecular association AB, is defined as

$$\Delta E = E_{AB} - (E_A + E_B) \tag{3.32}$$

where E_A, E_B, and E_{AB} denote the total energies (including the nuclear repulsion contributions) of A, B, and AB, respectively.

The problem to be faced in the evaluation of binding energies (independently of the practical difficulties, already mentioned) is one of precision. It is only in the case when very accurate energies would be known (say, from very complete CI calculations) that confidence could be placed on the values obtained for ΔE.

First of all there is the so-called basis set superposition error (BSSE) (Boys and Bernardi 1970), which will arise when different basis sets are used for A, B, and AB. That is, BSSE may only be avoided by using the same set for the evaluation of E_A, E_B, and E_{AB}.

But, even if BSSE does not exist, ΔE may still be affected by a considerable error when evaluated from *SCF* energies. This statement may be easily verified for the case of Hartree-Fock results. Let us denote by E_X, E_{HF}, and E_C the exact, Hartree-Fock, and correlation energies, respectively. These three energies are related (see Fig. 3.1) by

$$E_X = E_{HF} + E_C$$

Therefore, substituting in Eq. (3.32) one obtains

$$\Delta E = (E_{HF(AB)} + E_{C(AB)}) - [(E_{HF(A)} + E_{C(A)}) + (E_{HF(B)} + E_{C(B)})]$$

$$= \{E_{HF(AB)} - [E_{HF(A)} + E_{HF(B)}]\} + \{E_{C(AB)} - [E_{C(A)} + E_{C(B)}]\}$$

$$= \Delta E_{HF} + \Delta E_C$$

where ΔE_{HF} denotes the binding energy obtained from Hartree-Fock total energies and ΔE_C represents the correction arising from the correlation energies. Unless $\Delta E_C = 0$, it cannot be expected that ΔE_{HF} will represent necessarily a good approximation to ΔE. The situation will be even worse when other *SCF* results (not necessarily of Hartree-Fock accuracy) are used, although a coincidence may lead to a lucky cancellation of errors.

3.9 Summary

This chapter has served several purposes. On one hand it has become clear that, at present, it is not feasible to attempt a proper quantum chemical calculation of a large peptide or a protein, even though that would be the desirable approach to be adopted (with the qualifications to be presented in Chapter 4). It must be emphasized, however, that the difficulty is of a practical nature (i.e., availability of sufficiently large and extremely fast computers) as the formulation presented is equally appropriate for any system, independent of its size.

The main result for the purpose of this work has been the determination of average effective charges, to be used in a potential energy function. A detailed discussion has shown the deficiencies in both their determination and in their use.

4 Statistical Mechanics

In Chapter 3 we have outlined the basic concepts of Quantum Mechanics and developed the corresponding computational scheme. Therefore, supposedly, we should now be able to obtain the wave function and the energy of a given molecule, the association of two molecules, or even a collection of molecules (identical or not). We must realize, of course, that this statement, though correct in principle, does not reflect the reality, considering the many practical difficulties to be encountered in the implementation of those computational schemes.

Let us, however, consider a collection of particles and assume, for the sake of the discussion, that the corresponding quantum chemical information has been obtained. That information will describe the microstate of the collection of particles. Taking into account that each particle has access to a number of energy levels, it is evident that the collection of particles may exist in a variety of microstates.

These microstates do not correspond, individually, to the macrostate of the collection of particles. That collection of particles, in a given macrostate, can be in any one of a number of microstates, all of them with the same macroscopic properties.

The macroscopic behaviour of the collection of particles is provided by Thermodynamics and the problem to be faced is to establish the connection between Quantum Mechanics and Thermodynamics. That connection is provided by Statistical Mechanics which, ultimately, will allow us to calculate the macroscopic properties from the microscopic information.

4.1 Basic Concepts

In the preceding chapter we have used, as done customarily, the designation *system* for that collection of atoms (nuclei and electrons) which was the object of our study. That is, within the context of Chapter 3, *system* denoted either an atom, a molecule, or a collection of atoms and/or molecules. In this chapter, it will be necessary to refine our terminology.

We will use the designation *particle* to denote atoms or molecules, while *system* will correspond to a collection of particles. A collection of systems will be said to constitute an *ensemble*.

There are three main types of ensembles. A microcanonical ensemble (N, V, U) is a collection of adiabatic systems, each one having the same number of particles (N), the same volume (V), and the same energy (U), although the temperature (T) may fluctuate; this ensemble contains all the possible microstates. The canonical ensemble (N, V, T) is a collection of systems, each one having the same number of particles, the same volume, and the same temperature; each system can have any possible value of the energy, smaller than the total energy of the ensemble. The grand canonical ensemble (μ, V, T) is a collection of systems, each one having the same volume, the same temperature, and the same chemical potential (μ), although the number of particles in each system may fluctuate. [An additional ensemble, of interest in computer simulations, is the isobaric-isothermic ensemble (N, P, T).]

The thermodynamic properties are labelled as mechanical properties (P, pressure; V, volume; N, number of particles; U, internal energy; H, enthalpy), if their definition does not involve the concept of temperature, and non-mechanical properties (T, temperature; S, entropy; A, Helmholtz free energy; G, Gibbs free energy; μ, chemical potential), if their definition involves the concept of temperature.

4.2 The Canonical Ensemble

Let us denote by N the number of particles in each system and assume that there are n systems. We will also assume that, of those n systems, n_1 systems have energy U_1, n_2 systems have energy U_2, ..., and n_i systems have energy U_i. That is,

$$n = \sum_i n_i \tag{4.1a}$$

$$U = \sum_i n_i U_i \tag{4.1b}$$

where U is the total energy of the collection of particles. The probability of n_i systems having the energy U_i is then

$$p_i = \frac{n_i}{n} \tag{4.2}$$

The number of ways of distributing the n distinguishable systems so that n_1 systems have energy U_1, n_2 systems have energy U_2, etc., is

$$\Omega = \prod_i \frac{n!}{(n_i!)} \tag{4.3}$$

Among the many distributions that are possible, there will be a most probable distribution, with a corresponding Ω_{max}. Consequently, one can try and obtain the description of the collection of particles in terms of that most probable distribution. Therefore, the first step in this connection will consist of determining Ω_{max}.

4.2.1 The Partition Function

The determination of Ω_{max} implies finding the values of the corresponding n_i, subject to the constraints given by Eqs. (4.1). It can be performed by the method of Lagrangian multipliers (already mentioned in the preceding chapter, when developing the *SCF* theory).

The maximization of Ω is best performed for

$$\ln \Omega = \ln (n!) - \sum_i \ln(n_i!)$$

instead of Ω. Using Stirling's approximation, $\ln(n!) = n \ln(n) - n$ (for large n), the above expression may be rewritten as

$$\begin{aligned}
\ln \Omega &= \{n \ln (n) - n\} - \sum_i \{n_i \ln (n_i) - n_i\} \\
&= n \ln (n) - \sum_i n_i \ln (n_i) \\
&= -\sum_i n_i \ln \frac{n_i}{n} = -n \sum_i p_i \ln p_i
\end{aligned} \tag{4.4}$$

An arbitrary, infinitesimal variation, δn_i, in n_i results in

$$\delta \ln (\Omega) = -\sum_i \ln (n_i) \, \delta n_i$$

The necessary, though not sufficient, condition so that Ω may reach its absolute maximum is that $\delta\Omega = 0$ (or, equivalently, $\delta \ln \Omega = 0$) for any values of the variations δn_i, which are compatible with the restrictions arising from Eqs. (4.1), that is,

$$\sum_i \delta n_i = 0$$

$$\sum_i U_i \delta n_i = 0$$

The method of Lagrangian multipliers allows us to write

$$\sum_i \ell n(n_i)\delta n_i - \alpha \sum_i \delta n_i - \beta \sum_i U_i \delta n_i = 0$$

where α and β are the Lagrangian multipliers. This expression may be rewritten as

$$\sum_i \{\ell n(n_i) - \alpha - \beta U_i\} \delta n_i = 0$$

In order for this condition to be satisfied for any arbitrary, infinitesimal variation of the n_i it must be

$$\ell n(n_i) - \alpha - \beta U_i = 0$$

for every i. That is:

$$n_i = e^\alpha e^{\beta U_i}$$

One can reason that systems with large U_i will be less populated and therefore it is convenient to rewrite the above expression as

$$n_i = e^\alpha e^{-\beta U_i}$$

where β is now assumed to be positive. It can be proven (although the proof will be omitted here for simplicity) that $\beta = 1/k_B T$, where k_B is Boltzmann's constant (see Appendix 1). For simplicity, the subscript B will be omitted throughout this Chapter and the Boltzmann constant will be denoted as k.

Therefore, one can finally write

$$n_i = e^\alpha e^{-U_i/kT}$$

$$n = e^\alpha \sum_i e^{-U_i/kT}$$

$$p_i = \frac{n_i}{n} = \frac{e^{-U_i/kT}}{\sum_i e^{-U_i/kT}}$$

The expression in the denominator of the above expression receives the designation of partition function,

$$Q = \sum_i e^{-U_i/kT} \tag{4.5}$$

and therefore

$$p_i = \frac{1}{Q} e^{-U_i/kT}$$

4.2.2 Thermodynamic Properties

The total energy U of the collection of particles is then given by

$$U = \sum_i n_i U_i = n \sum_i p_i U_i = \frac{n}{Q} \sum_i U_i e^{-U_i/kT} = nkT^2 \left(\frac{\partial \ln Q}{\partial T} \right)_{N,V} \tag{4.6a}$$

so that the energy per system is

$$U = kT^2 \left(\frac{\partial \ln Q}{\partial T} \right)_{N,V} \tag{4.6b}$$

Taking into account the general postulate of Boltzmann and Eq. (4.4), the entropy S for the collection of particles may be written as

$$
\begin{aligned}
S = k \ln \Omega &= -nk \sum_i \left\{ \frac{e^{-U_i/kT}}{Q} \ln \frac{e^{-U_i/kT}}{Q} \right\} \\
&= nk \sum_i \left\{ \frac{U_i}{kT} e^{-U_i/kT} + \frac{e^{-U_i/kT}}{Q} \ln Q \right\} \\
&= \frac{U}{T} + n k \ln Q
\end{aligned}
\tag{4.7a}
$$

where U is the total energy. Therefore the entropy per system will be

$$S = \frac{U}{T} + k \ln Q \tag{4.7b}$$

where U stands now for the energy of the system.

The Helmholtz free energy A for the collection of particles may be obtained by comparing Eq. (4.6a) with the thermodynamic relationship

$$S = \frac{U}{T} - \frac{A}{T}$$

One obtains

$$A = -n k T \ln Q$$

so that the Helmholtz free energy per system will be

$$A = -k T \ln Q. \tag{4.8}$$

Then, taking into account the definitions of pressure (P), enthalpy (H), and Gibbs free energy (G), one obtains, per system,

$$P = kT \left(\frac{\partial \ln Q}{\partial V} \right)_T \tag{4.9}$$

$$H = kT \left\{ T \left(\frac{\partial \ln Q}{\partial T} \right)_V + V \left(\frac{\partial \ln Q}{\partial V} \right)_T \right\} \tag{4.10}$$

$$G = -kT \left\{ \ln Q - V \left(\frac{\partial \ln Q}{\partial V} \right)_T \right\} \tag{4.11}$$

The result is that one can calculate the energy, enthalpy, entropy, and the Helmholtz and Gibbs free energy of the system if the partition function, Eq. (4.5), is known. That, in turn, implies that the energies U_i must be known, with all the corresponding difficulties.

4.2.3 The Molecular Partition Function

Let us consider a system with energy U. This energy may be expressed as

$$U = \sum_p \mathscr{E}_p + \Delta E \tag{4.12}$$

where the summation over p extends to the N particles in the system, with energies $\mathscr{E}_p$, and ΔE represents the total interaction energy between the particles of the system. For simplicity in the development of the formulation it is convenient to consider first the case of non-interacting particles.

The energy of each particle will be one of the allowed energies, i.e., one of the eigenvalues, E_I, of the corresponding Hamiltonian operator (see Chapter 3). For a large collection of particles, the number of particles is given by

$$N_I = N \frac{g_I e^{-E_I/kT}}{\sum\limits_{J} g_J e^{-E_J/kT}}$$

where g_I and E_I denote the degeneracy and energy of the I-th state and the denominator

$$q = \sum_{J} g_J e^{-E_J/kT} \tag{4.13}$$

represents the molecular partition function.

The energy of the system will then be given by

$$U = \sum_{I} N_I E_I = \frac{N}{q} \sum_{I} g_I E_I e^{-E_I/kT}$$

which can be rewritten as

$$U = NkT^2 \left(\frac{\partial \ln q}{\partial T} \right)_{N,V} \tag{4.14}$$

Comparison of this expression with Eq. (4.6b) yields

$$\left(\frac{\partial \ln Q}{\partial T} \right)_{N,V} = N \left(\frac{\partial \ln q}{\partial T} \right)_{N,V}$$

or, equivalently

$$Q = q^N$$

This expression, valid for distinguishable particles, must be transformed into

$$Q = \frac{1}{N!} q^N \tag{4.15}$$

for indistinguishable particles. If several types of particles are present, this expression may be generalized to

$$Q = \frac{\prod\limits_{A} q_A^{N_A}}{\prod\limits_{A} (N_A!)}$$

where the products over A extend to all the types of particles.

These results show that, for non-interacting particles, the thermodynamic properties of the system may be evaluated from the corresponding molecular partition function(s). We already know of the difficulties associated with the evaluation of the eigenvalues of the Hamiltonian operator and, therefore, one can see the difficulties in the evaluation of the molecular partition function. In fact, the situation is far more complicated. For simplicity, in order not to confuse the issues, the discussion in Chapter 3 has been restricted to the eigenvalue equation for the electronic Hamiltonian operator, with neglect of all other motions.

Particles may translate and rotate in space and the nuclei may undergo vibrations. That is, the energies of the particles will consist of electronic, translational, rotational, and vibrational contributions. Within the Born-Oppenheimer approximation, the molecular partition function could then be expressed as

$$q = q_{el} \; q_{tr} \; q_{vib} \; q_{rot}$$

where q_{el}, q_{tr}, q_{vib}, and q_{rot} stand for the electronic, translational, vibrational, and rotational molecular partition functions. [There is also a contribution of the nuclear spins, not considered here, because of its irrelevance for the purpose of this work.]

One can now proceed to consider the case of interacting particles. The energy of the system will be given, taking into account Eq. (4.12) and the preceding discussion, by

$$U = \sum_{I} N_I E_I + \Delta E = NkT^2 \left(\frac{\partial \ln q}{\partial T} \right)_{N,V} + \Delta E$$

Therefore, one obtains

$$\left(\frac{\partial \ln Q}{\partial T} \right)_{N,V} = N \left(\frac{\partial \ln q}{\partial T} \right)_{N,V} + \frac{\Delta E}{NkT^2}$$

which yields

$$Q = q^N e^{-\Delta E / NkT}$$

for distinguishable particles. In the case of indistinguishable particles one will write, as above,

$$Q = \frac{1}{N!} q^N e^{-\Delta E / NkT} \qquad (4.16)$$

where the considerations presented above for the molecular partition function still hold. As all the thermodynamic properties are expressed in terms of $\ln Q$, one can write (using again Stirling's approximation)

$$\begin{aligned}
\ln Q &= -\ln(N!) + N \ln q - \frac{\Delta E}{NkT} \\
&= N \ln q - N \ln N + N - \frac{\Delta E}{NkT} \qquad (4.17) \\
&= N \ln (q / N) + N - \frac{\Delta E}{NkT}
\end{aligned}$$

where (q/N) is the molecular partition function per molecule. Using these expressions in the definitions of the thermodynamic properties, one can now obtain the corresponding expressions in terms of the molecular partition function. In this connection it is worthwhile to mention that the partition function depends not only on the number of particles in the system and the temperature [see, e.g., Eq. (4.16) and (4.17)] but also on the volume (mainly because of the translational contribution and the interaction energy).

4.3 Numerical Calculations

One could now proceed with the development of the corresponding formulation for the grand canonical ensemble (μ, V, T). This derivation is omitted here for simplicity, taking into account that it will be possible (see below) to obtain the information of interest from the canonical and the isobaric-isothermic ensembles.

We will, therefore, center the discussion on these two ensembles, considering separately the evaluation of mechanical and non-mechanical properties. The difficulties in the evaluation of the partition functions make it necessary to look for an alternate procedure for the evaluation of the thermodynamic properties, while taking into account the preceding considerations and theoretical developments. That is, the alternative should consist of simulating the canonical (as well as the isobaric-isothermic) ensemble, so that the properties of interest could be evaluated directly by averaging.

This simulation is achieved through the generation of a trajectory in the configuration space, that will sample the configurations according to the canonical Boltzmann distribution of configurations. Each configuration is characterized by a different distribution of the particles in space (i.e., different inter-particle separations and orientations) and, consequently, by a different interaction energy. The averages over the trajectory will then correspond to a canonical ensemble average. For example, the average value of a mechanical property, F, which depends only on the configuration, will be given by

$$<F> = \frac{\int F e^{-\Delta E/kT} d\tau}{\int e^{-\Delta E/kT} d\tau} \qquad (4.18)$$

where integration is performed over the configuration space.

The simulation may be achieved through the use of the Monte Carlo technique (Metropolis *et al.* 1953), which builds up a Markov chain through the selection of successive configurations, under certain constraints [see, e.g., the work of Lluch *et al.* (1992)]. This technique, however, will not allow for the evaluation, in particular, of free energies unless the grand canonical ensemble is considered (Yao *et al.* 1982), with the corresponding increase in complexity. This problem may be solved, for example, through the use of statistical perturbation theory (see, e.g., the work of Zwanzig 1954, Jorgensen *et al.* 1985, 1987, 1989abc, Brooks 1986, Brooks and Fleischmann 1990, Migliore *et al.* 1988, Rao and Singh 1989, and Gonzalez-Lafont *et al.* 1990). In this formulation, the difference in free energy between an (unperturbed) reference system and the perturbed system may be evaluated from a canonical ensemble average (for A) or from an isobaric-isothermic ensemble average (for G) over the unperturbed system.

4.3.1 The Monte Carlo Method

We will, consequently, concentrate on the details of the Monte Carlo (*MC*) method, under the assumption that, directly or through the use of some additional approximations, the resulting trajectory of configurations will permit us to evaluate the properties of interest.

The steps in a basic *M C* procedure are illustrated in Fig. 4.1 (In this connection, the reader is referred to the initial work of Metropolis *et al.* 1953, the early application of Barker and Watts 1969 and Abraham 1974, and the review of

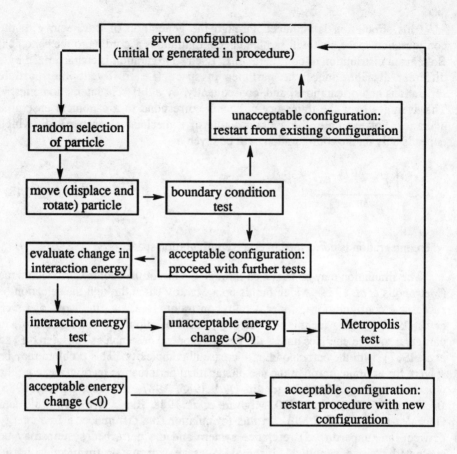

Figure 4.1. Schematic representation of the flow in a Monte Carlo procedure

Lluch *et al.* 1992, where extensive references as well as details of the practical improvements, introduced over the years, may be found.) These steps are as follows:

(a)   Define the boundaries of the system (e.g., cubic or spherical boundary).

(b)   Choose the initial configuration of the collection of particles.

(c)   Select at random a particle, its displacement, and its rotation around an axis and move the particle.

(d)   Test for the boundary condition. [For example: If the move takes the particle outside the cube, assume that it has reentered it from the opposite side; if the move has brought the particle outside the spherical boundary, reject the configuration and restart the process from the existing one.]

(e)   If the generated configuration has passed the test, the interaction energy ΔE is evaluated (see below) and its change, $\delta(\Delta E)$, with respect to the previous configuration, is obtained.

(f)   If the interaction energy has decreased, the new configuration is accepted and the process is restarted.

(g)   If the interaction energy has increased, the Metropolis test $e^{-\delta(\Delta E)/kT} > h$ (h: random number, $0 \leq h \leq 1$) is applied. If the *lhs* is greater than h, the configuration is accepted and the procedure is restarted with the new configuration; otherwise the configuration is rejected and the procedure is restarted using the old configuration.

The *MC* procedure, as such, is extremely simple and its implementation in a computer code offers no difficulties. The practical problems that may arise are related to the selection of the initial configuration and the computing time and storage needs of the calculation. In order to avoid any bias in the results, due to a poor choice of the initial configuration, the procedure should be executed for an appropriate number of moves (say 500,000 to 1,000,000 moves), until statistical equilibration has been reached. It is only after that point that the results should be saved, as they will be needed at the end of the simulation for the averaging calculation., The considerable storage needs arise precisely from the fact that all the configurations, obtained after equilibration has been reached, must be stored; in a typical calculation, there may be, say, 1,000,000 configurations to be saved.

The bottleneck in a *MC* simulation, however, is to be found in the evaluation of the interaction energies for each newly generated configuration. Of course, at each point, one has to evaluate only the interaction energy of the particle, which has been moved, with all the remaining particles, which have remained fixed; but it must be pointed out that the number of particles must be large (say, several hundreds) in order for the simulation to have any statistical meaning. Ideally, the evaluation of the interaction energy should be performed as described in Chapter 3, in which case (assuming such calculation to be feasible) the computer time requirement would be staggering. That is why a fundamental requisite for a *MC* calculation is the availability of a potential energy function, which will allow for an extremely fast computation while yielding satisfactory results. This subject will be discussed in detail in Chapter 5.

The *MC* method is amenable to a dynamic interpretation by introducting a *timescale*, which labels the sequential order of the configurations (Müller-Krumbhaar and Binder 1973, Binder 1979, 1987). With an appropriate definition of the relationship between the *MC* and physical times, the *MC* is equivalent to solving numerically the diffusion equation for Brownian particles (Kikuchi *et al*.

1991). This procedure, equivalent to a Brownian dynamics simulation (Ermak and McCammon 1978; see Section 12.2.3), has been extended to include the hydrodynamic interactions mediated by the solvent (Kikuchi *et al*. 1992).

4.4 Summary

The formulation for the Statistical Mechanics prediction of the thermodynamic properties of a given system is straightforward but its application is hindered by the fact that it requires a knowledge of the (molecular) partition function, with all the corresponding difficulties. A practical solution, however, may be found in the use of the *MC* method.

This simulation method has been summarized in a rather brief manner for the following reason. As it has already been mentioned, its practical implementation is rather simple, once an appropriate potential energy function is available. And that is the stumbling block. Once that block has been removed (see Chapter 5), the reader may easily incorporate the potential energy function into a *MC* program, readily available from different sources, and proceed with the simulation.

The two pieces of information we have gathered so far are that effective charges are available from quantum-chemical calculations and that statistical results may be simulated through *MC* calculations. Now the only piece missing is the development of the potential energy function.

An additional comment, however, is appropriate at this point. Such statistical calculations are not performed, as a rule, for proteins. And the designation *Monte Carlo*, within the context of simulations of protein structures, has evolved to indicate simply a stochastic procedure for a random walk on an energy hypersurface (see Section 6.1).

5 Molecular Mechanics: The Potential Energy Function

It should have become evident by now that an accurate treatment of the modeling of protein structures is not feasible in practice, even though the necessary formulations are available. There is, therefore, no recourse but to introduce approximations, some of which, in fact, may seem very drastic. Judgment, however, should be passed only after an evaluation of the correctness of the results that the corresponding treatments may yield.

In this Chapter we will proceed to develop the transition from Quantum to Molecular Mechanics, on the basis of formal, historical, and practical considerations. There are certain characteristics of Molecular Mechanics treatments that should be emphasized right from the start.

The basic requirement for a Molecular Mechanics treatment is the availability of a proper potential energy function (*PEF*). Such a *PEF* could be developed from experimental information or from purely theoretical considerations, as presented below. Once the *PEF* has been obtained, the Molecular Mechanics treatment is performed with almost complete neglect of the principles of Quantum Mechanics. The system(s) under study is (are) considered to consist of a number of particles (the nuclei) which are allowed to move in space, perhaps with some restrictions (see below), as determined by the *PEF*, under a criterion of minimum energy. Such a criterion is certainly a left-over from the original quantum-mechanical formulation, as embodied in the variational principle (discussed in Chapter 3). The other link to quantum-chemical calculations is the use of effective charges, in the case of a theoretically-derived *PEF*, and we have already mentioned the practical approximations in this connection. The most striking characteristic in Molecular Mechanics treatments is the fact that no mention is made at all of electrons and that bonds do not exist, except as the result of a mathematical constraint introduced in the *PEF*.

5.1 Forces in Protein Folding

Electric, magnetic, and gravitational forces (Buckingham 1967) are at play in the folding process of proteins and in their interactions with other systems. Gravitational and magnetic interactions are negligible (as already mentioned in Chapter 3, regarding the latter) and therefore the only interactions to be considered are those embodied in the corresponding, electronic Hamiltonian operator. That is, the electron kinetic energy, the electron-nucleus attraction, and the electron-electron and nucleus-nucleus repulsion terms.

Protein folding and interactions have been under study long before the possibility of a theoretical approach was considered and consequently a parallel terminology has evolved over the years. As an example, as it will prove useful later on, let us summarize how one could describe the folding process (King 1989) and the interactions which guide it and stabilize the final tertiary structure (Weisman and Kim 1991) using macroscopic terminology. One might postulate that local, weak interactions between hydrophobic residues, adjacent in the sequence, lead to the generation of secondary structure units. These structural units may be stabilized by association, with formation of H-bonds and salt-bridges, and by additional hydrophobic interactions as well as interactions between the main-chain dipole moments of the structural units (Hol *et al.* 1981). On the other hand, however, one could postulate as well that a hydrophobic collapse takes place first, followed by the generation of secondary structure units (Dill 1990). And, finally, there is the possibility that the generation of the structural units and the hydrophobic collapse are strongly coupled (Honig *et al.* 1993), that is, that they take place simultaneously.

This simple summary emphasizes the difficulty in explaining what is a rather complex process in terms of macroscopic terminology. But because this language is widely used, it is convenient to continue with further examples, after which we may try and link it with the microscopic language. The best example in this connection is the designation of hydrophobic interactions (Kauzman 1959, Muller 1992). Current thinking, for example, is that internal residues involved in cluster interactions are conserved during evolution and that this confirms the role of the hydrophobic domains in the folding process (Plochocka *et al.* 1988). Then the folding process may be interpreted in terms of only close packing (Ngo and Marks 1992, Rose and Wolfenden 1993), once the existence of close-packed hydrophobic cores (Murzin and Finkelstein 1988) is established, with the result that the side-chain conformation in the protein core may be determined on the basis of that requirement (Lee and Subbiah 1991); examples of helix-to-helix (Chothia *et al.* 1981) and β-sheet (Cohen *et al.* 1981, Lesk and Chothia 1982) packing have been observed. In this connection it is interesting to note that the free energy difference between the folded and unfolded states is proportional to the number of buried non-polar residues (Koehl and Delarue 1994).

All the interactions, mentioned in the macroscopic language, arise from the terms in the electronic Hamiltonian operator. The macroscopic terminology tries to reflect their magnitude, ranging from the strong Coulombic interactions (Davis and McCammon 1990, Oberoi and Allewell 1993) present in salt-bridges to the weaker interactions denoted as dipole-interactions and van der Waals forces (Buckingham *et al.* 1988), consisting of the dipole-dipole (Keesom), dipole-induced dipole (Debye), and induced dipole-induced dipole (London) interactions. In an alternate way, as discussed below, it would be more convenient to talk simply about short, intermediate, and long-range interactions, as long as one would clearly define what are the corresponding distance ranges.

There is, however, no direct correspondence between the so-called *H*-bonding, of predominantly electrostatic character (Baker and Hubbard 1984), and hydrophobic forces and the terms in the Hamiltonian operator and therefore some comments are in order. The designation *H*-bonding describes an experimental observation and it should be considered as a result of the appropriate, microscopic interactions. That is, there is no specific term in the electronic Hamiltonian operator associated with *H*-bonding. In other words: *H*-bonding is not a distinct interaction and therefore it needs not be considered independently. If it should exist in a given system, its existence should be confirmed in an accurate calculation without need of imposing its formation as an additional constraint.

The same comments apply to the disulfide bridges (Hazes and Dijkstra 1988, Sowdhamini *et al.* 1989, Benham and Jafri 1993) and the concept of hydrophobic forces. The latter cannot be ascribed either to any particular microscopic interaction. This designation, useful as it may be, should be taken only as to represent the competition between internal forces and the interactions with the solvating, aqueous environment.

5.2 Theoretical Formulation of Molecular Interactions

The task ahead, therefore, consists of developing the formal expression of a *PEF*, as a basis on which to introduce whatever approximations may be needed or are convenient for practical reasons. As mentioned above, the system under study will be considered to consist of a set of charged particles, with a given spatial configuration. Therefore, the first step is to establish the basic characteristics associated with such a set of charges. [See also Section 3.7.1].

5.2.1 Charges, Moments, and Fields

The charge distribution under consideration will be assumed to consist of N particles, with charges q_i, at positions (X_i, Y_i, Z_i). Each particle may then be described in terms of its charge, q_i, and its position vector (referred to the origin of coordinates) r_i, with magnitude

$$r_i = (x_i^2 + x_i^2 + x_i^2)^{1/2} \tag{5.1}$$

The charge distribution is characterized by its corresponding total charge,

$$Q = \sum_i q_i \tag{5.2}$$

(where the summation extends to the N particles), and its moments.

Those moments, which are denoted as dipole, quadrupole, octupole, hexadecapole, etc., are tensors of rank 1, 2, 3, ..., respectively, and they are characterized by their components. Thus, the dipole moment, which is a tensor of rank 1 (i.e., a vector) has components

$$\mu_x = \sum_i q_i\, x_i, \quad \mu_y = \sum_i q_i\, y_i, \quad \mu_z = \sum_i q_i\, z_i, \tag{5.3}$$

with the summations over the N particles. Its magnitude is

$$\mu = (\mu_x^2 + \mu_y^2 + \mu_z^2)^{1/2} \tag{5.4}$$

and in vector form it may be expressed as

$$\mu = \sum_i q_i\, r_i \tag{5.5}$$

The quadrupole moment, $\mathbf{Q}$, is a tensor of rank 2, with components

$$
\begin{aligned}
Q_{xx} &= \sum_i q_i x_i^2, & Q_{xy} = Q_{yx} &= \sum_i q_i x_i y_i, & Q_{xz} = Q_{zx} &= \sum_i q_i x_i z_i \\
Q_{yy} &= \sum_i q_i y_i^2, & Q_{yz} = Q_{zy} &= \sum_i q_i y_i z_i, & Q_{zz} &= \sum_i q_i z_i^2
\end{aligned}
\tag{5.6}
$$

and its trace (i.e., the sum of its diagonal elements Q_{xx}, Q_{yy}, Q_{zz}), with the value $\sum_i q_i r_i^2$, is invariant under a rotation of the coordinate axes.

We could present the corresponding definitions for the octupole, hexadecapole, and higher moments, but there is no need because our purpose is to develop the formal expression of the *PEF* and we will see it from the terms obtained from a consideration of the dipole and quadrupole moments.

The definitions of the moments in terms of Cartesian coordinates have the advantage of an intuitive simplicity but it is possible to obtain a formally simpler description in terms of spherical harmonics. The normalized spherical harmonics are defined

$$Y_{\ell m}(\theta,\phi) = (-1)^m \left[\frac{2\ell+1}{4\pi}\right]^{1/2} \left[\frac{(\ell-m)!}{(\ell+m)!}\right]^{1/2} P_{\ell m}(\cos\theta)\, e^{im\phi} \tag{5.7a}$$

$$Y_{\ell -m}(\theta,\phi) = \left[\frac{2\ell+1}{4}\right]^{1/2} \left[\frac{(\ell-m)!}{(\ell+m)!}\right]^{1/2} P_{\ell m}(\cos\theta)\, e^{-im\phi} \tag{5.7b}$$

(for positive integer m) in terms of the unnormalized associated Legendre polynomials

$$P_{\ell m}(\cos\theta) = (1-\cos^2\theta)\frac{d^m P_\ell(\cos\theta)}{d\cos\theta^m}$$

where

$$P_\ell(\cos\theta) = \frac{1}{2^\ell \ell!}\frac{d^\ell(\cos^2\theta-1)^\ell}{d\cos\theta^\ell}$$

is a Legendre polynomial. In these expressions θ and ϕ are the spherical polar coordinates, related to the Cartesian coordinates by

$$x = r\sin\theta\cos\phi \quad y = r\sin\theta\sin\phi, \quad z = r\cos\theta \tag{5.8}$$

To second-order, the spherical harmonics in explicit form are

$$Y_{00} = \left(\frac{1}{4\pi}\right)^{1/2} \tag{5.9a}$$

$$Y_{10} = \left(\frac{3}{4\pi}\right)^{1/2}\cos\theta \tag{5.9b}$$

$$Y_{1\pm 1} = \frac{-}{+}\left(\frac{3}{4\pi}\right)^{1/2}\left(\frac{1}{2}\right)^{1/2}\sin\theta\, e^{\pm i\phi} \tag{5.9c}$$

$$Y_{20} = \left(\frac{5}{4\pi}\right)^{1/2}\frac{1}{2}(3\cos^2\theta-1) \tag{5.9d}$$

$$Y_{2\pm1} = \frac{-}{+}\left(\frac{5}{4\pi}\right)^{1/2}\left(\frac{3}{2}\right)^{1/2} \sin\theta\cos\theta\, e^{\pm i\phi} \tag{5.9e}$$

$$Y_{2\pm2} = \left(\frac{5}{4\pi}\right)^{1/2}\left(\frac{3}{8}\right)^{1/2} \sin^2\theta\, e^{\pm i2\phi} \tag{5.9f}$$

The generalized definition of the moments of a charge distribution in terms of spherical harmonics is

$$M_{\ell m} = \left[\frac{4\pi}{2\ell+1}\right]^{1/2} \sum_i q_i r_i^\ell\, Y_{\ell m}(\theta,\phi) \tag{5.10}$$

and the relationship between this general definition and the one given above in terms of Cartesian coordinates may easily be obtained taking into account the definitions of the spherical harmonics, Eqs. (5.9), as well as

$$\cos\phi = \frac{1}{2}(e^{i\phi} + e^{-i\phi}), \quad \sin\phi = \frac{1}{2i}(e^{i\phi} - e^{-i\phi})$$

or, conversely,

$$e^{i\phi} = \cos\phi + i\sin\phi, \quad e^{-i\phi} = \cos\phi - i\sin\phi$$

For example, for the components of the moment of rank 1 one can write

$$M_{1+1} = \left[\frac{4\pi}{3}\right]^{1/2} \sum_i q_i r_i Y_{1+1}(\theta_i,\phi_i) = -\left(\frac{1}{2}\right)^{1/2} \sum_i q_i r_i \sin\theta_i(\cos\phi_i + i\sin\phi_i)$$

$$= -\left(\frac{1}{2}\right)^{1/2} \sum_i q_i(x_i + i\, y_i) = -\left(\frac{1}{2}\right)^{1/2}(\mu_x + i\,\mu_y)$$

$$M_{10} = \left[\frac{4\pi}{3}\right]^{1/2} \sum_i q_i r_i Y_{10}(\theta_i,\phi_i) = \sum_i q_i r_i \cos\theta_i = \mu_z \tag{5.11}$$

$$M_{1-1} = \left[\frac{4\pi}{3}\right]^{1/2} \sum_i q_i r_i Y_{1-1}(\theta_i,\phi_i) = \left(\frac{1}{2}\right)^{1/2} \sum_i q_i r_i \sin\theta_i(\cos\phi_i - i\sin\phi_i)$$

$$= \left(\frac{1}{2}\right)^{1/2} \sum_i q_i(x_i - i\, y_i) = \left(\frac{1}{2}\right)^{1/2}(\mu_x - i\,\mu_y)$$

or, conversely,

$$\mu_x = \sqrt{2}\,(M_{1-1} - M_{1+1}), \quad \mu_y = -\sqrt{2}\,(M_{1+1} + M_{1-1}), \quad \mu_z = M_{10}$$

Proceeding in a similar fashion one can obtain the corresponding relationships for the moments of higher rank.

Let us now consider a single charge, q, at a point (X_p, Y_p, Z_p), with position vector **R**, and evaluate the potential, electric field, and electric field gradient at another point (x_0, y_0, z_0), with position vector r_0. The separation between the two points may be expressed as $|r_0\text{-R}|$ and it will be denoted as r_{op}.

The potential V at the point (x_0, y_0, z_0) due to the charge q is a tensor of rank 0 (i.e., a scalar). It is defined as

$$V = q/r_{op} = q[(x_0\text{-}X_p)^2 + (y_0\text{-}Y_p)^2 + (z_0\text{-}Z_p)^2]^{-1/2} \tag{5.12}$$

The electric field **F**, at the same point, due to the charge q, is a tensor of rank 1 (i.e., a vector), with components

$$F_x = -\frac{\partial V}{\partial x_0} = q(x_0 - X_p)/r_{op}^3$$

$$F_y = -\frac{\partial V}{\partial y_0} = q(y_0 - Y_p)/r_{op}^3$$

$$F_z = -\frac{\partial V}{\partial z_0} = q(z_0 - Z_p)/r_{op}^3$$

and a magnitude

$$F = (F_x^2 + F_y^2 + F_z^2)^{1/2} = q/r_{op}^2$$

The electric field gradient **F'**, at the same point, due to the charge q, is a tensor of rank 2, with components

$$F'_{xx} = \frac{\partial^2 V}{\partial x_0^2} = q[3(x_0 - X_p)^2 - r_{op}^2]/r_{op}^5$$

$$F'_{xy} = F'_{yx} = \frac{\partial^2 V}{\partial x_0 \partial y_0} = 3q(x_0 - X_p)(y_0 - Y_p)/r_{op}^5$$

$$F'_{xz} = F'_{zx} = \frac{\partial^2 V}{\partial x_0 \partial z_0} = 3q(x_0 - X_p)(z_0 - Z_p)/r_{op}^5$$

$$F_{yy}' = \frac{\partial^2 V}{\partial y_0^2} = q[3(y_0 - Y_p)^2 - r_{op}^2]/r_{op}^5$$

$$F_{yz}' = F_{zy}' = 3q(y_0 - Y_p)(z_0 - Z_p)/r_{op}^5$$

$$F_{zz}' = \frac{\partial^2 V}{\partial z_0^2} = q[3(z_0 - Z_p)^2 - r_{op}^2]/r_{op}^5$$

and a trace (i.e., sum of the diagonal elements)

$$\text{tr } F' = F_{xx}' + F_{yy}' + F_{zz}' = 0$$

which is invariant under a rotation of the coordinate axes.

In the particular case when the point (x_0, y_0, z_0) is the centre of coordinates, then in the above expressions r_{op} becomes R, i.e., the distance between the position of charge q and the origin. Furthermore, if, for example, the charge q is on the z-axis, that is, at point $(0, 0, R)$ then the above expressions become

$$F_x = F_y = 0, \quad F_z = -q/R^2 \tag{5.13}$$

$$F_{xx}' = F_{yy}' = -q/R^3, \quad F_{zz}' = 2q/R^3, F_{xy}' = F_{yz}' = F_{zx}' = 0 \tag{5.14}$$

As it was the case for the electric moments, one can give a generalized definition of the components of the electric field, the electric field gradient, etc. in terms of spherical harmonics. One can write

$$F_{\ell m}(R) = \left(\frac{4\pi}{2\ell + 1}\right)^{1/2} (q/R^{\ell+1}) Y_{\ell m}(\theta, \phi) \tag{5.15}$$

for the m-th component of the electric field ($\ell = 1$), the electric field gradient ($\ell = 2$), etc., at the origin, due to a charge q with a position vector **R**. Taking into account the definitions of the spherical harmonics one can easily obtain the relationships between the generalized definitions and those given in terms of Cartesian coordinates. For example, for the case when the charge q is at position $(0, 0, R)$ one would obtain $F_{1+1} = F_{1-1} = 0$, $F_{10} = -q/R^2$ for the components of the electric field.

5.2.2 The Multipole Expansion

The electrostatic potential interaction energy between two electric charges is given by the product of the first charge times the potential due to the second charge at the position occupied by the first charge, or *vice versa*. Let us consider a charge q_i, with position vector r_i, and a charge q, with position vector **R**. Their interaction may then be expressed as the product of q_i times the potential due to charge q at the position of q_i, i.e.,

$$\Delta E = qq_i/|r_i\text{-}R|$$

and, therefore, for a set of charges, one can write

$$\Delta E = q\sum_i q_i/|r_i\text{-}R|$$

This expression may be rewritten, using the Neumann expansion, as

$$\Delta E = q \sum_i q_i \sum_k (r_<^k/r_>^{k+1}) P_k(\cos \omega_i) \tag{5.16}$$

where the summation over *k* runs from 0 to ∞, $r_<$ and $r_>$ denote the lesser and the greater, respectively, of **R** and r_i, ω_i is the angle between the two position vectors **R** and r_i, and $P_k(\cos \omega_i)$ is a Legendre polynomial.

In the case when R is larger than all the r_i, Eq. (5.16) may be rewritten as

$$\Delta E = q \sum_i q_i \sum_k (r_i^k/R^{k+1}) P_k(\cos \omega_i) \tag{5.17}$$

Taking into account (Condon and Shortley 1964) that

$$P_k(\cos \omega_i) = \left(\frac{4\pi}{2k+1}\right) \sum_m Y_{km}(\theta,\phi) Y_{km}(\theta_i,\phi_i)$$

where the summation extends over $-k \leq m \leq k$ and θ,ϕ denote again the polar spherical coordinates, then one obtains

$$\Delta E = q \sum_i q_i \sum_k \left(\frac{4\pi}{2k+1}\right)(r_i^k/R^{k+1}) \sum_m Y_{km}(\theta,\phi) Y_{km}(\theta_i,\phi_i)$$

$$= \sum_k \sum_m \left\{\left(\frac{4\pi}{2k+1}\right)^{1/2} (q/R^{k+1}) Y_{km}(\theta,\phi)\right\}\left\{\left(\frac{4\pi}{2k+1}\right)^{1/2} \sum_i q_i r_i^k Y_{km}(\theta_i,\phi_i)\right\}$$

Taking into account Eqs. (5.10) and (5.16) one has finally

$$\Delta E = \sum_k \sum_m F_{km}(R) M_{km}(r_i) \tag{5.18}$$

so that the electrostatic potential interaction energy is given in terms of the products of the moments associated with the charge distribution and the potential, electric field, electric field gradient, etc., at the origin, due to charge q.

This expression would allow us to calculate the value of the interaction energy between the charge q and the charge distribution, under the above assumption that $R > r_i$ (for any i). Here, however, we are interested in examining the R-dependence of this expression.

Let us assume, for simplicity, that the charge q is on the z-axis. The first term in the expansion will be qQ/R. The second term, taking into account Eqs. (5.11) and (5.14) will be $-q\mu_z/R^2$. Similarly, taking into account Eq. (5.15), one can see that the third term will be proportional to $1/R^3$, involving the components of the quadrupole moment of the charge distribution. If we were to continue examining higher terms we would observe the same $1/R^n$ dependence, so that in a general form we could express the interaction energy as

$$\Delta E = C^{(1)}/R + C^{(2)}/R^2 + C^{(3)}/R^3 + ... \tag{5.19}$$

where $C^{(1)}$, $C^{(2)}$, $C^{(3)}$, ..., etc., depend on q and the moments of the charge distribution.

5.2.3 Long-range Electrostatic Interaction Between Rigid Molecules

The above formulation has been developed on the basis, exclusively, of classical electrostatics, without any reference to quantum-mechanical concepts. The only assumption introduced is that R must be greater than every r_i; that is, it is only applicable to the case when the charge distribution is centred around the origin of coordinates and the interacting charge q is positioned at a larger distance. The question that we must consider next is whether this result is of any use in connection with the problem of interest, that is, the evaluation of the interaction energy between two molecules.

We already know that the corresponding calculation should be performed by application of the appropriate quantum-mechanical formulations, as described in Chapter 3. Therefore, when considering the possible application of the preceding formulation, we must determine first what are the necessary conditions for a satisfactory result.

At short ranges, when the overlap between the electron clouds of the two molecules is considerable, with a corresponding electron exchange between them, there is no possibility of any approximation: a quantum-mechanical calculation is needed for satisfactory results. However, as the separation between the two molecules increases, the penetration of the two electron clouds decreases, until it may be assumed to be negligible at an appropriately large separation.

It is in such a case that the formulation developed above may be used, within the context of perturbation theory, with satisfactory results (Buckingham 1967). The corresponding formulation is straightforward but laborious, particularly because of the complicated notation needed, as one is dealing with the components of tensor operators. The final results are summarized below for two reasons: on one hand because they will illustrate very clearly the complexity of the calculations, requiring the evaluation of rather complicated expectation values (assuming, of course, that the unperturbed functions of the two molecules were known, which would not be the case for two large peptides); second, and this is the important point for this work, because they will show clearly the R-dependence of the interaction energy.

The formulation may be summarized as follows [from A.D. Buckingham, *Permanent and Induced Molecular Moments and Long-Range Intermolecular Forces*, in *Intermolecular Forces*, edited by J.O. Hirschfelder, Advances in Chemical Physics, vol. 12, pp.107-142 (1967). Copyright © 1967 Interscience Publishers, New York, U.S.A. Reprinted by permission of John Wiley & Sons, Inc.] The two molecules will be denoted as A and B and the quantities associated with each one will be affected by the corresponding subscript. Their unperturbed eigenfunctions are $\Psi_{I(A)}^{(O)}$ and $\Psi_{J(B)}^{(O)}$, with $I = 0, 1, 2, ..., J = 0, 1, 2, ...,$ labelling the various eigenstates. The permanent electric moments of the molecules are represented by q (total charge), μ (electric dipole moment), θ (electric quadrupole moment), Ω (electric octupole moment), etc. The subscripts $\alpha, \beta, \gamma, ...,$ denote tensor components and can be equal to x, y, z, and a repeated Greek subscript indicates a summation over all three Cartesian coordinates. The R-dependence is given by the tensors $T_{(B)}$, with components

$$T_{(B)} = R^{-1}$$

$$T_{\alpha(B)} = \nabla_\alpha R^{-1} = -R_\alpha R^{-3}$$

$$T_{\alpha\beta(B)} = \nabla_\alpha\nabla_\beta R^{-1} = (3R_\alpha R_\beta - R^2\delta_{\alpha\beta})R^{-5}$$

$$T_{\alpha\beta\gamma(B)} = \nabla_\alpha\nabla_\beta\nabla_\gamma R^{-1} = -3[5R_\alpha R_\beta R_\gamma - R^2(R_\alpha\delta_{\beta\gamma} + R_\beta\delta_{\gamma\alpha} + R_\gamma\delta_{\alpha\beta})]R^{-7}$$

where R is the vector from the origin of A to B and δ represents a Kronecker delta (i.e., $\delta_{\alpha\alpha} = 1, \delta_{\alpha\beta} = 0$). As the vector from A to B is equal to the vector from B to A, with sign changed, then $T_{(A)} = (-1)^n T_{(B)}$, where n is the order of the tensor.

These tensors are symmetric and therefore a repeated Greek subscript reduces any $\mathbf{T}$ to zero. It can be seen from the above expressions that the tensor $\mathbf{T}$ of order n is proportional to $R^{-(n+1)}$, with the result (as further discussed below) that the interaction between the two molecules may be evaluated satisfactorily when R is large compared to the molecular dimensions.

The interaction energy, when the two free molecules are in their ground states, may be expressed [see Eq. (3.17c)] as

$$\Delta E = < \Psi_{O(A)}^{(O)} \Psi_{O(B)}^{(O)} \left| \mathcal{H}^{(1)} \right| \Psi_{O(A)}^{(O)} \Psi_{O(B)}^{(O)} >$$

$$- {\sum}' \frac{| < \Psi_{O(A)}^{(O)} \Psi_{O(B)}^{(O)} \left| \mathcal{H}^{(1)} \right| \Psi_{I(A)}^{(O)} \Psi_{J(B)}^{(O)} > |^2}{(E_{I(A)}^{(O)} - E_{O(A)}^{(O)}) + (E_{J(B)}^{(O)} - E_{O(B)}^{(O)})} + ... \tag{5.20}$$

where the summation extends over all the unperturbed states $\Psi_{I(A)}^{(O)} \Psi_{J(B)}^{(O)}$, except $\Psi_{O(A)}^{(O)} \Psi_{O(B)}^{(O)}$.

The first term in Eq. (5.20) represents the first-order correction and it is usually denoted as the *electronic interaction energy*, ΔE_{elect}. For non-degenerate unperturbed eigenstates $\Psi_{O(A)}^{(O)}$ and $\Psi_{O(B)}^{(O)}$ it may be written as

$$\Delta E_{elect} = T_{(B)} q_A q_B + T_{\alpha(B)} (q_A \mu_{\alpha(A)} - q_B \mu_{\alpha(A)})$$

$$+ T_{\alpha\beta B)} (\frac{1}{3} q_A \, \theta_{\alpha\beta(B)} + \frac{1}{3} q_B \, \theta_{\alpha\beta(B)} - \mu_{\alpha(A)} \mu_{\beta(B)})$$

$$+ T_{\alpha\beta\gamma(B)} (\frac{1}{15} q_A \, \Omega_{\alpha\beta\gamma(B)} - \frac{1}{15} q_B \, \Omega_{\alpha\beta\gamma(A)} - \frac{1}{3} \mu_{\alpha(B)} \theta_{\beta\gamma(B)} + \frac{1}{3} \mu_{\alpha(B)} \theta_{\beta\gamma(A)})$$

$$+ ... \tag{5.21}$$

The second term in Eq. (5.20) includes the so-called *induction* and *dispersion* interaction energies, ΔE_{ind} and ΔE_{dis}. The sets of excited unperturbed states $\Psi_{I(A)}^{(O)} \Psi_{O(B)}^{(O)}$ and $\Psi_{O(A)}^{(O)} \Psi_{J(B)}^{(O)}$, for which the matrix elements of $\mathcal{H}^{(1)}$ are diagonal in the ground states of the free molecules, are responsible for the induction energy,

$$\Delta E_{ind} = -\sum_{I \neq 0} \frac{|<\Psi_{O(A)}^{(O)} \Psi_{O(B)}^{(O)}|\mathcal{H}^{(1)}|\Psi_{I(A)}^{(O)} \Psi_{O(B)}^{(O)}>|^2}{E_{I(A)}^{(O)} - E_{O(A)}^{(O)}}$$

$$-\sum_{J \neq 0} \frac{|<\Psi_{O(A)}^{(O)} \Psi_{O(B)}^{(O)}|\mathcal{H}^{(1)}|\Psi_{O(A)}^{(O)} \Psi_{J(B)}^{(O)}>|^2}{E_{J(B)}^{(O)} - E_{O(B)}^{(O)}}$$

$$= \Delta E_{ind(A)} + \Delta E_{ind(B)} \tag{5.22}$$

where

$$\Delta E_{ind(A)} = -\frac{1}{2} \alpha_{\alpha\beta(A)} F_{\alpha(A)} F_{\beta(A)} - \frac{1}{3} A_{\alpha,\beta\gamma(A)} F_{\alpha(A)} F_{\beta\gamma(A)}$$

$$-\frac{1}{6} C_{\alpha\beta,\gamma\delta(A)} F_{\alpha\beta(A)} F_{\gamma\delta(A)} - \cdots \tag{5.23}$$

(with a similar expression for $\Delta E_{ind(B)}$). In this expression, α, $\mathbf{A}$, $\mathbf{C}$ denote the (tensor) molecular polarizabilities while the tensors $\mathbf{F}$, with components

$$F_{\alpha(A)} = -T_{\alpha(A)} q_B + T_{\alpha\beta(A)} \mu_{\beta(B)} - \frac{1}{3} T_{\alpha\beta\gamma(A)} \theta_{\beta\gamma(B)} + \cdots \tag{5.24a}$$

$$F_{\alpha\beta(A)} = -T_{\alpha\beta(A)} q_B + T_{\alpha\beta\gamma(A)} \mu_{\gamma(B)} - \frac{1}{3} T_{\alpha\beta\gamma\delta(A)} \theta_{\gamma\delta(B)} + \cdots \tag{5.24b}$$

are the electric field and electric field gradient at the origin of molecule A arising from the permanent moments of molecule B in its unperturbed state $\Psi_{O(B)}^{(O)}$. The expressions for the corresponding quantities for molecule B are obtained by exchanging the subscripts A and B in the preceding equations.

The remainder of the second-order correction in Eq. (5.20) is the dispersion energy

$$\Delta E_{dis} = -\sum_{\substack{I \neq 0 \\ J \neq 0}} \frac{|<\Psi_{O(A)}^{(O)} \Psi_{O(B)}^{(O)}|\mathcal{H}^{(1)}|\Psi_{I(A)}^{(O)} \Psi_{J(B)}^{(O)}>|^2}{(E_{I(A)}^{(O)} - E_{O(A)}^{(O)}) + (E_{J(B)}^{(O)} - E_{O(B)}^{(O)})}$$

which can only be simplified after introducing some approximations (Buckingham 1967). The resulting expression becomes

$$\Delta E_{dis} = -K\{T_{\alpha\beta(B)} T_{\gamma\delta(B)} \alpha_{\alpha\gamma(A)} \alpha_{\beta\delta(B)}$$

$$+ \frac{2}{3} \ T_{\alpha\beta(B)} \ T_{\gamma\delta\epsilon(B)} \ (\alpha_{\alpha\gamma(A)} \ A_{\beta,\delta\epsilon(B)} - \alpha_{\alpha\gamma(B)} \ A_{\beta,\delta\epsilon(A)})$$

$$+ T_{\alpha\beta\gamma(B)} \ T_{\delta\epsilon\phi(B)} \ (\frac{1}{3} \ \alpha_{\alpha\delta(A)} \ C_{\beta\gamma,\epsilon\phi(B)} + \frac{1}{3} \ \alpha_{\alpha\delta(B)} \ C_{\beta\gamma,\epsilon\phi(A)}$$

$$- \frac{2}{3} \ A_{\alpha,\epsilon\phi(A)} \ A_{\delta,\beta\gamma(B)})$$

$$- \frac{2}{9} \ T_{\alpha\beta(B)} \ T_{\gamma\delta\epsilon\phi(B)} \ A_{\alpha,\gamma\delta(A)} \ A_{\beta,\epsilon\phi(B)} + ... \}$$ (5.25)

where

$$K = \frac{U_A U_B}{4(U_A + U_B)}$$

may be evaluated from the quantities U, which may be taken to be either the first ionization potentials or the lowest allowed excitation energies.

The formulation could be extended to higher orders but it is not necessary to do so for the purposes of this discussion. As mentioned above, the numerical calculation would converge rapidly for large separations and therefore it can be said that this formulation provides an excellent tool for the evaluation of interaction energies between rigid molecules at large separations. The main drawbacks are the need of a knowledge of the unperturbed eigenfunctions for the isolated molecules and the time-consuming computation of the expectation values. For the purpose of this work, the first difficulty is all that matters as those unperturbed functions will not be available for large peptides and therefore one will not be able to consider the practical application of the formulation to the study of the interaction between proteins.

It is interesting, however, to analyze the above expressions regarding two points. First, one can see that the interaction is ultimately expressed in terms of (permanent) moment-moment (such as charge-charge, charge-dipole, charge-quadrupole, ..., dipole-dipole, dipole-quadrupole, ...) interactions as well as interactions involving the various electric polarizabilities of the two molecules. That is, these expressions establish the connection between the macroscopic language and the microscopic formulation. On the other hand, one can see from the expressions of the tensors T that the complete expression represents an expansion of $1/R^n$ powers, starting with n=1 (corresponding to the Coulombic interaction, which will be the dominant term for the interaction between charged systems). That is, we have obtained an expression of the type of Eq. (5.19), with coefficients that, in principle, could be evaluated accurately but which will not be available in the cases of practical interest.

5.2.4 The Atom-Pair Potential Approximation

A practical application may be found, however, for the expressions obtained in the preceding section, albeit if one is ready to introduce some very drastic approximations.

Let us apply the formulation to the evaluation of the interaction between two charged particles, which do not possess any moments but have electric polarizabilities. In such a case the expression for the interaction energy [see e.g., Eq. (5.19)] becomes

$$\Delta E = C^{(1)}/R + C^{(4)}/R^4 + C^{(6)}/R^6 + C^{(7)}/R^7 + C^{(8)}/R^8 + ... \qquad (5.26)$$

where

$$C^{(1)} = q_A q_B \qquad (5.27a)$$

$$C^{(4)} = -\frac{1}{2} (\alpha_A q_B^2 + \alpha_B q_A^2) \qquad (5.27b)$$

$$C^{(6)} = -(C_A q_B^2 + C_B q_A^2) \qquad (5.27c)$$

$$C^{(7)} = -\frac{1}{2} (B_A q_B^3 + B_B q_A^3) \qquad (5.27d)$$

$$C^{(8)} = -\frac{1}{24} (\gamma_A q_B^4 + \gamma_B q_A^4) \qquad (5.27e)$$

...

where the standard notation has been used for the electric dipole polarizability (α), dipole hyperpolarizability (γ), uniform field quadrupole polarizability (B), and field gradient quadrupole polarizability (C).

The first approximation to be introduced, in order to be able to apply this result to the study of the interaction between two molecules, is to consider that the molecules consist of a set of charged particles, characterized by the corresponding charges and polarizabilities. At this point the molecule, as an entity with certain properties, is being replaced with that collection of particles. Neither bonds nor electrons are the subject of our consideration.

Within the context of this approximation, the interaction energy for the two molecules will be approximated with a summation over the interactions between all the pairs of particles, each pair consisting of a particle of molecule A and a particle of molecule B. That is, the interaction is approximated as a summation of atom-pair potentials. Equation (5.26), applied to this case, may be written as

$$\Delta E = \sum_a \sum_b \{c_{ab}^{(1)}/R_{ab} + c_{ab}^{(4)}/R_{ab}^4 + c_{ab}^{(6)}/R_{ab}^6 + c_{ab}^{(7)}/R_{ab}^7 + c_{ab}^{(8)}/R_{ab}^8 + ...\} \quad (5.28)$$

where the summation over a extends to the particles in molecule A and the summation over b extends to the particles in molecule B. The expansion coefficients, $c_{ab}^{(n)}$, are defined by Eqs. (5.27) and R_{ab} denotes the separation between the two particles being considered.

In spite of the approximation just introduced, the evaluation of the interaction energy is still hindered by insurmountable difficulties. As discussed in Chapter 3, we may assume that effective charges are available for the molecules under consideration and electric dipole polarizabilities may be obtained from existing theoretical tabulations (Fraga *et al.* 1976, 1981). The higher-order electric polarizabilities, however, are not so easily evaluated and therefore there is a scarcity of values.

Several approaches are possible in order to break this bottleneck and they will be examined elsewhere in this work. At this point, for consistency in the development, we will consider an approximation of a theoretical nature. The solution is to disregard the higher terms, keeping only the first three terms. That is, Eq. (5.28) becomes

$$\Delta E = \sum_a \sum_b \{c_{ab}^{(1)}/R_{ab} + c_{ab}^{(4)}/R_{ab}^4 + c_{ab}^{(6)}/R_{ab}^6\} \quad (5.29)$$

Unfortunately, there is a scarcity of numerical values for the field gradient quadrupole polarizability, C, and therefore it is not possible to evaluate $c_{ab}^{(6)}$ [see Eq. (5.27c)]. Usually, this parameter is approximated as

$$c_{ab}^{(6)} = -\frac{3}{2}\alpha_a\alpha_b/[(\alpha_a/n_a)^{1/2} + (\alpha_b/n_b)^{1/2}] \quad (5.30)$$

where n_a and n_b denote the effective numbers of valence electrons of atoms a and b, respectively (Pitzer 1959).

5.2.5 Short-range Interaction Between Rigid Molecules

Let us assume, for a moment, that the above result, Eq. (5.29), is satisfactory, in spite of the various approximations needed to obtain it. It might happen that, when studying the interaction between two particular systems, their approach (towards a molecular association with lowest energy) could take place with such relative orientations that the Coulombic interaction would be highly attractive. As the contributions of the two other terms are always attractive, the result would be that the approach of the systems would continue until they would collide. Such a situation is not physically acceptable and would not appear in a proper quantum-mechanical calculation. The reason that it exists here is because the formulation to this point has been developed from the treatment for long-range interactions, where the overlap of the electronic clouds was negligible.

The only possible solution to this problem is to be found in the introduction of an additional term in Eq. (5.29), of repulsive nature. That is, the interaction energy will be approximated as

$$\Delta E = \sum_a \sum_b \{ c_{ab}^{(1)} / R_{ab} + c_{ab}^{(4)} / R_{ab}^4 + c_{ab}^{(6)} / R_{ab}^6 + c_{ab}^{(12)} / R_{ab}^{12} \} \tag{5.31}$$

where the term $1/R^{12}$ is adopted for historical reasons (Lennard-Jones) although a term $1/R^{10}$ could be just as acceptable. The coefficient $c_{ab}^{(12)}$ will always be positive but its value cannot be obtained from theoretical considerations.

5.3   Determination of a Theoretical Potential Energy Function

The uncertainty regarding the coefficients $c_{ab}^{(12)}$ makes it necessary to look for a procedure whereby all the coefficients $c_{ab}^{(n)}$, n = 1, 4, 6, 12, in Eq. (5.31) will be determined in a consistent fashion. Throughout this work reference will be made to various *PEF*, but, for simplicity in the conceptual development, at this point it is more appropriate to present the approach adopted by Clementi for the determination of an actual potential energy function. [For specific details and applications the reader is referred to the review work of Clementi (1980) as well as to the specific papers where the calculations were reported (Clementi *et al*. 1977, Scordamaglia *et al*. 1977, and Sordo *et al*. 1986, 1987). See also the related work of Iglesias *et al*. 1991.]

5.3.1 The Method of Clementi

The approach adopted by Clementi is conceptually simple but the calculations are time-consuming and the results are dependent on a good fitting procedure. The original calculations were concerned with the interaction of amino acids, as well as the *DNA* puric and pyrimidinic bases, with water (Clementi *et al*. 1977,

Scordamaglia *et al.* 1977) but they were extended later on to the interactions between amino acids (Sordo *et al.* 1986, 1987).

The method relies on accurate *SCF* calculations, using appropriate basis sets (see Section 3.5.1). For the interaction between two molecules *A* and *B*, the procedure may be summarized as follows:

(a) First, calculations are performed for the independent, non-interacting systems, yielding the effective charges and accurate values of the corresponding molecular energies, denoted as E_A and E_B.

(b) Then calculations are performed for the interacting pair *A-B*, for different relative positions (separation and orientation) of the two molecules, yielding the corresponding molecular energies, denoted as E_{AB}, for the molecular association. [The calculations will also yield the effective charges for the two molecules. These charges will be different from those obtained for the independent molecules and they will also be different for the various relative positions of the two molecules. These new effective charges are of no use, because in order to obtain them it is necessary to perform the corresponding calculation for the molecular association, which would not be possible for two large peptides. Consequently, as discussed in Chapter 3, use will be made only of the effective charges for the independent molecules.]

(c) The corresponding interaction energies, for the various relative positions of the two molecules, are then obtained as $\Delta E_{AB} = E_{AB} - (E_A + E_B)$.

(d) The final step consists of the determination of the coefficients $c_{ab}^{(n)}$ from these values of the interaction energies.

The original calculations were performed for the interaction of twenty-five systems (the twenty natural amino acids, hydroxyproline, the two puric bases, and the two pyridiminic bases) with water. In each case, the evaluation of the interaction energy between any of those systems and a water molecule was carried out for 200 relative positions of the two molecules: that is, at 100 different positions of the water molecule with respect to the other system, with the water molecule with a certain orientation (say, with the Hydrogen atoms pointing away from the interacting system), and then repeating the calculation at each point after having flipped the water molecule (which would now be oriented with the Hydrogen atoms pointing towards the interacting system). This procedure will reduce the risk of a bias in the results (see below).

The calculations yielded, therefore, 10,000 values of interaction energies, which are to be used for the determination of the expansion coefficients, $c_{ab}^{(n)}$. This determination poses a considerable problem, which may be simplified by the consideration of the atom classes (see Chapter 3). Altogether there are 30 atom classes: the Hydrogen and the Oxygen in water and the 28 atom classes in the amino acids, puric, and pyrimidinic bases. Within the atom-pair potential approximation, using Eq. (5.31) as a basis, one can then write the interaction energy between any of these systems and water as

$$\Delta E = \sum_{a(X)} \sum_{b(W)} \{c_{ab}^{(1)}/R_{ab} + c_{ab}^{(4)}/R_{ab}^4 + c_{ab}^{(6)}/R_{ab}^6 + c_{ab}^{(12)}/R_{ab}^{12}\} \tag{5.32}$$

where the summation over a (i.e., the atoms of the interacting system) is partitioned according to their atom classes (denoted by X) and the summation over the three atoms b of the water molecule is partitioned into a term for the Oxygen atom and two terms for the two Hydrogen atoms. The calculations were performed with Eq. (5.32) simplified by omission of the term in $1/R^4$ and, therefore, the 10,000 interaction energy values were to be fitted in terms of 56 parameters $c_{ab}^{(n)}$: that is, 56 parameters $c_{ab}^{(1)}$, 56 parameters $c_{ab}^{(6)}$, and 56 parameters $c_{ab}^{(12)}$. For example, for the amino Hydrogen atom of the amino acids one will obtain the coefficients $c_{H(1)H(W)}^{(1)}$, $c_{H(1)O(W)}^{(1)}$, $c_{H(1)H(W)}^{(6)}$, $c_{H(1)O(W)}^{(6)}$, $c_{H(1)H(W)}^{(12)}$, and $c_{H(1)O(W)}^{(12)}$, where the subscripts H(1) indicate a class 1-Hydrogen atom and H(W) and O(W) denote the water Hydrogen and Oxygen atoms. As the coefficients $c_{ab}^{(1)}$ will be related to the effective charges in the two atoms, it was found convenient to express them as

$$c_{ab}^{(1)} = k_{ab}\, q_a\, q_b$$

An inspection of the results obtained in the calculations brings out certain points of interest. First, it can be observed that the values of k_{ab} are very close to unity, which shows the Coulombic interaction may be well approximated in terms of the effective charges for the independent molecules. The values of $c_{ab}^{(6)}$ are extremely small, which is in agreement with the fact that SCF energies do not account for the dispersion energy. A rather interesting point is the range of values of the coefficients $c_{ab}^{(12)}$ for the various classes of a given type of atom, without any possibility of a theoretical justification. An approximate interpretation may be offered, however, as follows.

The fitting procedure tries to reproduce the values of the total interaction energies in terms of atom-pair interactions. This is a mathematical procedure, without any consideration of chemical characteristics. For an atom on the periphery of the interacting system, the fitting procedure will find that the water molecule may approach it rather closely and correspondingly the coefficient $c_{ab}^{(12)}$ will reflect, or will be related to, the actual size of the two interacting atoms. For an atom buried inside the interacting system, however, the fitting procedure will sense that the water molecule may not approach it so much, as it is repelled by the other atoms which surround the one under consideration. Therefore the fitting procedure will decide that such an atom has a larger size. These peculiarities do not affect the quality of the results to be obtained using these coefficients, as those relative situations will persist in actual calculations and the above comments are presented only in order to point out that no physical interpretation should be ascribed to those coefficients, which are simply the product of a fitting. It will be interesting, however, to see below how well they may be interpreted in terms of semiempirical parameters.

This set of parameters could be used only for the evaluation of the interaction energy of a system, built up from the ones considered above (say, a peptide) and water. Later on, when the parameters for the interaction between amino acids were determined, calculations could be performed also for the interaction between two peptides. It is interesting, however, because of the comments made above, to consider the transformation of the above parameters into a new set, which could be used for the evaluation of the interaction between atoms of the various classes.

5.3.2 Transformation of the Expansion Parameters

The expansion parameters obtained by Clementi are specific for pairs of atoms. Thus, in the example given above, the coefficients $c^{(n)}_{H(1)O(W)}$ are to be used exclusively for the evaluation of the interaction between an amino Hydrogen atom of an amino acid and the Oxygen of water. If one were able, however, to separate them into terms specific for the amino Hydrogen, $H(1)$, and the water Oxygen, $O(W)$, then the resulting coefficients could be used for the evaluation of the interaction between atoms of any of the classes under consideration. This separation may be achieved with the help of some semiempirical considerations. [See Section 11.2 for further comments on this factorization and its relevance for the fast evaluation of interaction energies.]

The procedure involves a new fitting, based on Eq. (5.31) (for a pair of atoms), with the first three coefficients given by

$$c^{(1)}_{ab} = q_a q_b$$

$$c^{(4)}_{ab} = -\frac{1}{2}(f_a \alpha_a q_b + f_b \alpha_b q_a)$$

$$c^{(6)}_{ab} = -\frac{3}{2}\frac{f_a \alpha_a f_b \alpha_b}{(f_a \alpha_a / n_a)^{1/2} + (f_b \alpha_b / n_b)^{1/2}}$$

where f_a and f_b are optimization factors to be determined. The remaining quantities have the same meaning as above and may be chosen as follows: The effective charges are the average values obtained by Clementi; the polarizabilities may be obtained by interpolation, according to the effective charge, from existing tabulations (Fraga *et al.* 1976, 1979; Fraga and Muszynska 1981); and the effective number of electrons may be defined as $n = Z-q$ (where Z is the nuclear charge and q is the effective charge).

The interaction energy, over a given separation interval, is evaluated for an atom of a given class and, say, the Hydrogen of water. These results are then fitted with an expression of the form of Eq. (5.31), with the coefficients $c_{ab}^{(1)}$, $c_{ab}^{(4)}$, and $c_{ab}^{(6)}$ defined as above, so that the coefficient $c_{ab}^{(12)}$ may be determined by a least-squares procedure. The calculations are then repeated with different values of the optimization parameters, f, until the best fitting (with a minimum absolute value of the maximum error) is obtained.

This procedure is repeated for every class of atoms, so that one obtains 28 new coefficients $c_{a(X)H(W)}^{(12)}$, for the interaction of each atom class with the Hydrogen of water. A similar set of calculations will then yield the corresponding values of $c_{a(X)O(W)}^{(12)}$, for the interaction with the Oxygen of water. The subscript (X) denotes any of the 28 atom classes.

The problem has not been solved yet, as those coefficients still correspond to pairs of atoms. A possible solution to this problem may be obtained through the approximation

$$c_{a(X)H(W)}^{(12)} = c_{a(X)}^{(12)} c_{H(W)}^{(12)}$$

$$c_{a(X)O(W)}^{(12)} = c_{a(X)}^{(12)} c_{O(W)}^{(12)}$$

whereby each coefficient is now expressed as the product of two coefficients, which are now ascribed to the individual atoms (Minicozzi and Bradley 1969). These expressions may then be used, within the context of the following considerations, for the determination of the parameters for the individual atoms:

(a) When considering the interaction between the hydroxylic Oxygen of the carboxyl group of an amino acid and the Oxygen of water, that is, O(9) and O(W), it may be assumed that both atom classes are, if not identical, at least very similar. Therefore one will have

$$c_{O(9)}^{(12)} \simeq c_{O(W)}^{(12)} \simeq \left[c_{O(9)O(W)}^{(12)} \right]^{1/2} = 552 \text{ kcal}^{1/2} \text{ Å}^6 \text{ mol}^{-1/2}$$

which compares rather well with the value 534 kcal$^{1/2}$ Å^6 mol$^{-1/2}$ obtained by Minicozzi and Bradley (1969) in optimization calculations which reproduced the experimental results for the dimer of the formic acid. Then, using the above value

$c_{O(W)}^{(12)}$, one can obtain $c_{O(10)}^{(12)} = 473$ kcal$^{1/2}$ Å^6 mol$^{-1/2}$ from the value of $c_{O(10)O(W)}^{(12)}$. Again, this value agrees satisfactorily with the value 496 kcal$^{1/2}$ Å^6 mol$^{-1/2}$ obtained by Minicozzi and Bradley (1969).

(b) Alternatively, using the value $c_{O(9)}^{(12)} = 534$ kcal$^{1/2}$ Å^6 mol$^{-1/2}$ (Minicozzi and Bradley 1969) one obtains, from $c_{O(9)O(W)}^{(12)}$, the value $c_{O(W)}^{(12)} = 570$ kcal$^{1/2}$ Å^6 mol$^{-1/2}$. Similarly, from $c_{O(10)O(W)}^{(12)}$, using $c_{O(10)}^{(12)} = 496$ kcal$^{1/2}$ Å^6 mol$^{-1/2}$ (Minicozzi and Bradley 1969), one obtains $c_{O(W)}^{(12)} = 526$ kcal$^{1/2}$ Å^6 mol$^{-1/2}$.

(c) Finally, taking into account the preceding results, one may use an average value $c_{O(W)}^{(12)} = 548$ kcal$^{1/2}$ Å^6 mol$^{-1/2}$, which will lead to the values $c_{O(9)}^{(12)} = 555$ kcal$^{1/2}$ Å^6 mol$^{-1/2}$ and $c_{O(10)}^{(12)} = 476$ kcal$^{1/2}$ Å^6 mol$^{-1/2}$, in better agreement with the results of Minicozzi and Bradley (1969).

(d) Then, using the above value for $c_{O(9)}^{(12)}$ one will obtain $c_{H(W)}^{(12)} = 5.47$ kcal$^{1/2}$ Å^6 mol$^{-1/2}$ from $c_{O(9)H(W)}^{(12)}$. Now one can proceed to the determination of all the remaining $c_{a(X)}^{(12)}$ parameters. [In this connection it must be mentioned that the calculations will yield two independent sets of parameters $c_{a(X)}^{(12)}$, depending on whether they are obtained from the coefficients $c_{a(X)O(W)}^{(12)}$ or from the coefficients $c_{a(X)H(W)}^{(12)}$. The values finally adopted are those, with a few exceptions, that give a minimum repulsion contribution at very short distances. See the work of Fraga (1982) for complete details.]

This procedure yields, therefore, a set of coefficients which may be used for the evaluation of the interaction between two atoms of any of the 28 classes originally considered. An important fact to be emphasized is that the results obtained, as discussed above, show a consistency between the semiempirical values of Minicozzi and Bradley (1969) and the theoretical results of Clementi, which is rather striking taking into account all the approximations involved in the complete procedure.

The coefficients obtained by this transformation suffer from the same basic defects as the original parameters of Clementi. In particular, one must mention that they will not account for dispersion contributions. The proper way in which one could proceed in order to include those contributions would be to repeat the whole procedure, starting from the original theoretical calculations of Clementi, but at a higher level of sophistication (say, within the context of a CI treatment). The cost of such calculations would be staggering and therefore it may be appropriate to use an alternative. For example, one might introduce the dispersion contribution through consideration of additional $1/R^6$ terms, with coefficients obtained independently (Hobza 1985).

As examples of the correctness of the coefficients one could mention their satisfactory application to the study of the growth mechanism of benzene clusters and crystalline benzene (Oikawa *et al.* 1985), the recognition of amino acids in solution (Seijo *et al.* 1986), the ion channels in Southern bean mosaic virus capsid (Silva *et al.* 1987), the adsorption of hydrocarbons on graphite (Sordo *et al.* 1990), and the evaluation of spectral shifts (Gorse and Pesquer 1994). [Additional details and improvements may be found in the work of Fraga (1983), Torrens *et al.* (1988), and Bidacovich *et al.* (1990).]

5.3.3 Additional Terms for Non-rigid Molecules

The above *PEF* (or an equivalent one, obtained by some other approach) may be used for the study of the interaction between two (or more) rigid molecules (including the case of solvation) as well as for the evaluation of the non-bonded interactions within a molecule, also considered as rigid.

In order to conform more closely, however, to physical reality it is necessary to consider the possibility of distortions in the molecular geometry. Those distortions may result, in particular, as a consequence of changes in the bond lengths, the bond angles, and the dihedral (torsion) angles. The terms to be included in the *PEF*, in order to account for the contributions due to these changes, are discussed below.

[In this discussion, the notation recommended by the IUPAC-IUB Commission on Biochemical Nomenclature is used: bond lengths, bond angles, and torsion angles are denoted by b, τ, and θ, respectively. In this context, the latter symbol should not be confused with the polar spherical coordinated used elsewhere in this Chapter. This notation has not been used consistently in the literature and the reader should get acquainted with the notation used by different authors.]

(a) Changes in bond lengths

The contribution to the potential energy, arising from changes in bond lengths, is usually approximated by harmonic oscillator terms,

$$\Delta E_b = \frac{1}{2} \sum_i K_{i(b)} (b_i - b_{i(O)})^2 \tag{5.33}$$

where the summation extends to all the bonds in the molecule, $b_{i(0)}$ denotes the reference length of the i-th bond, and $K_{I(b)}$ are appropriate parameters, usually determined from experimental information.

This term always makes a positive contribution to the potential energy; that is, if alone, it will always favor the standard bond length but changes may occur due to the effect of the other contributions in the *PEF*, particularly the non-bonded interactions.

The effect of this contribution may be controlled through an appropriate choice of the parameters $K_{i(b)}$. Thus, small values of the $K_{i(b)}$ will allow for appreciable changes in the bond lengths while large values of the $K_{i(b)}$ will result in stiff potential barriers which will restrict the changes in bond lengths.

[The out-of-plane torsions may be handled in an analogous way. See the work of Lifson (1982) for details.]

(b) Changes in bond angles

In a similar fashion, the changes in the potential energy due to changes in the bond angles are expressed as

$$\Delta E_\tau = \frac{1}{2} \sum_i K_{i(\tau)} (\tau_i - \tau_{i(0)})^2 \tag{5.34}$$

where the summation extends to all the bond angles, $\tau_{i(0)}$ denotes the i-th reference angle, and $K_{i(\tau)}$ are appropriate parameters, also to be determined from experimental information.

Because of the analogy between the expressions for ΔE_b and ΔE_τ, the comments made above for ΔE_b apply as well for ΔE_τ.

(c) Changes in dihedral (torsion) angles

The contribution to the potential energy due to changes in the dihedral angles reflects the existence of potential barriers to rotation around the corresponding central bond (associated with the dihedral angle). It may be expressed as

$$\Delta E_\theta = \frac{1}{2} \sum_i K_{i(\theta)} [1 - \cos n (\theta_i - \theta_{i(0)})] \tag{5.35}$$

where the summation extends to all the dihedral angles in the molecule, $\theta_{i(0)}$ denotes the reference value of the i-th dihedral angle, n is related to the periodicity of the potential barrier, and $K_{i(\theta)}$ is a parameter to be determined from experimental information. [See the work of Momany *et al.* (1975) for further details and the work of Gelin and Karplus (1975) and McCammon *et al.* (1979) for variants of this expression.]

As for bond lengths and bond angles, this contribution is always positive and its effect may again be controlled by an appropriate choice of the parameters $K_{i(\theta)}$. In this connection it should be mentioned, in particular, that variation of the peptide dihedral angle ω is usually restricted through the use of an appropriately large value of $K_{i(\theta)}$. In addition, for simplicity, usually the side-chain dihedral angles are kept unchanged, except for the first one.

The development in this Chapter has been focused on those points, which will provide an understanding of the most common approaches used at present. For that reason, certain topics [such as, e.g., the three-body interactions (Lifson 1982)] have been disregarded.

5.3.4 Other Interactions

A *PEF* consisting of Eqs. (5.32)-(5.35), with appropriate coefficients, should account for all the significant contributions to the potential energy and therefore predict correctly the possible existence of *H*-bonds and disulfide bridges.

The inclusion of the terms, Eqs. (5.33)-(5.35), needed in order to take into account the possibility of distortions of the molecular geometry, may result in the need of a parameterization of the *PEF*, in which case it may be possible to include in it additional terms. The two additional terms most often included are those that try to improve the probability of formation of *H*-bonds and disulfide bridges.

(a) *H*-bonding

The additional terms for *H*-bonding interactions are usually (McCammon *et al.* 1979) of the form

$$\Delta E_{HB} = \sum_a \sum_b \left\{ c_{ab(HB)}^{(10)} / R_{ab}^{10} + c_{ab(HB)}^{(12)} / R_{ab}^{12} \right\} \qquad (5.36)$$

where the summation extends only to those atoms capable of being involved in *H*-bonding. The terms in $1/R_{ab}^{10}$ are attractive while the terms in $1/R_{ab}^{12}$ are repulsive. This expression may be further modified by consideration of a directional character, corresponding to the geometrical peculiarity of a *H*-bond (interpreted in terms of the hybridization scheme usually assumed for the atoms involved in the bond). [The development of the *H*-bonding potential energy terms may be followed through the work of Lippincott and Schroeder (1955), Moulton and Kromhout (1956), Schroeder and Lippincott (1957), Scott and Scheraga (1966), Ooi *et al.* (1967), and McGuire *et al.* (1971).]

(b) Disulfide Bridges

Similarly one might reinforce the possibility of formation of disulfide bridges through inclusion of a term (Coghlan and Fraga 1985) of the form

$$\Delta E_{SS} = 215.0\{[1-e^{-2.64(R_{SS}-2.04)}]^2 - 1.0\}$$

(in kJ mol^{-1}), where R_{SS} denotes the separation between the two sulfur atoms that might be involved in the disulfide bridge. This term, which corresponds to a Morse potential (Herzberg 1957) could also be parameterized with consideration of a directional character.

5.3.5 The Use of a Dielectric Constant

The first term in Eq. (5.32), for the non-bonded interactions, corresponds to the Coulomb interaction between two charges and should, therefore, include (in the denominator) the dielectric constant of the medium. Consideration of such a constant will change the relative weights of the Coulomb term *versus* the other contributions (dispersion, induction, and repulsion) and affect the results of the simulation.

One must, therefore, view the need of the dielectric constant as another difficulty to be overcome, insofar as to what is the appropriate value to be used (Harvey 1989). A way out of this dilemma consists of repeating the calculations with different values of the dielectric constant and in this regard it may be illustrative to mention the argument of McKelvey *et al.* (1991), in the study of the pentapeptide Tyr-Ile-Gly-Ser-Arg from the $B1$ chain of laminin, a major component of the basement membrane. Taking into account that simple esters of carboxylic acids have dielectric constants in the range 3-7 (e.g., methyl oleate, 3.2; dibutyl sebacate, 4.5; ethyl acetate, 6.0) and given the lipid nature of all membranes, McKelvey *et al.* (1991) performed the calculations with dielectric constant values of 1 (for comparison purposes), 4, and 10 (a value which already gives a reduction of 90% in the Coulomb contribution).

A positive aspect of the consideration of the dielectric constant is that, when properly chosen, it may help in simulating the effect of the solvent (Zimmerman 1985).

5.4   Summary

In this Chapter we have attempted to offer a theoretical justification for the general form of the *PEF* most often used in computer simulations of protein structures and interactions within the framework of Molecular Mechanics. The justification is rather reasonable for the non-bonded interactions but even in that case the approximations and possible sources of error have become quite evident. Accounting for the possibility of distortion of the molecular geometry is done in an *ad hoc* manner and parameterization of the *PEF* becomes a prerequisite.

For simplicity, no attempt has been made to analyze the distinctions between the most important *PEF* in use, which differ only in the manner (theoretical, semiempirical, or combination thereof) in which their parameters are determined. That summary will be presented in Chapter 7, with additional comments in other Chapters, as appropriate.

6 Molecular Mechanics: Computer Simulations

Once the potential energy function (*PEF*), as a whole, has been chosen, one may proceed to its practical application in a computer simulation. At this point new decisions must be made, based on additional considerations regarding the purpose of the calculations, the computing facilities available and, consequently, the computing time required for the simulations. These considerations are not independent of each other and the final approach adopted may imply the inclusion of additional approximations, as explained below.

6.1 General Considerations

The two main factors with a bearing on the approach to be adopted for the computer simulation are the temperature (T) and the time (t) dependence. In order to be completely safe, insofar as a comparison with experimental results is concerned, one should proceed, without doubt, with consideration of both dependences. Such is the approach embodied in the Molecular Dynamics (*MD*) treatments.

A *MD* simulation should, when using a satisfactory *PEF*, yield correct results and provide a prediction/interpretation of both the static and dynamic properties of the system under study, as well as of their *T*-dependence. The only problem, of a practical nature, of any such a simulation is that it may be extremely expensive in terms of computing time, especially for large peptide chains.

New approximations may then be introduced in order to reduce those computing-time requirements. The first approximation that comes to mind is to disregard the *t*-dependence, which will result in an appreciable savings of computing time. The corresponding technique, denoted as Monte Carlo (*MC*), will then be suitable for the study of the *T*-dependence of static properties of proteins. Although the distinction will become clear later in this Chapter, it is convenient, however, to offer some comments at this point regarding the use of the designation Monte Carlo given to this procedure. As discussed in Chapter 4, the original *MC* method was developed for the simulation of a canonical ensemble and as such it would presuppose the availability of an appropriate structure for the protein under study, such a structure remaining unchanged during the *MC* simulation. Within the context of Molecular Mechanics, however, the designation *MC* is used to denote a different approach, concerned exclusively with the optimization of the protein structure. The simulation is applied, not to a collection of protein molecules, but rather to the particles which constitute the protein. Another difference is that the complete trajectory (after equilibration) is required in the original *MC* method while the goal of the Molecular Mechanics simulation is the conformation with the lowest energy. What remains unchanged is the algorithm and, particularly, the use of the Metropolis test.

Even so, *MC* calculations may still be rather expensive and therefore, at a still lower level of approximation, one could decide to remove also the *T*-dependence. The results to be obtained in a *T*- and *t*-independent simulation will represent only a very rough representation of the true characteristics of the system under study. The simple energy minimization procedures are to be placed, consequently, at the low end of the scale of computer simulations. Their usefulness, however, is not to be discarded lightly, but one should be aware of their limitations. Energy minimization simulations have at least, especially in the absence of the corresponding crystallographic data, a very important application: They serve as an auxiliary tool, their results serving as a starting point for both *MC* and *MD* simulations.

Energy minimization calculations, with a complete *PEF*, may still be rather expensive. For most purposes, once the *T*- and *t*-dependence has already been removed, it may not be necessary to include all the terms in the *PEF*, as presented in the preceding Chapter. Thus, it is customary to omit the terms corresponding to the changes in bond lengths and bond angles as well as the terms for the out-of-plane distortions, keeping only the terms corresponding to the non-bonded interactions as well as those for the changes in the dihedral (torsion) angles (and the *H*-bonding terms, if originally present in the *PEF*). These simulations are denoted as constrained, in the sense that the molecular geometry is maintained (except for the variation of the torsion angles).

Therefore, in order of increasing sophistication and correctness, the various possible approaches are constrained energy minimizations, unconstrained energy minimization (usually not considered), *MC* simulations, and *MD* modeling. There are still other possibilities, such as constrained *MC* and *MD* simulations, but they should not be included in the above scale, because imposing certain constraints may have additional implications regarding the character of the results, as discussed below.

6.2 Energy Minimization

Conceptually, an energy minimization procedure is extremely simple: the simulation strives to find the geometric conformation, with lowest energy, for the system under study. That system may be either an isolated peptide structure or a molecular association and the discussion is best presented if these two cases are considered separately.

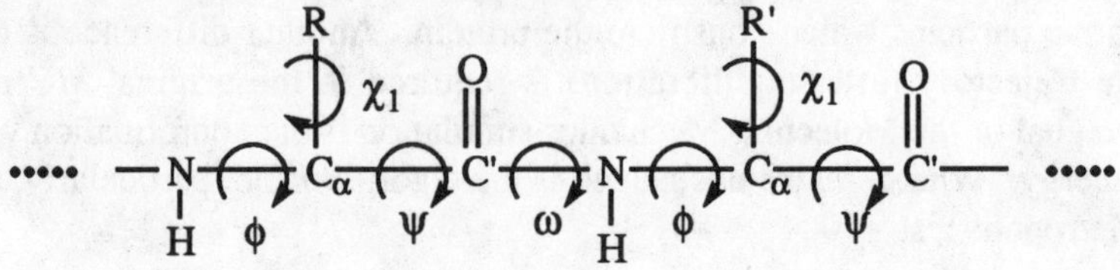

Figure 6.1   Schematic representation of part of a peptide chain showing the dihedral angles to be varied during an optimization procedure. R and R' denote side chains, with additional torsion angles, which may also be optimized.

6.2.1 Modeling of Isolated Peptide Structures

The simulation may be performed in a brute force manner: that is, by variation of the torsion angles until no change in the energy is observed. Those torsion angles are the ϕ, ψ, and χ angles of all the residues in the peptide and the ω angles for the peptide linkages (see Fig. 6.1). The large number of degrees of freedom for large peptide chains has two rather serious implications.

On one hand, the energy hypersurface (giving the dependence of the energy on the dihedral angles) may be extremely complicated, with a considerable number of local minima in addition to a global minimum. The consequence is that the calculation may yield a conformation corresponding to a local minimum instead of the conformation corresponding to the global minimum. The starting geometry will have a crucial bearing on the final result and therefore it will be appropriate to repeat the calculations for different initial geometries. In addition, once a minimum energy has been reached it will be necessary to subject the resulting geometry to an abrupt, random change, after which the simulation is restarted; hopefully, this procedure may allow the conformation to escape from a local minimum, but there will be no certainty that such a goal has been achieved.

This problem may be somewhat reduced, but not much, by a reduction in the number of degrees of freedom to be considered. The usual candidates are the ω angles and the χ angles (except for χ_1). The ω angles may be maintained fixed at 180°, corresponding to the *trans* conformation of the O-C'-N-H atoms of the peptide linkage. [An exception to this rule is to be found in the case of proline, for which the values 0° and 180° are possible.] The χ angles, if desired, may be optimized later, as a fine-tuning operation, once the main features of the conformation (as determined by the backbone) have been established. A far more appreciable reduction in the number of degrees of freedom may be achieved through the use of fragments with fixed geometry, as discussed in Section 10.3.3.

The number of degrees of freedom has also a direct bearing on the length of the calculations. Each time a torsion angle is varied, it is necessary to evaluate the energy corresponding to the resulting conformation. The evaluation of the non-bonded interaction involves a double summation (see Chapter 5): in the present case one of the summations extends to all the atoms on the left of the bond around which the rotation is being performed, while the second summation extends to all the atoms on the right of that bond. For a large peptide, the cost of the calculation may be staggering, which is one of the reasons for the use of 1/R-expansions for the non-bonded interactions.

Energy minimization calculations constitute the basic tool for the *de novo* computer-synthesis of the peptide structure with a given amino acid sequence. But before we proceed to summarize the corresponding procedure it is convenient to discuss the relaxation of crystallographic structures.

Given the availability of the crystal structure of the peptide to be studied, one could proceed directly with a *MC* or a *MD* simulation. One could, however, perform first an energy minimization of that structure, as a test of the correctness of the *PEF* adopted for the calculations. Such a procedure should yield a relaxed structure, which should still maintain most of the characteristics of the original crystal structure. This relaxation results from the difference in the conditions present in the crystal structure and those in the computer simulation: the crystal structure reflects the equilibrium between the internal forces and the interactions with neighboring molecules while the energy minimization is carried out for an isolated molecule. For large, globular proteins the internal interactions should make the predominant contribution and therefore the energy minimization should produce only a certain relaxation of the structure. A strong distortion of the structure, with elimination of some of its important features, would indicate a poor *PEF*.

The procedure for the *de novo* synthesis of a protein structure, corresponding to a given amino acid sequence, is formally simple. It consists of two basic steps, requiring as input data the structures (i.e., the Cartesian coordinates) of the individual amino acids. These three points (input data, preliminary build-up of the peptide chain, and final energy minimization) are discussed separately below:

(a) Input data

The peptide chain may be built-up from *L*- or *D*-amino acids. Most of the simulations have been performed with naturally-occurring amino acids but studies with non-natural amino acids are now proliferating, in order to design new peptides with new/enhanced characteristics. In any case, experimental data (e.g., crystallographic data; see Appendix 2.1) are the best source for the corresponding Cartesian coordinates. Some comments are needed in this connection.

On one hand, attention must be given to the fact that, within the context of the approximation of linear combination of atomic orbitals, the Nitrogen atom of the terminal amino groups is supposed to be in an sp^3 hybridization state in the free amino acid and in an sp^2 hybridization state when involved in the peptide bond. Consequently, the coordinates of the Hydrogen atoms of the terminal amino group should be transformed accordingly.

Depending on the conditions of the simulation, it may be necessary to introduce some additional changes in the original (crystallographic) coordinates, as well as in the corresponding effective charges. The crystallographic coordinates, and the effective charges discussed in Chapter 3, are appropriate for the modeling of protein structures in vacuum. However, when the purpose of the simulation is to model a peptide chain, say, in aqueous solution, it is necessary to take into account the pH of the (simulated) solution. For calculations at a neutral pH, the carboxylic groups in the side chains of the aspartic and glutamic acids will be assumed to be ionized while the N atoms in the side chains of arginine and lysine will be protonated. The effective charges may be changed as follows: For the ionized aspartic and glutamic acids, the same effective charge may be ascribed to both Oxygen atoms in the side chain carboxylic group; the corresponding value may be obtained by adding -0.5 to the average value of their effective charges in the neutral forms. For the protonated residues, the excess of positive charge may be distributed equally among all the atoms in the protonated group (assigning the same charge to the Hydrogen atoms).

When the possibility of disulfide bridge formation exists, the input data for the cysteine residues will be obtained by removal of the Hydrogen atom in the HS-group, with an appropriate renormalization of the effective charges.

(b) Preliminary build-up of the peptide chain

The peptide chain is built-up by adding one amino acid at a time and forming the corresponding peptide linkage. The sequence of steps may be as follows:

The OH-group of the terminal carboxylic group of the first residue and one of the Hydrogen atoms of the terminal amino group of the second residue are removed.

The coordinates of the atoms of the second residue are transformed so that its terminal Nitrogen atom will be at the appropriate bond distance from the terminal Carbon atom of the first residue (say, at 1.34 Å, which is the standard bond length of a peptide bond) and so that the atoms O-C'-N-H of the peptide bond are on the same plane. [Having transformed appropriately, as discussed above, the coordinates of the terminal amino group, the approach of the two residues is performed so that the directions of the original bonds C'-OH and H-NH coincide.]

If the *PEF* to be used in the calculations makes use of the effective charges on the atoms, at this point it is necessary to proceed with a renormalization of those charges. The effective charge of the OH-group, removed from the first residue, is not equal (in absolute value) to the charge of the Hydrogen atom, removed from the second residue; therefore, if such a renormalization is not carried out, the dipeptide just formed will have a net effective charge (different from zero), a situation which would affect severely the simulated folding process. The manner in which the renormalization is performed will also have a bearing on the results of the energy minimization procedure. The procedure described in Chapter 3 (Section 3.7.2) will not produce major changes in the effective charges, as it distributes the charge change over the complete dipeptide.

At this point, and if desired, one can proceed to a (partial) optimization of the structure just obtained, but a word of caution must be issued in this connection. In general, it may be more appropriate to only allow the structure to relax slightly, without trying to optimize it fully. A complete optimization may lead to difficulties at later stages in the build-up procedure, if the chain has evolved prematurely towards a globular structure. As already mentioned above, this relaxation may be carried out for all the dihedral angles in the dipeptide or just, say, for ϕ, χ_1, and ψ torsion angles.

The complete procedure, just summarized, is now repeated for the dipeptide formed and the third residue in the sequence, and so on for all the amino acids, until the complete chain has been constructed.

(c) Protection of the termini

In the modeling of proteins, one can then proceed with the full optimization of the preliminary structure obtained in (b) above. It may be, however, that the purpose of the simulation is to model a (short) peptide, which has been (or will be) studied experimentally in solution. Usually, such experimental studies (e.g., by NMR) are performed on peptides, whose termini have been protected in order to eliminate the possible interference due to the protonated terminal amino group or the ionized terminal carboxylic group.

Therefore it will be necessary to protect the terminal groups in the computer simulation. If possible (i.e., if the corresponding Cartesian coordinates and all other needed data are available) one should use the same protecting groups as in the experimental study but, if such is not the case, any protecting group will at least reduce the effect of the terminal groups as well as introduce part of the corresponding steric hindrance. Possible protecting groups are

$$H_3C-C\overset{O}{\diagdown} \qquad H_3C-\overset{CH_3}{\underset{CH_3}{\overset{|}{\underset{|}{C}}}}-O-C\overset{O}{\diagdown}$$

$$-HN-CH_3 \qquad -HN-\overset{CH_3}{\underset{CH_3}{\overset{|}{\underset{|}{C}}}}-CH_3$$

The addition of the protecting groups is carried out as described above in (b).

(d) Optimization of the peptide structure

Optimization, by energy minimization, may now be performed, either for all the torsion angles or in a constrained manner (with variation only of the ϕ and ψ angles of the backbone and the first side-chain angles χ_1), with full optimization later on once a minimum has been reached. In either case, the optimization procedure is full of risks, as the starting point and the path followed will have a significant influence on the final result.

In a brute force minimization it may be worthwhile to try all the possibilities that come to mind. For example, one could proceed in a straight fashion, starting the optimization from the H_2N-terminus and proceeding towards the HOOC-terminus, or *vice versa*. In an alternate procedure one could proceed simultaneously from both termini or start from the middle of the chain. In all cases a bias is introduced and that is why it is considered advisable to introduce a random distortion once a minimum has been reached and then restart the optimization. Alternatively, one could perform the optimization through random variation of the torsion angles (see below). Some of these difficulties disappear when the peptide chain is constructed from fragments, whose geometry is maintained fixed throughout the optimization procedure. This variant is discussed in detail in Chapter 10.

The procedure for the determination of energy minima may be rationalized and speeded up. There is an extensive literature [see, e.g., the work of Fraga *et al.* (1978)], as well as a wealth of corresponding software packages, on a variety of methods for the determination of minima and therefore there is no need of repeating here the corresponding formulation. In order to illustrate the general characteristics of the problem, it will be sufficient to discuss some of the simple procedures (e.g., the steepest-descent and the Newton-Raphson methods), which may be sufficient for most applications.

Let us consider a constrained minimization procedure, that is, with fixed bond lengths and bond angles. The *PEF* will consist of the non-bonded interaction terms and the contributions from the changes in the torsion angles. We will consider the rotation around a given bond and denote by A that part of the molecule on the left of that bond and by B the part of the molecule on the right of that bond; A will remain fixed in space while B moves under the rotation. The coordinates of the atoms in the complete system are referred to a system fixed in space (e.g., with its origin at one of the Hydrogen atoms of the *N*-terminus) but in addition it is convenient to consider another system of coordinates with its origin at the right-hand atom of the bond around which the rotation is being performed. The relative position of the second system of coordinates with respect to the first one is defined by the polar coordinates R, ϕ, θ (see Fig. 6.2).

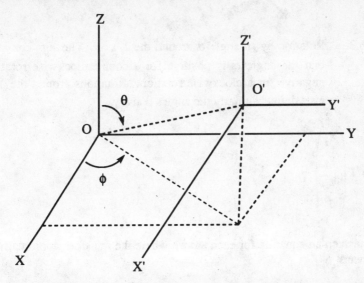

Figure 6.2.   Relative position of the two systems of Cartesian coordinates, (XYZ) and (X'Y'Z'), associated with the interacting systems A and B. The origins of the two systems, O and O', are at a distance R.

When the rotation is performed, the coordinates of the atoms of B in the coordinate system (XYZ) change correspondingly. The transformation of the coordinates may be expressed very easily in the following manner: We will assume that the coordinates of the atoms of B remain fixed in the coordinate system (X'Y'Z'), which undergoes the transformation due to the rotation. The net result of the transformation may then be reproduced through the following procedure: (a) bring the system (X'Y'Z') into coincidence with the system (XYZ) by a translation R; (b) rotate system (X'Y'Z') by an angle α about the X-axis (Fig. 6.3), then by an angle β about the Y-axis, and finally by an angle γ about the Z-axis; (c) translate the system (X'Y'Z') from O to O'. The new coordinates of the atoms of B, with respect to system (XYZ) are then given by

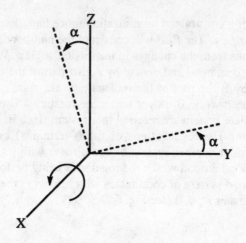

Figure 6.3. Rotation by an angle α around the X axis. The sign convention is that the angle α is positive for a counterclockwise rotation and negative for a clockwise rotation. Rotations around the Y and Z axes define similarly the angles β and γ.

$$\begin{pmatrix} x'_i \\ y'_i \\ z'_i \\ 1 \end{pmatrix} = T \begin{pmatrix} x_i \\ y_i \\ z_i \\ 1 \end{pmatrix}$$

(with one such an equation for each atom), where the transformation matrix T has the elements

$T_{11} = \cos \gamma \cos \beta$

$T_{12} = \cos \gamma \sin \beta \sin \alpha - \sin \gamma \cos \alpha$

$T_{13} = \cos \gamma \sin \beta \cos \alpha + \sin \gamma \sin \alpha$

$T_{14} = R \sin \theta \cos \phi$

$T_{21} = \sin \gamma \cos \beta$

$T_{22} = \sin \gamma \sin \beta \sin \alpha + \cos \gamma \cos \alpha$

$T_{23} = \sin \gamma \sin \beta \cos \alpha - \cos \gamma \sin \alpha$

$T_{24} = R \sin \theta \sin \phi$

$T_{31} = - \sin \beta$

$T_{32} = \cos \beta \sin \alpha$

$T_{33} = \cos \beta \cos \alpha$

$T_{34} = R \cos \theta$

$T_{41} = 0$

$T_{42} = 0$

$T_{43} = 0$

$T_{44} = 1$

Once the new coordinates of the atoms of B, resulting from the rotation, have been obtained one can then evaluate the new interatomic distances R_{ab} and, consequently, the non-bonded interactions. That is, the non-bonded interactions are a function of the coordinates of the atoms of B, which are the variables of the problem. If the function ΔE exists in a closed region, all its partial derivatives (to order n) are continuous in that region, and the first (n+1) partial derivatives exist in the open region, then it is possible to express ΔE by a Taylor expansion. For example, illustrating the formulation for a single atom, one would write

$$\Delta E(x'_i, y'_i, z'_i) \equiv \Delta E(x_i + \Delta x_i, y_i + \Delta y_i, z_i + \Delta z_i)$$

$$= \Delta E_\theta + \Delta E(x_i, y_i, z_i) + \frac{\partial \Delta E}{\partial x_i} \Delta x_i + \frac{\partial \Delta E}{\partial y_i} \Delta y_i + \frac{\partial \Delta E}{\partial z_i} \Delta z_i$$

$$+ \frac{1}{2} \left\{ \frac{\partial^2 \Delta E}{\partial x_i^2} (\Delta x_i)^2 + \frac{\partial^2 \Delta E}{\partial x_i \partial y_i} \Delta x_i \Delta y_i + \frac{\partial^2 \Delta E}{\partial x_i \partial z_i} \Delta x_i \Delta z_i + ... \right\}$$

+ higher terms

where ΔE_θ represents the direct contribution from the change in the torsion angle [see Chapter 5; θ is the subscript denoting the change in the dihedral angle and should not be confused with the polar coordinate used above in order to refer to the position of (X'Y'Z') to (XYZ)].

Denoting by x the column vector of coordinates $(x_1, y_1, z_1, x_2, y_2, z_2, ..., z_n)$, where n denotes the number of atoms in B, the above expression may be generalized (to second order) as

$$\Delta E(x + \Delta x) = \Delta E_\theta + \Delta E(x) + \Delta x^\dagger \cdot G(x) + (\Delta x^\dagger \cdot H(x)) \cdot \Delta x \qquad (6.1)$$

where $G(x)$ is the gradient vector of first derivatives and $H(x)$ is the Hessian matrix of second derivatives.

This equation may now be used for the automatic search of the minimum in the energy. The two basic methods (steepest descent and Newton-Raphson) differ on whether only the first-order term or both the first- as well as the second-order terms are considered. The computation, using matrices, is efficient but the real bottleneck (especially for large proteins) is to be found in the evaluation of the matrices of derivatives, particularly the Hessian matrix; the algorithm of Noguti and Go (1983), however, reduces the corresponding number of necessary operations from n^4 to n^2, where n is the number of torsion angles.

The steepest-descent method assumes linear behavior. The change in the interaction energy is taken to be given by

$$\Delta(\Delta E(x)) = \Delta E(x + \Delta x) - \Delta E(x) = \Delta E_\theta + \Delta x^\dagger \cdot G(x) = \Delta E_\theta + \Delta x G(x) \cos \omega \qquad (6.2)$$

where Δx and $G(x)$ denote the magnitudes of the vectors Δx and $G(x)$. For a given change Δx, the maximum change (in absolute value) in the interaction energy corresponds to $\cos \omega = -1$, which defines the direction of steepest descent (opposite to that of the gradient).

The procedure will then consist of the following steps. Starting at a given point (i.e., a given conformation), one evaluates the gradient, system B is transformed (by the appropriate rotations) in the direction of steepest descent, the gradient is evaluated at the new position, etc., until $\Delta(\Delta E(x))$ (in absolute value) becomes smaller than a chosen threshold (depending on the accuracy needed). This method is only approximately valid near the minimum and suffers, in addition, of slow convergence.

In the Newton-Raphson method the interaction energy is approximated as

$$\Delta(\Delta E(x)) = \Delta E_\theta + \Delta x^\dagger \cdot G(x) + (\Delta x^\dagger \cdot H(x)) \cdot x$$

$$= \Delta E_\theta + \Delta x^\dagger \cdot (G(x) + H(x) \cdot \Delta x) \qquad (6.3)$$

Disregarding the ΔE_θ contribution, one can see that if $x + \Delta x$ corresponds to the minimum, that is, $\Delta(\Delta E(x)) = 0$, then it must be

$$\Delta x^\dagger \cdot (G(x) + H(x) \cdot \Delta x) = 0$$

or, equivalently,

$$G(x) + H(x) \cdot \Delta x = 0$$

If $(H(x))^{-1}$ exists, then

$$(H(x))^{-1} \cdot G(x) + (H(x))^{-1} \cdot (H(x) \cdot \Delta x) = (H(x))^{-1} \cdot G(x) + \Delta x = 0$$

so that

$$\Delta x = -(H(x))^{-1} \cdot G(x) \qquad (6.4)$$

That is, the position of the minimum could be obtained in a single step.

This derivation will hold only (in the absence of ΔE_θ) if the energy hypersurface may be approximated by the expansion to second order (which will be true, as a rule, if the initial point is close to the minimum). Therefore, in the real case, it is necessary to follow an iterative procedure.

Given a starting point, one evaluates the gradient vector, $(H(x))^{-1}$, and Δx, according to Eq. (6.4). $\Delta(\Delta E(x))$, evaluated from Eq. (6.3), will not be zero, because of the approximation mentioned above and the existence of ΔE_θ, but the new point $x+\Delta x$ may be closer to the minimum (unless the initial guess was extremely poor). The procedure is restarted at that new point and the calculations are continued until the minimum has been reached (to the desired degree of accuracy). As the calculation proceeds towards the minimum, the contribution of ΔE_θ will become smaller and smaller.

The difficulties associated with a poor starting point are not eliminated in the automatic search for the minimum, as Fig. 6.4 illustrates in a schematic manner. This is called the multiple-minima problem (see Section 10.2).

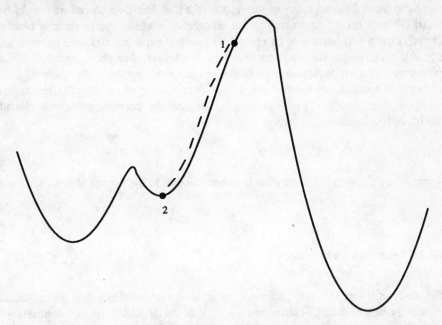

Figure 6.4   Schematic representation of a simplified energy hypersurface illustrating the difficulties in reaching the global minimum. Starting from point 1, a normal minimization procedure (e.g., by the steepest-descent method) will lead to point 2, which is a local minimum.

(e) Results

The build-up/optimization procedures yield the Cartesian coordinates of the final (optimized) structure. For a protein, with an appreciable number of residues (and, therefore, a considerable number of atoms), the analysis of the results should start with a graphic representation.

Numerical information, however, may also be easily obtained, regarding specific features of the peptide chain, through a geometric analysis of the Cartesian coordinates. For example, evaluation of the distances between the C_α atoms of the residues and the center of mass of the protein will identify the residues lying on its surface, thus pointing out where surface turns are found. In addition, the distances between pertinent atoms (say, C_α atoms, atoms in charged groups, heteroatoms that might be involved in H-bonding, S atoms, etc.) will point out the possible existence of α-helices, β-chains, and turns as well as salt-bridges, H-bonds, and disulfide bridges. [In this connection the reader is referred to the work of Rose *et al.* (1985), and references therein, for a summary of the working definitions used for the identification of turns.]

Judgment of the accuracy of the results of a minimization procedure should be based (though taking into account the comments made at the beginning of this Chapter) on a comparison with the corresponding experimental results, such as, e.g., the crystallographic structure (for large proteins). For short peptides, *NMR* experiments constitute an excellent source of comparative values. For example, the accuracy of a simulation may be ascertained from a comparison of the experimental values of the $d_{NN}(i,i+1)$ *nOe* connectivities and the corresponding theoretical results, approximated as

$$d_{NN}(i,i+1) = 10^3 \, (1.75/R_{i,i+1})^6 \tag{6.5}$$

where $R_{i,i+1}$ denotes the separation between the amido hydrogens of residues i and $i+1$.

6.2.2 Molecular Associations

The study of molecular associations, through an energy minimization procedure, is essentially free of the difficulties associated with the *de novo* computer synthesis of peptide structures. Given two interacting systems, the problem consists of finding the relative position (separation and orientation) of the two systems, which is characterized by the lowest energy.

The comments and derivations presented in the preceding section apply in this case, with the difference that (except as discussed below) the *PEF* consists only of non-bonded interactions and that system B (which, in this case, is a separate molecule) undergoes both rotations and translations. Therefore, here it will suffice to comment on the nature of the possible interacting systems as well as on the variants of the manner in which the molecular association may be studied.

Being interested in protein structures, it is evident that at least one of the interacting systems must be a peptide chain. The other system may be another peptide, an organic/inorganic molecule, a metallic cation, a simulated surface, etc.

The only restriction will arise from the lack of the corresponding data, such as coordinates, effective charges, expansion parameters for the *PEF*, etc. The last point is especially important in the case of the interaction of a peptide with a metallic cation for which, as a rule, the corresponding $c^{(12)}$ coefficient will not be available; the solution is to be found in a preliminary parameterization, with comparison with experimental results.

In the simplest simulation procedure, the two interacting systems will be let to approach each other, with rigid geometry, until a stable association has been formed. In a more sophisticated simulation, in better correspondence with reality, each interacting system should be allowed (if pertinent) to undergo whatever structural changes may be induced by the presence of the other system. Naturally, this procedure will be more expensive in terms of computing time but the results will be more rewarding. A properly performed docking process should yield valuable information on, say, the active centers of peptide chains.

The only possible source of conflict, which in fact may be easily corrected, is to be found in the selection of the starting positions of the two systems, as they will predetermine somewhat the final association. Therefore, the calculations should be repeated for a number of different starting positions, covering as much as possible all possible directions of approach. The most stable of all the resulting associations may then be accepted as the one representing the true molecular association.

Solvation constitutes a very important case of molecular associations, considering that in biological systems as well as in industrial processes, proteins are found in solution. Solutions are usually aqueous, but the practical details discussed below apply as well to any other solvent or to mixtures of solvents (such as, for example, the water/methanol solutions used in *NMR* experiments on short peptides).

The modeling of a solvated peptide/protein should be performed by either a *MC* or a *MD* simulation (see below). A *T*- and *t*-independent simulation, however, may yield preliminary results, which may provide useful information, such as, e.g., a prediction of the hydrophilic/hydrophobic character of the surface regions of the protein. Furthermore, the results thus obtained for a solvated peptide may be taken as starting point for a subsequent *MC* or *MD* simulation.

The simplest study will involve a single water molecule, in two different modalities. For example, one can move the water molecule over the surface of the protein, thus obtaining its contour, which yields information on the accessibility of the various regions of the surface (see Chapter 8) as well as an outline of the shape of the protein (which may prove useful in the design of drugs). This procedure does not involve energy minimization, being simply based on the contact of the van der Waals spheres.

On the other hand, one can proceed with the prediction of (all) the possible stable molecular associations of the peptide and a single water molecule, as described above in general terms. The ordering of these associations, according to increasing energies (i.e., from more to less stable), will already give a feeling for the hydrophilicity/hydrophobicity of the surface regions of the protein.

It is also a very simple matter to create a solvation shell around a peptide through the successive addition of solvent molecules. Starting with the unsolvated peptide, one first adds one solvent molecule, as described above; that is, one forms the most stable peptide-solvent molecule association. The process is then repeated for the addition of one more solvent molecule to the association just obtained, and so on, until the desired number of solvent molecules has been incorporated into the association. Usually, the geometry of the peptide is maintained fixed throughout this process. Although the simulation is simple, it may offer one difficulty. As the solvation shell grows, it may happen that the solvent molecules will tend to cluster together, yielding a rather biased picture of the solvated peptide. That is, one is facing a competition between the solvent-solvent and solvent-peptide interactions and therefore care must be exercised in order to ensure a homogeneous growth of the solvation shell. Although the bias may still be present, its effects will be diminished when an appreciable number of solvent molecules is added and the peptide is completely surrounded by solvent molecules.

When such a shell has been obtained, a spatial analysis of the distribution of the solvent molecules may show, depending on the particular peptide, some interesting features. In the simulation of an aqueous environment, the solvation shell will be found to have approached closer to the hydrophilic surface regions of the protein than to the hydrophobic regions; a *bubble* (that is, an empty space) will be observed between the solvent shell and a hydrophobic surface region (Seijo *et al.* 1986).

There is no foolproof procedure by which this simple simulation may be improved, although it is possible to think of a variety of approaches. One such procedure could be as follows. One proceeds first with the simulation of the association of the peptide and one water molecule, as described above, but allowing for the optimization of the peptide structure under the interaction with the water molecule. Once the stable association has been obtained, the water molecule is incorporated into the structure of the peptide, as a component of the side chain of the nearest residue. The procedure is then repeated with this solvated peptide and a new water molecule, and so on, until all the water molecules have been added; the procedure should be limited to a small number of water molecules, say, one per residue, as if one were in fact just creating the first solvation shell. The rationale behind such a procedure is that the first solvation shell will be tightly bound to the peptide, forming a single structural unit, so that the water molecules will follow the side chains in their motion towards an optimized structure. When, for whatever reason, *MC* or *MD* simulations are not feasible, this simple approach may be used in order to obtain information on the stability of internal *H*-bonds and salt-bridges in short peptides in solution (Thornton and Fraga 1991, Fraga and Thornton 1993).

6.3 Monte Carlo Simulations

The energy minimization procedures suffer from the neglect of the T-dependence, which should be considered for a more realistic simulation. The Monte Carlo (MC) technique constitutes an appropriate tool by which such a dependence may be introduced.

6.3.1 The Basic Approach

The MC method has been presented above (Chapter 4) as a technique for the simulation of the canonical ensemble description of a collection of particles. The MC procedure satisfies the ergodicity requirement (that any local minimum may be reached from any other after a finite number of sampling steps) and generates a Markov walk through the energy hypersurface, with Boltzmann transition probabilities.

Within the framework of a Molecular Mechanics simulation, the MC procedure is not applied to a collection of peptide chains but rather to a collection of particles, which are the atoms of the peptide chain under study. Their possible displacements are dependent on the PEF used in the calculations, with the usual controversy regarding the use of unconstrained/constrained geometry. The calculations are faster and more economical if the displacements are generated by changes in the dihedral angles but it is more realistic if changes are allowed in the bond angles and bond lengths (with the possibility of a compromise, as described below).

The dihedral angles to be varied must be chosen at random, as discussed in general in Chapter 4, and two points must be emphasized in this connection. On one hand, simultaneous random changes in more flexible dihedral angles may be required in order to escape from local minima, which are energetically favored (Caflisch *et al.* 1992). The second comment pertains to the random selection of the torsion angles, which is a basic requirement of the MC technique, and close attention should be paid to the random number generator used. It may be that the random number generator available in a computing facility is rather poor, with the possibility of leading to disastrous results. As the subject of random number generators is outside the scope of this work, the reader is referred to the work of Park and Miller (1988), and references therein, for an understanding of what constitutes the acceptable minimal standard and its possible improvement. Here it will suffice to mention that the minimal standard random number generator is a multiplicative linear congruential generator with multiplier 16807 and prime modulus $2^{31}-1$ (Park and Miller 1988).

A fundamental defect of the *MC* procedure, when applied in a straightforward manner, arises from the large number of degrees of freedom of the peptide chain, the consequent large number of local minima, and the existence of large potential energy barriers. As a consequence, it may happen that the sampling will be restricted to a limited region of the hypersurface, with a correspondingly biased description.

6.3.2 Monte Carlo Minimization with/without Thermalization/Annealing

Modifications may be introduced in order to correct this defect and, in fact, search for the global minimum, which is not the usual purpose of the *MC* technique when applied within the context of Statistical Mechanics.

The Monte Carlo Minimization (*MCM*) procedure, combining the *MC* technique with an energy minimization (Li and Sheraga 1987, 1988), is based on a sampling of minima instead of points in the complete energy hypersurface. It may be easily summarized as follows. A randomly generated conformation is subjected to an energy minimization procedure and the resulting conformation is used as starting point for the *MCM* iterative cycles. Each *MCM* cycle consists of a random change of (a) dihedral angles(s), an energy minimization procedure, and a *MC* test of the resulting minimum by comparison with the previously accepted local minimum.

This procedure may be further improved by addition of a thermalization step (Caflisch *et al.* 1992), appropriate in those cases when the conformation cannot escape, during a given number of cycles, from a set of local minima with similar energies. When thermalization is included, the procedure is performed as follows [where the specific values, given in parentheses, for the numbers of cycles, iterations, and T are those used by Caflisch *et al.* (1992)].

An arbitrarily chosen starting conformation is subjected to a number (50,000) of *MCM* cycles with thermalization (*MCMT*). Each *MCMT* cycle involves, first, random changes in n torsion angles, followed by a given number (100) of energy minimization iterations, with unconstrained geometry, and then a Metropolis test of the minimized conformation at the chosen temperature (310 K). If the energy is not lowered by a given amount (1 kcal/mole) after a number (100) of *MCM* cycles, then several (5) thermalization cycles are carried out, whereby a very high value of T is used for the Metropolis test.

Repetition of the procedure for a number of starting conformations will yield a corresponding number of final structures. Of those, the one with the lowest energy is chosen and subjected to a number (20,000) of *MCM* cycles. Thermalization is not performed at this stage.

The rationale for the thermalization may be understood as follows. The Metropolis test (Chapter 4) will lead to acceptance of the conformation if $e^{-\Delta E/kT} > h$, where ΔE is the energy difference between the new and the old conformation and h is a random number. Using a very high T in this test ensures acceptance of a conformation, which otherwise would have been rejected at the lower temperature of 310 K. In this way, thermalization will expand the search beyond the region of clustered local minima with very similar energies.

This idea of temperature variation is also found in MCM procedures with simulated annealing (Snow 1992). The simulation starts from a randomly-chosen conformation, with the temperature, the step-size ($\Delta\theta$) for the change in the angles, and the number (n) of sweeps over the angles given as input. The annealing procedure, consisting of a number of iterations, is performed with semiconstrained geometry, with changes in both the bond and the dihedral angles, while the energy minimization procedure is carried (in Cartesian coordinates) with unconstrained geometry, allowing also for changes in the bond lengths.

A MC iteration, with n sweeps over all the angles, is carried out, with the Metropolis test applied after having changed each angle. At the end of the iteration, the acceptance rate of trial conformation is evaluated and a decision is made on whether T and $\Delta\theta$ should be adjusted for the next iteration:

(a) if the acceptance rate is lower than 40% or greater than 60%, the temperature is multiplied by a factor in the ranges 1.0-2.0 and 0.5-1.0, respectively;

(b) if the acceptance rate is acceptable (in the range 40-60%), the change in the best-observed energy is examined. If the energy has been lowered, T and $\Delta\theta$ are maintained unchanged for the following iteration, but if the energy has dropped, T and $\Delta\theta$ are multiplied by a factor in the range 0.5-1.0 (preferably, 0.6 and 0.75, respectively, for a better performance).

Iterations are repeated until a predetermined number has been performed or lower limits of either T or $\Delta\theta$ have been reached. At this point the simulation proceeds with an energy minimization step.

6.4 Molecular Dynamics

The procedure just described represents a considerable improvement over the simple energy minimization simulations and seems to yield very satisfactory results for isolated peptides/proteins, predicting the conformation corresponding to the global minimum. Such is not the case, however, when the simulation is performed for a short peptide in solution, in order to try to reproduce the experimental observations in, say, circular dichroism (CD) or nuclear magnetic resonance (NMR) studies. The MC simulation may yield a number of different conformations with comparable energies (Li and Scheraga 1987) but they cannot reproduce the *t*-dependence implicit in the experiment. When the interpretation of experimental results is to be carried out at this level, molecular dynamics (*MD*) constitutes the appropriate tool to be used.

6.4.1 The Equations of Motion

We will be considering a system of N particles (i.e., the atoms of the peptide), characterized at time t by the corresponding position, velocity, and acceleration vectors $r_i(t)$, $\dot{r}_i(t)$, and $\ddot{r}_i(t)$, respectively, where the subscript i labels the particle and the usual notation has been used to indicate differentiation with respect to time; alternatively, when convenient, the notation $v_i(t)$ and $a_i(t)$ will be used for the velocity and the acceleration, respectively.

This system is also characterized by a potential energy, $V(r_1, r_2, ..., r_n)$, and a kinetic energy, $T(\dot{r}_1, \dot{r}_2, ..., \dot{r}_n)$. The potential energy is defined by the *PEF* to be used in the calculations and the kinetic energy is given by

$$T(\dot{r}_1, \dot{r}_2, ..., \dot{r}_n) = \frac{1}{2} \sum_i m_i \dot{r}_i^2(t) \equiv \frac{1}{2} \sum_i m_i v_i^2(t)$$

where the summation extends to all the particles in the system, with corresponding masses m_i. In Cartesian coordinates, taking into account that

$$r_i = x_i \mathbf{i} + y_i \mathbf{j} + z_i \mathbf{k}$$

$$\dot{r}_1 = \dot{x}_i \mathbf{i} + \dot{y}_i \mathbf{j} + \dot{z}_i \mathbf{k}$$

$$\dot{r}_i^2 = \dot{r}_i \cdot \dot{r}_i = \dot{x}_i^2 + \dot{y}_i^2 + \dot{z}_i^2$$

(where $\mathbf{i}$, $\mathbf{j}$, and $\mathbf{k}$ denote the unit vectors along the x-, y-, and z-axes, respectively) the above expression becomes

$$T(\dot{r}_1, \dot{r}_2, ..., \dot{r}_n) = \frac{1}{2} \sum_i m_i(\dot{x}_i^2 + \dot{y}_i^2 + \dot{z}_i^2)$$

The Lagrangian function is defined as

$$L = T - V$$

and the equations of motion in Lagrangian form are

$$\frac{d}{dt}\frac{\partial L}{\partial \dot{q}_j} - \frac{\partial L}{\partial q_j} = 0$$

with one equation for each coordinate q_j (i.e., x_j, y_j, or z_j) and associated $\dot{q}_j$; that is, in Cartesian coordinates there are $3N$ equations of motion.

Taking into account the above dependence of the potential and kinetic energy functions, for the i-th particle the above equation yields

$$\frac{d}{dt}\frac{\partial T}{\partial \dot{x}_i} + \frac{\partial V}{\partial x_i} = \frac{d}{dt}(m_i\dot{x}_i) + \frac{\partial V}{\partial x_i} = m_i\ddot{x}_i + \frac{\partial V}{\partial x_i} = 0$$

$$\frac{d}{dt}\frac{\partial T}{\partial \dot{y}_i} + \frac{\partial V}{\partial y_i} = \frac{d}{dt}(m_i\dot{y}_i) + \frac{\partial V}{\partial y_i} = m_i\ddot{y}_i + \frac{\partial V}{\partial y_i} = 0$$

$$\frac{d}{dt}\frac{\partial T}{\partial \dot{z}_i} + \frac{\partial V}{\partial z_i} = \frac{d}{dt}(m_i\dot{z}_i) + \frac{\partial V}{\partial z_i} = m_i\ddot{z}_i + \frac{\partial V}{\partial z_i} = 0$$

In vector form these three equations may be collected into a single one,

$$m_i(\ddot{x}_i\mathbf{i} + \ddot{y}_i\mathbf{j} + \ddot{z}_i\mathbf{k}) = m_i\ddot{\mathbf{r}}_i = -(\frac{\partial V}{\partial x_i}\mathbf{i} + \frac{\partial V}{\partial y_i}\mathbf{j} + \frac{\partial V}{\partial z_i}\mathbf{k})$$

which can be rewritten as

$$\mathbf{F}_i(\mathbf{r}_1, \mathbf{r}_2, ..., \mathbf{r}_n,) = m_i\frac{d^2\mathbf{r}_i}{dt^2} = -\nabla_i V$$

which shows that the force, $\mathbf{F}_i$, acting on the i-th particle, is derived from the potential, V. That is, at any time t it must be

$$\frac{d^2\mathbf{r}_i}{dt^2} = -\frac{1}{m_i}\nabla_i V \tag{6.6}$$

with one such equation for each particle in the system. In other words: the (classical) behavior of the system of particles is expressed in terms of a set of N coupled, second-order differential equations; the designation *coupled* simply states the fact that the equation for each particle depends on the positions of all the particles in the system (as implicit in the expression of the potential).

6.4.2 Solutions of the Equations of Motion

Solution of the above set of equations, subject to some initial conditions, will yield the positions, velocities, and accelerations of the particles in the system as a function of time. For simplicity in the following formulation, Eqs. (6.6) may be written (van Gunsteren and Berendsen 1977) as

$$y'' = f(y) \tag{6.7a}$$

with the initial conditions given as

$$y_0 = y(t_0) \tag{6.7b}$$

$$y_0' = y'(t_0) \tag{6.7c}$$

Van Gunsteren and Berendsen (1977) have examined this problem in detail and have concluded that the solution, by a step-by-step method, is best obtained with a k-value predictor-corrector algorithm, written in the so-called N-representation, using predictor parameters obtained by a Taylor expansion and with the corrector parameters obtained from stability and accuracy requirements (Gear 1966, 1967, 1971). The corresponding formulation is summarized below and the reader is referred to the work of van Gunsteren and Berendsen (1977) for additional details.

Let us assume that $y(t)$ and its successive derivatives (with respect to t), denoted by $y^{(k)}(t)$, $k = 1, 2, ...,$ are known at time t. For simplicity in the formulation, they will be represented as y_n and $y_n^{(k)}$ (where n is the value defining t in terms of the spacing h). We want to find now the values of $y(t+h)$ and $y^{(k)}(t+h)$, which in the notation just introduced will be represented as y_{n+1} and $y_{n+1}^{(k)}$, respectively. The solution may be obtained through the use of a Taylor expansion, as described below.

Given a function $F(t)$, for which the values $F(a)$, $F^{(1)}(a)$, $F^{(2)}(a)$, etc., are known, the corresponding Taylor expansion

$$F(t) = F(a) + \frac{(t-a)}{1!} F^{(1)}(a) + \frac{(t-a)^2}{2!} F^{(2)}(a) + \frac{(t-a)^3}{3!} F^{(3)}(a) + \frac{(t-a)^4}{4!} F^{(4)}(a) \, ...$$

will allow us (assuming that all the mathematical conditions are met) to evaluate the value $F(b)$ at $t=b$. Differentiation of the above equation with respect to t yields

$$F^{(1)}(t) = F^{(1)}(a) + \frac{2(t-a)}{2!} F^{(2)}(a) + \frac{3(t-a)}{3!} F^{(3)}(a) + \frac{4(t-a)}{4!} F^{(4)}(a) + \dots$$

$$= F^{(1)}(a) + \frac{(t-a)}{1!} F^{(2)}(a) + \frac{(t-a)^2}{2!} F^{(3)}(a) + \frac{(t-a)^3}{3!} F^{(4)}(a) + \dots$$

$$F^{(2)}(t) = F^{(2)}(a) + \frac{(t-a)}{1!} F^{(3)}(a) + \frac{(t-a)^2}{2!} F^{(4)}(a) + \dots$$

....

which will allow us to evaluate the values $F^{(1)}(b)$, $F^{(2)}(b)$, ..., at t=b. In the present case, with $F \equiv y_n$ and (b-a) = h, the above equations become

$$y_{n+1} = y_n + \frac{h}{1!} y_n^{(1)} + \frac{h^2}{2!} y_n^{(2)} + \frac{h^3}{3!} y_n^{(3)} + \frac{h^4}{4!} y_n^{(4)} + \dots$$

$$y_{n+1}^{(1)} = y_n^{(1)} + \frac{h}{1!} y_n^{(2)} + \frac{h^2}{2!} y_n^{(3)} + \frac{h^3}{3!} y_n^{(4)} + \dots$$

$$y_{n+1}^{(2)} = y_n^{(2)} + \frac{h}{1!} y_n^{(3)} + \frac{h^2}{2!} y_n^{(4)} + \dots$$

....

which allows us to evaluate y_{n+1} and $y_{n+1}^{(k)}$ from the values of y_n and $y_n^{(k)}$.

These equations may be collected together in a single equation in matrix form:

$$\begin{pmatrix} y_{n+1} \\ y_{n+1}^{(1)} \\ y_{n+1}^{(2)} \\ y_{n+1}^{(3)} \\ y_{n+1}^{(4)} \\ \dots \end{pmatrix} = \begin{pmatrix} 1 & 1 & 1 & 1 & 1 & 1 & 1 & \dots \\ & 1 & 2 & 3 & 4 & 5 & 6 & \dots \\ & & 1 & 3 & 6 & 10 & 15 & \dots \\ & & & 1 & 4 & 10 & 20 & \dots \\ & & & & 1 & 5 & 15 & \dots \\ & & & & & \dots & \dots \end{pmatrix} \begin{pmatrix} y_n \\ hy_n^{(1)}/1! \\ h^2 y_n^{(2)}/2! \\ h^3 y_n^{(3)}/3! \\ h^4 y_n^{(4)}/4! \\ \dots \end{pmatrix}$$

which may be rewritten as

$$y_{n+1} = Ty_n$$

where y_n is the column vector with elements y_n, $hy_n^{(1)}/1!$, $h^2 y_n^{(2)}/2!$, ... (and similarly for y_{n+1}). The transformation matrix $\mathbf{T}$ is an upper-triangle matrix, formed by column vectors t_k, whose elements are the binomial coefficients $_kC_m \equiv \begin{pmatrix} k \\ m \end{pmatrix}$, where k labels the column and $m = 1, ..., k$.

As only a limited number of terms are used in the Taylor expansion above, y_{n+1} and $y_{n+1}^{(k)}$ will not be correct. That is, when substituting in Eq. (6.7a) it will be

$$f(y_{n+1}) - y_{n+1}^{(2)} \neq 0$$

In order to indicate that the values obtained represent only a predicted estimate, based on a given size of the Taylor expansion, the notation should be changed to $y_{n+1(p)}$ and $y_{n+1(p)}$, where the subscript p indicates the *predictive* character. The corrected values, such that Eq. (6.7a) will be satisfied, may be obtained through an iterative procedure, based on the corrector equation

$$y_{n+1} = y_{n+1(p)} + \frac{ch^2}{2!} [f(y_{n+1(p)}) - y_{n+1(p)}''] \tag{6.8}$$

where the column vector $\mathbf{c}$ is the corrector vector. In principle, the iterative procedure should be continued until no change is observed (to the desired accuracy) in y_{n+1}, but the calculation may be expensive for large proteins, because of the evaluation of $f(y_{n+1(p)})$. The values of the corrector parameters (i.e., the elements of $\mathbf{c}$) are determined so that the stability requirements are satisfied and the accuracy of the algorithm is optimized; those values may be found in the work of Gear (1966, 1967, 1971).

The optimum values of the step size, h, and of k (i.e., the number of terms in the Taylor expansion) are to be chosen depending on the system under study.

6.4.3 Practical Details

Although the formulation is straightforward, a feeling for what the calculations entail, the possible difficulties to be faced, and the problems that may arise is best attained from an illustration of an actual calculation. [The specific data, presented in the following summary, are from the work of McCammon *et al.* (1979); see also the work of Karplus and McCammon (1983).]

Before the calculations are started, a decision must be made regarding the *PEF*, predictor-corrector algorithm, k and h to be used. For example, the *PEF* may include the terms ΔE_b, ΔE_τ, ΔE_θ, and ΔE_{HB} in addition to the non-bonded interactions (see below) and the predictor-corrector algorithm, described above, may be adopted, with k = 5 and h = 9.78•10⁻¹⁶ s. That is: with X denoting any of the Cartesian coordinates (x, y, z) of any of the atoms in the system and maintaining the same meaning as above for the superscripts and subscripts, the working equations become

$$X_{n+1(p)} = X_n + \frac{h}{1!} X_n^{(1)} + \frac{h^2}{2!} X_n^{(2)} + \frac{h^3}{3!} X_n^{(3)} + \frac{h^4}{4!} X_n^{(4)} + \frac{h^5}{5!} X_n^{(5)}$$

$$X_{n+1(p)}^{(1)} = X_n^{(1)} + \frac{h}{1!} X_n^{(2)} + \frac{h^2}{2!} X_n^{(3)} + \frac{h^3}{3!} X_n^{(4)} + \frac{h^4}{4!} X_n^{(5)}$$

$$X_{n+1(p)}^{(2)} = X_n^{(2)} + \frac{h}{1!} X_n^{(3)} + \frac{h^2}{2!} X_n^{(4)} + \frac{h^3}{3!} X_n^{(5)}$$

$$X_{n+1(p)}^{(3)} = X_n^{(3)} + \frac{h}{1!} X_n^{(4)} + \frac{h^2}{2!} X_n^{(5)}$$

$$X_{n+1(p)}^{(4)} = X_n^{(4)} + \frac{h}{1!} X_n^{(5)}$$

$$X_{n+1(p)}^{(5)} = X_n^{(5)}$$

and

$$X_{n+1} = X_{n+1(p)} + \frac{3}{20} \Delta f$$

$$X_{n+1}^{(1)} = X_{n+1(p)}^{(1)} + \frac{251}{360} \Delta f$$

$$X_{n+1}^{(2)} = X_{n+1(p)}^{(2)} + \Delta f$$

$$X_{n+1}^{(3)} = X_{n+1(p)}^{(3)} + \frac{11}{18} \Delta f$$

$$X_{n+1}^{(4)} = X_{n+1(p)}^{(4)} + \frac{1}{6} \Delta f$$

$$X_{n+1}^{(5)} = X_{n+1(p)}^{(5)} + \frac{1}{60} \Delta f$$

where

$$\Delta f = \frac{h^2}{2!} [f_{n+1(p)} - X_{n+1(p)}^{(2)}]$$

These equations give clearly an idea of the considerable number of calculations to be performed for a large protein.

The X-ray structure of the protein under study may be used as starting point of the calculations. Preferably it should be subjected to a prior relaxation, through an energy minimization procedure, in order to relieve the stresses that may be present in the crystal structure, as already mentioned above (arising from the different situation existing in a crystal and in the consideration of an isolated molecule). In the absence of the X-ray structure, recourse may be made of a *de novo* structure, obtained as described earlier in this Chapter.

The initial velocities, with equal magnitude for all the atoms but with random directions may be evaluated, at a low temperature, as

$$v = (3N \, k_B \, T/M_{av})^{1/2}$$

where M_{av} is the average atomic mass, and the initial accelerations may be obtained from the forces acting on each atom in the crystal structure. The remaining higher derivatives are set initially equal to zero.

The simulation is performed for a few picoseconds (2.0 ps), during which period the atomic velocities are modified as required in order to maintain the temperature (at 300 K) of the protein and to prevent localized heating. The overall temperature may be maintained by multiplying all the atomic velocities by an appropriate factor (1.005 or 0.995) after each step in which the average temperature is less than $T-\Delta T$ or greater than $T+\Delta T$ (with $\Delta T = 5$ K). If after any step, the temperature of any atom exceeds a chosen upper limit of temperature (670 K), its velocity is multiplied by a factor slightly less than unity. [This treatment for the elimination of localized heating may result in a net linear and angular velocity of the protein.] The equilibration treatment is continued for an additional period (2.9 ps), without any modification of the atomic velocities, in order to allow for a complete relaxation of the protein. [The equilibration treatment is considered complete when no systematic changes are observed in the temperature over a given period (10 ps).] At the end of the equilibration period, the temperature will have reached a certain value (308 K), different from the initial one. Throughout the equilibration period it is necessary to check that the atomic momenta conform to a Maxwell distribution and that all the regions of the protein are at the same average temperature.

The proper dynamical simulation is then carried out over a given period (9.8 ps) and the complete trajectory used for the statistical analysis. [The complete calculation, for a 14.7-ps trajectory took over 2 hours on an IBM 360/91 computer.]

The considerable computing time required for a *MD* simulation of a large protein suggests the consideration of constraints, which might reduce it. In the above development, mention has been made of the fact that the calculations may be expensive because of the evaluation of terms dependent on the *PEF*. Manipulation with the corresponding parameters (especially those in ΔE_b and ΔE_τ) will enhance or reduce their contributions, but such a manipulation will not affect the computing time. The evaluation of those terms will always be present in the calculations and therefore one has to look elsewhere for a possible reduction in the computing time.

Because of the assumed generality of the *PEF*, the formulation has been presented in terms of the Cartesian coordinates of the atoms of the system. A simple calculation will immediately give an idea of the magnitude of the problem to be faced. Let us consider as an example the case of a protein with, say, 300 residues, which could have of the order of 5,000 atoms. With k=5, as in the above example, it means that one must deal with 12 equations per (x, y, z) coordinate, for a total of $5,000 \times 3 \times 12 = 180,000$ values to be calculated at each point of the trajectory (and this under the assumption that the corrector step is performed only once).

Within the framework of a constrained formulation, with fixed bond angles and bond lengths, the number of variables is reduced considerably as one can operate with only the torsion angles. Let us assume, for simplicity, that only the angles ϕ, ψ, and χ_1 of each residue, as well as the peptide ω angles are to be varied. That is, there are 1,197 coordinates, for a total of $1,197 \times 12 = 14,364$ quantities to be evaluated at each point of the trajectory. In other words, there is a reduction of an order of magnitude.

This approach, used by Go and Scheraga (Noguti and Go 1983, 1985; Gibson and Scheraga 1990ab), was rejected earlier (van Gunsteren and Berendsen 1977, van Gunsteren and Karplus 1982) as leading to misleading results: the constraint of fixed bond lengths seems to be acceptable, which is not the case for fixed bond angles.

An acceptable constraint, of interest in the building of protein structures by homology (see Chapter 10), consists of restricting the simulation to only the side chains (Holm and Sander 1992).

A more detailed examination, at a practical level, of the difficulties and sources of error in *MD* simulations is presented in Section 10.4.2.

6.4.4 Analysis of the Results

A *MD* calculation yields the trajectory for the period of the proper simulation, after equilibration has been reached: that is, one has collected the positions (as well as the velocities and accelerations) of all the particles of the system as a function of time, at the chosen temperature. Analysis of these results will provide information on the internal motions of the protein under study: there are local (atom fluctuations, side chain oscillations, and loop displacements), rigid-body (helices, domains, and subunits), large-scale (opening fluctuations, and folding and unfolding), and collective (elastic-body modes, coupled atom fluctuations, soliton and other non-linear contributions) internal motions, with amplitudes between 0.01 and 100 Å and energies between 0.1 to 100 kcal/mole, over periods of 10^{-15} to $10^3 \, s$ (Karplus and McCammon 1983).

A simple representation of the behavior of the position of chosen atoms as a function of time will already give some feeling for what is happening during the simulation. Naturally, a full pictorial representation, in movie form, will show clearly the main features of the evolution of the protein structure, helping towards an understanding of the (biological) function of the protein. The structural analysis for the identification of the relevant conformations in the *MD* trajectory may be performed by *clustering*, based on the use of *similarity matrices* for a selected quantity, such as coordinates, intramolecular distances, sum of squared distances, dihedral angles, ... (Levitt 1983, Unger *et al.* 1989, Rooman *et al.* 1990, Gordon and Somorjai 1992, Karpen *et al.* 1993, Shenking and McDonald 1994, Torda and van Gunsteren 1994). The clustering, however, may be somewhat fuzzy (Gordon and Somorjai 1992), as a conformation may well belong to two clusters. In addition, *Fourier transforms* (see Section 11.2) of given variables (atomic coordinates, torsion angles, combinations of internal coordinates, variables for relative geometries, ...) or of their correlation functions (see below) have been used to obtain the frequencies of oscillation of those variables; in particular, Dauber-Osguthorpe and Osguthorpe (1993) extract, from the atomic trajectories, the vectors, analogous to those obtained from a normal mode analysis, which define the characteristic motion for each frequency and provide a graphic description of the motion.

At a quantitative level, the results may be analyzed through the evaluation of the mean-square fluctuations as well as through the so-called correlation functions.

The mean-square fluctuations of the atoms from their average positions are of interest because of their relationship to the atomic temperature factors, B, obtained in X-ray diffraction studies of crystal structures. That relationship is given (Karplus and McCammon 1983) by

$$\left\langle \Delta r^2 \right\rangle_{MD} = \frac{3B}{8\pi^2} - \left\langle \Delta r^2 \right\rangle_{disorder}$$

where the last term represents a correction arising from lattice disorders, etc. An estimate of this contribution may then be obtained from the value of $<\Delta r^2>_{MD}$, obtained in the simulation, and the experimental result for B.

Correlation functions are used for the interpretation of the experimental results obtained from time-dependent techniques (such as, in particular, spectroscopic experiments). In general, the correlation function

$$C(t) = <A(0)B(t)>$$

defines, as a canonical ensemble average, the relationship between the dynamic property A(0), at a given time, and the dynamic property B(t), at some later time. In the particular case when A and B refer to the same dynamic property,

C(t) = <A(0)A(t)>

defines the autocorrelation function of that property (Zwanzig 1965, Berne and Harp 1970). The canonical ensemble average of a single particle property may be replaced, in the case of ergodic systems, by a (time averaged) infinite-limit time integral (Borstnik et al. 1980, McCammon et al. 1979)

$$C(t) = \lim_{t_u \to \infty} \frac{1}{t_u} \int_0^{t_u} d\tau \, A(t)A(t+\tau)$$

which, in *MD* simulations, may be approximated by the time average over discrete points

$$C(t_n) = \frac{1}{m_u} \sum_{m=1}^{m_u} A(t_m)A(t_{m+n})$$

(where the subscript *u* has been used to denote upper limits). [See the work of Borstnik et al. (1980) for references of work on the characteristics of, and alternatives to, these calculations.]

Of particular interest are the orientational autocorrelation functions

$$C_\ell(t) \equiv \left\langle P_\ell(\mathbf{u}(0) \cdot \mathbf{u}(t)) \right\rangle$$

where P_ℓ is a Legendre polynomial and $\mathbf{u}$ is a unit vector. Their rates of decay (relaxation times) may be related to experimental, spectroscopic infrared ($\ell = 1$) and Raman ($\ell = 2$) results. An interpretation of both infrared and Raman spectra may be attempted on the basis of the autocorrelation functions $C_1(t)$ and $C_2(t)$, respectively, but only under the assumption of a classical description of the electric dipole moment and the electric dipole polarizability. In such treatments, the vector $\mathbf{u}$ is either a unit vector parallel to the transition dipole moment or a unit vector along the bond which contributes to the polarizability tensor. In addition, correlation times, obtained from integration of the autocorrelation function $C_2(t)$, may provide an estimate of *NMR* relaxation times. [For more details see the work of Clementi et al. (1990) and references therein.]

Correlation functions may be also used in the analysis of torsional fluctuations of subunits of the system (Levy and Karplus 1979). Denoting by <A> the mean value (i.e., the canonical ensemble average) of an internal coordinate A and by $\Delta A(t)$ the fluctuation at time *t* from that mean value, the corresponding autocorrelation function will be <$\Delta A(0)\Delta A(t)$>, which gives the time decay of the fluctuation. Within the framework of the harmonic model, this correlation function, as well as the related velocity correlation function $< \Delta\dot{A}(0)\,\Delta\dot{A}(t) >$, may be evaluated from the sets of normal coordinates, $\{Q_i\}$, and normal angular velocities, $\{\omega_i\}$. Expanding $\Delta A(t)$ in terms of the normal coordinates

$$\Delta A(t) = \sum_{i=1}^{3N-6} \alpha_i Q_i(t)$$

where α_i denotes the component of Q_i in the direction of A, and taking into account that the correlation function for normal coordinates is given (Wang and Uhlenbeck 1945) by

$$<Q_i(t_1)Q_j(t_2)> = \delta_{ij}\frac{k_BT}{\omega_i^2}\cos\omega_i(t_1-t_2)$$

one obtains

$$< \Delta A(0)\Delta A(t) > = k_BT\sum_{i=1}^{3N-6}\frac{\alpha_i^2}{\omega_i^2}\cos\omega_i t$$

$$< \Delta\dot{A}(0)\Delta\dot{A}(t) > = k_BT\sum_{i=1}^{3N-6}\alpha_i^2\cos\omega_i t$$

[which can be expressed in terms of frequencies, ν_i, and force constants, k_i, taking into account the relationships $k_i = \omega_i^2 m_i = 4\pi^2\nu_i^2 m_i$ (Herzberg 1960)]. The normal frequencies and normal-mode eigenvectors are obtained in the solution of the secular equation (Wilson *et al.* 1955)

$$|F - \omega^2 T| = 0$$

where F is the Cartesian second-derivative potential energy matrix and T is a diagonal matrix with elements $T_{ij} = \delta_{ij}m_i$. (3N-6) of the 3N eigenvalues of this secular equation correspond to the normal frequencies; the components of the associated eigenvectors give the relative amplitudes of the contributing atomic displacements (in terms of mass-scaled Cartesian coordinates).

6.5 Monte Carlo/Molecular Dynamics Simulations

It is also possible to perform hybrid *MD/MC* simulations. At this point, however, only the procedure of Morley *et al.* (1992), not applied yet to proteins, will be examined because of its peculiarities. The discussion of hybrid *MD/MC* simulations, as actually applied to the study of protein structures, is postponed to Section 10.7.

A trial conformation is adopted and its energy, E_{old}, is evaluated. A bond is selected and initial velocities are assigned in order to produce a rotation around that bond. The overall translation and rotation kinetic energy of the whole molecule and the velocities are renormalized to the desired temperature, T_{MD}, at which a *MD* trajectory will be obtained. A 1 fs *MD* trajectory generates a new conformation, whose energy, E_{new}, is calculated. The new conformation is accepted if the energy has been lowered or the Metropolis test, at a temperature T_{MC}, is passed, in which case a new *MD* trajectory is started. Otherwise, the simulation restarts from the previously accepted conformation. The steps in the search may be controlled by a manipulation of both T_{MD} and T_{MC} (see the section on *MC* for a discussion on the effect of the value adopted for T_{MC}).

It must be mentioned, however, that this procedure, as applied to organic molecules, depends on the definition of so-called spinnable bonds. This definition includes all the non-terminal, saturated bonds as well as those π-bonds with a bond order smaller than a given threshold value. Therefore, for the application of this algorithm to protein modeling, it will be necessary to prepare, first of all, a master table of bond orders for the unsaturated bonds present in proteins. The corresponding values could be obtained, say, by a Mulliken population analysis (which was discussed in Chapter 3, in connection with the evaluation of effective charges; the corresponding formulation for bond orders was not presented because of the gradual disappearance of bond orders in Computational Chemistry work).

7 Practical Overview

At this point it is necessary to pause and look at the overall picture, lest a distorted conclusion is reached. Let us analyze the dangers, objectives, and available means in the modeling of proteins and of their interactions.

The preceding chapters have offered an outline of the theoretical foundations for the majority of procedures in use for the computer simulations of proteins. It has not been possible to cover in detail all the pertinent formulations and a limited purpose has been adopted: to *acquaint* the non-theoretical reader with the general ideas, characteristics, limitations, and, particularly, the possible deficiencies arising from the many approximations introduced when proceeding from the original Schrödinger equation to the Molecular Mechanics techniques used in actual simulations. But the following chapters will focus on experimental information and its use in semiempirical approaches, which may prove to be rather useful when applied intelligently.

This two-fold approach illustrates the dangers in work in this field. By necessity, the researcher must be knowledgeable in such distinct fields as Quantum and Statistical Mechanics, on one hand, and Biochemistry, Biology, Immunology, and Pharmacology, on the other. This problem is compounded by the proliferation of software packages, which may produce a wealth of numbers and pictures, valuable only when interpreted with awareness of the *reality* of the problem under study and of the possible sources of error in the treatment. In other words: work in this field will be particularly successful and rewarding when carried out with close collaboration of experimental and theoretical researchers.

The proliferation of formulations and techniques may blur the picture, hiding what the real objectives are. Development work (with new parameterizations and algorithms) will continue and will be tested in the refinement of crystallographic structures, the *de novo* prediction of conformations, and the interpretation of physicochemical results [see, e.g., the review of Nemethy and Scheraga (1990)]. But the real objective is the prediction/interpretation of the function of proteins, whether in biological or industrial processes, and, ultimately, the rational design of enzymes and drugs. When that stage is reached, computer simulations will be fully integrated with experimental work.

Are all the necessary tools for those simulations already available? Since the pioneering work of Scheraga and co-workers [see, e.g., Momany *et al.* (1975), Tanaka and Scheraga 1975, Burgess and Scheraga 1975, Dunfield *et al.* (1978), and Nemethy *et al.* (1983)] the growth in this field has been most impressive [see the review of Fraga and Parker (1994)]. One should mention, at least, the following potential energy functions, auxiliary programs, and software packages: AMBER (Weiner and Kollman 1981) and the related ORAL (Zimmerman 1991) and OPLS (Jorgensen and Tirado-Rives 1988); CHARM (Brooks *et al.* 1983) and the related QUANTA 3.2/CHARMm (Momany and Rose 1992); *CVFF* (Dinur and Hagler 1991); DISCOVER (Dauber-Osguthorpe *et al.* 1988) and the related work of Dauber-Osguthorpe and Osguthorpe (1993); ECEPP (Scheraga); maPSI (Coghlan and Fraga 1985; Thornton *et al.* 1991); *MM*2 and *MM*3 (Allinger 1977, Allinger *et al.* 1989, and Lii and Allinger 1989) and the related *MM*EPP (Wolfe *et al.* 1988ab); RATTLE (Andersen 1983), SHAKE (Ryckaert *et al.* 1977), and SETTLE (Miyamoto and Kollman 1992); SYBYL and ALCHEMY (Clark *et al.* 1989); and the work of Robson and Platt (1986) and Snow (1993). [See also Section 12.3.5.] Before proceeding to answer the original question above, it may be convenient to pose a new one: Is it possible to discriminate among the many approaches and packages? First, we would like to refer the reader to the work of Roterman *et al.* (1989) and Kollman and Dill (1991) for a sampling of arguments in this connection. Our opinion, however, is that all may be useful in one way or another, as long as the user is aware of the possible deficiencies but especially if the interpretation of the results is carried out in a competent manner, always with final recourse to a rationalization on the basis of past experience and whatever experimental information is available. That is why all the available information must be used (Thornton 1988).

The answer to the first question is *'no'*. The reason is to be found in the limitations of the *MM* approach, with neglect of the electronic structure. Such a neglect may have important consequences, because of the strong charge fluctuations that may result from the changes in the molecular geometry in the course of a simulation (Merz 1992): that is, in a proper quantum-chemical calculation, with geometry optimization, the effective charges will change as the molecular geometry changes. Thus, the use of fixed effective charges represents a serious deficiency and, in due time, it will be necessary to complement *MM* simultions with a quantum-chemical component (Field *et al.* 1990, Thery *et al.* 1994). The need for such a quantum-chemical component is even more crucial, one might say, in conjunction with a rational drug design (as discussed in Chapter 12).

Experimental and Theoretical Data

We have just stressed that, when simulating protein structures, all the available information should be used. This statement was made, fundamentally, in connection with the results obtained from different approaches and techniques. But, ultimately, the quality of the simulation must be tested against experimental information. For this purpose, the structural data in the Brookhaven Protein Data Bank and the Cambridge Structural Database is of paramount importance.

This fact is emphasized here because the short space dedicated to them in Chapter 8 might seem to suggest otherwise. In this work, whose purpose is to bring all the pertinent information to the attention of the reader, it should suffice to point out the existence of such data banks, without a detailed description.

The primary information in these data banks is the tertiary and secondary structures but it has been complemented in a variety of databases with additional information on angles, patterns, function, ..., and analyzed for homologies, sequence conflicts, variants, residue conservation, ... This wealth of information is of extreme usefulness for simulation procedures, as discussed in the following section.

At a different level, there is also a variety of physicochemical data, of interest in conjunction with semiempirical methods for diverse predictions. The determination and use of the hydrophilicity, accessibility, flexibility, and recognition factors is discussed in Chapter 8, but their application to actual predictions is postponed to Chapters 9 and 12.

8 Databases

This summary of experimental information on amino acids, peptides, and proteins is best accomplished by considering separately the structural data, and related processed information, and the physico-chemical data. This decision is based, fundamentally, in the different usage of both types of data: the structural data constitute both a benchmark and a source of information for the simplification/improvement/interpretation of simulation treatments while the physico-chemical data play a role in semiempirical predictions.

The terminology in use in the literature (*databanks*, *libraries*, *databases*, and *programs*) requires some comments, in order to acquaint the reader with the possible meanings that may be associated with each term. In principle, the designation *databank*, *library*, and *database* should be equivalent and be used to refer to a collection of data. In many cases, however, the designation *database* implies a data storage and retrieval system, which incorporates a program for the analysis of the data. On the other hand, a so-called *program* may include databanks. Consequently, the term *database* will be used below in a general fashion, to refer to databanks, libraries, databases, retrieval systems, and programs.

8.1 Structural Data

8.1.1 Brookhaven Protein Data Bank and Cambridge Structural Database

X-ray crystallography has constituted the main source of structural data, providing the Cartesian coordinates of the constituent atoms. This raw information is collected, in an on-going project, in the Brookhaven Protein Data Bank (*PDB*) (Bernstein *et al*. 1977, Abola *et al*. 1987) and the Cambridge Structural Database (*CSD*) (Allen *et al*. 1979). Unfortunately, there are far more sequences known than tertiary structures but, nevertheless, consultation of these data banks should be a prerequisite to a simulation work. In this connection it must be mentioned that the PDB contain redundant entries, for very similar structures, and therefore it may be more appropriate to use limited, representative sets (Boberg *et al.* 1992, Hobohm *et al.* 1992).

8.1.2 Additional Databases

Related databases, containing additional information, have proliferated in recent years and the purpose of this work will be best attained by a simple listing of them. The reader is referred to the general work of Parsaye *et al.* (1989) as well as the original sources for more detailed information on their contents and how to access them. For simplicity, they will be separated into two groups: those with given acronyms are given first, while the remaining ones are lumped together according to their main characteristics.

GenBank (Benson *et al.* 1993)
Data from the major protein sequence and structural databases.

HSSP (Sander and Schneider 1993)
Homology-derived structures of proteins, useful for the analysis of residue conservation, sequence patterns, etc.

IPSA [inductive protein structure analysis (Schulze-Kremer and King 1992)]
Geometrical, topological, and physico-chemical information on secondary structure and relationships between secondary structures in connection with biological function.

OODBS [object-oriented database: *P/FDM* (Gray *et al.* 1990)]
Primary, secondary, and tertiary structure information for the modeling of homologous protein structures.

PIR - International databases (Barker *et al.* 1993)
Sequence, and sequence-related, databases, using sequence homology.

PROSITE (Bairoch 1991, 1993)
Sites and patterns in protein sequences
[see also the work of Barker *et al.* (1988), Hodgman (1989), Bork (1989), Smith *et al.* (1990), and Taylor and Jones (1991)].

RDBS [relational databases: *BIPED* (Islam and Sternberg 1989)], *SESAM* (Huysmans *et al.* 1991), *SCAN3D* (Vriend *et al.* 1994).
Coordinates, salt-bridges, *H*-bonds, disulfide bridges, close tertiary contacts, ...

REBASE (Robers and Macelis 1993)
Restriction enzyme database.

SBASE (Pongor *et al.* 1993)
Annotated protein sequence segments (with information on structure, function, binding specificity, and similarity to other proteins).

SEQSEE (Wishart *et al.* 1994)
Multipurpose, menu-driven software for the analysis and display of protein sequences and databases; it includes the databanks *SEQBANK*, *SEQSITE*, *SEQMOTIF1*, and *SEQMOTIF2*.

SWISS-PROT (Bairoch and Boeckmann 1993)
Protein sequence database containing information on function, domains and sites, secondary structure, quaternary structure, similarities, sequence conflicts, and variants.

Other databases and programs of interest, listed according to their main emphasis, are:

β-turn databases (Wilmot & Thornton 1990).

Fragment peptide library (Seto *et al.* 1990)
Fragments (5-7 residues) characterizing specific superfamilies.

Homology database (Pascarella and Argos 1992; see also Gribskov *et al.* 1987, Bowie *et al.* 1991)
Data for structurally superposed proteins with similar main-chain folds.

Homology pattern database (Parker and Hodges 1994)
Database of homologous patterns derived from families of proteins.

NMR chemical shift databanks (Seavey *et al.* 1991)
Chemical shifts and sequence information of approximately 200 peptides and proteins.

Rotamer libraries (McGregor *et al.* 1987, Ponder and Richards 1987, Dunbrack and Karplus 1993, Schrauber *et al.* 1993)
Side-chain conformations.

Secondary-structure related databases (Kabsch and Sander 1983, Kneller *et al.* 1990, Boberg *et al.* 1992)
Secondary structure, solvent exposure, disulfide bridges, and function.

Protein surfaces (Bauman *et al.* 1989)

Signature sequences and sequence motifs (Smith *et al.* 1990)

Tertiary structures (Schulz 1988, Bazan 1990, Niermann and Kirschner 1990, Sali *et al.* 1990)

8.2 Physico-chemical Data

We have already discussed the Cartesian coordinates of individual amino acids and the effective charges of their constituent atoms. In this section we turn our attention to those parameters - hydrophilicity, flexibility, accessibility, and recognition - that are of use in semiempirical predictions. We will examine separately their definition/determination and the preparation of the corresponding sequence profiles.

8.2.1 Hydrophilicity, Flexibility, Accessibility, and Recognition Parameters for Individual Amino Acids

The hydrophilicity (*h*), accessibility (*a*), flexibility (*f*), and recognition (*r*) factors, assigned to each amino acid, are parameters related to a particular characteristic, identifiable from their designation, of the amino acid and/or the peptide chain of which it forms part. The *h*-, *a*-, and *f*-factors are of experimental origin while the *r*-factors were derived theoretically; they are transportable except for the *f*-factors, which show a limited sequence dependence (see below).

(a) Hydrophobicity scales

The importance attached to the hydrophobic forces in protein folding (see Section 5.1) has been the driving force behind the innumerable efforts dedicated throughout the years to the derivation of hydrophobicity scales. From the early work of Levitt (1976) to the recent theoretical prediction of Iglesias *et al.* (1994), more than forty hydrophobicity scales have been proposed. Mention must be made, in particular, of the hydrophobicity (Bull and Breese 1974), bulk hydrophobic character (Manavalan and Ponnuswamy 1978), hydration potential (Wolfenden 1978; Wolfenden *et al.* 1979, 1981; Radzicka and Wolfenden 1988), hydropathy (Kyte and Doolittle 1982), and π (Fauchere and Pliska 1983) scales. [See also the work of Cornette *et al.* (1987) for a comparison of the prediction of amphipathic character using different hydrophobic scales.] In this work, however, we will only discuss the hydrophilicity scale of Parker *et al.* (1986) which, based exclusively on directly-determined experimental information, should reflect accurately the relative hydrophilic/hydrophobic character of the individual amino acids.

This hydrophilicity scale (Parker *et al.* 1986, Parker and Hodges 1991abc) was derived from the contribution in high-performance liquid chromatography of each amino acid side chain to the retention time of model synthetic peptides, Ac-Gly-X-X-(Leu)₃-(Lys)₃-amide, where *X* was substituted by the 20 natural amino acids. The amino acid (Trp) with the maximum retention coefficient was assigned a hydrophobic value of -10.0 and the amino acid (Asp) showing the minimum retention coefficient a hydrophilic value of +10.0. The values for the remaining amino acids were then scaled proportionally to their retention coefficients. The values of these *h*-factors are presented in Table 8.1.

(b) Accessible surface areas

The accessibility scale (Lee and Richards 1971, Janin 1979; see also Chothia 1976) was obtained from a study of the accessible surface areas of residues in a sample of 22 globular proteins. It was found that the number, n_B, of buried residues (i.e., with an accessible surface smaller than a threshold value, A_m) varies with the total number of residues, *n*, in the peptide chain according to

$$n^{1/3} - n_B^{1/3} = k$$

where *k* depends only on A_m. The results of the analysis yielded, for each amino acid, the molar fractions of buried and accessible residues and, consequently, the corresponding partition coefficient, from which one could evaluate the change in free energy, ΔG, for the transfer from inside the protein to its surface. Both the partition coefficient and the associated ΔG may then be taken as representing a characteristic of each amino acid in globular proteins. The values of the *a*-factors are presented in Table 8.1.

The relationship between accessible surface area and hydrophobicity (see, e.g., the work of Rose *et al.* 1985) will be discussed in conjunction with the interpretation of the corresponding profiles in the next section.

(c) Flexibility parameters

The *f*-factors (Karplus and Schulz 1985) were defined from the temperature factors (i.e., the B-values; see Section 6.4.4) of the C_α-atoms in an analysis of 31 proteins, normalized according to

$$B_{norm} = (B + D_p)/(<B>_p + D_p)$$

where $<B>_p$ is the average B-value of all the C_α-atoms in protein *p* (omitting the 3-*N*- and 3-*C*-terminal residues); the value of D_p for a given protein *p* was chosen so that the root-mean-square deviation of the B_{norm}-values would be 0.3. These results were then refined through a nearest-neighbor analysis, yielding three sets of *f*-factors: first, the natural amino acids were separated into two groups, rigid and flexible, depending on whether their B_{norm}-values were smaller or greater than 1.0, respectively; then the B_{norm}-values for each amino acid were determined for residues with no rigid neighbor, one rigid neighbor, and with both neighbors rigid. A striking nearest-neighbor effect was observed and it is for this reason that, as mentioned above, the *f*-factors are not fully transportable, except insofar as the existence of rigid neighbors is taken into account in the construction of the corresponding profile (see below). Those values are presented in Table 8.1.

Table 8.1 Hydrophilicity, accessibility, flexibility, and recognition factors for the individual amino acids.

	Hydrophilicity[a]	Accessibility[a]	Flexibility[a,b]			Recognition
Ala	2.1	2.7	-0.5	-2.6	-3.0	78
Arg	4.2	9.8	-0.7	4.6	-2.3	95
Asn	7.0	8.4	5.7	2.7	0.0	94
Asp	10.0	8.4	-1.1	10.0	0.1	81
Cys	1.4	-10.0	-7.1	-8.6	-0.4	89
Glu	7.8	8.9	3.8	5.3	0.7	87
Gln	6.0	8.9	9.6	4.6	-3.5	78
Gly	5.7	2.3	7.8	5.8	-0.5	84
His	2.1	6.7	-5.3	-2.1	-2.8	84
Ile	-8.0	-3.4	-3.7	-7.3	-4.6	88
Leu	-9.2	-0.3	-6.5	-1.3	-0.7	85
Lys	5.7	10.0	3.8	9.4	10.0	87
Met	-4.2	1.9	-8.2	-10.0	-9.9	80
Phe	-9.2	0.5	-9.6	-5.6	-1.2	81
Pro	2.1	7.5	0.6	9.6	0.1	91
Ser	6.5	6.7	10.0	6.4	-0.5	107
Thr	5.2	7.1	2.1	6.6	0.3	93
Trp	-10.0	3.2	-10.0	-5.1	-10.0	104
Tyr	-1.9	8.0	-7.0	-4.0	-7.3	84
Val	-3.7	-2.5	-5.3	-4.2	-1.3	89

[a]As given by Parker *et al.* (1986). See the text for the original references.

[b]See the text for the explanation of the three sets of values.

(d) Recognition factors

The *r*-factors were determined (Fraga 1982,1983) from theoretical results. Calculations were first carried out for all the possible molecular associations of pairs of the natural amino acids (see Sections 3.8 and 6.2.2). The procedure yielded, for each pair of amino acids, a number of stable associations, characterized by different relative positions (separation and orientation) and interaction energies. Each pair of amino acids was then assigned the value of the interaction energy corresponding to the most stable association found. The r-factor for a given amino acid was then approximated as the average of the absolute values of such interaction energies for all the pairs formed by that amino acid with all the others as well as with itself. Those values are presented in Table 8.1.

8.2.2 Profiles

The h-, a-, f-, and r-factors are of use in the construction of the corresponding profiles (to be denoted as H-, A-, F-, and R-profiles, respectively) for peptide chains, taking into account their amino acid sequences. We will discuss here the construction and characteristics of those profiles, postponing the consideration of their application in predictive schemes to following chapters (see Chapter 9, 10 and 12).

The values of the f-factors of Karplus and Schulz (1985), collected in Table 8.1, have proved to be satisfactory (see above) but the reader may consider also using the new values obtained by Vihinen et al. (1994) from an analysis of 92 proteins, in an extension of the work of Karplus and Schulz.

Construction of the profiles

We will denote by x the factor under consideration, so that x_i will represent the factor of the residue at the i-th sequence position of the protein under consideration. The corresponding value for the protein, with consideration of its sequence, is then evaluated (see, e.g., Hopp and Woods 1981 and Hopp 1985) as

$$X_i = \sum_j x_j w_j$$

where the summation runs from $j = i\text{-}3$ to $j = i + 3$, in steps of one. For the H-, A-, and R-profiles, the weight factors, w_j, are given the values 1.0, except for $w_{i-3} = w_{i+3} = 0.50$. In the case of the F-profile, the weights are 0.25, 0.50, 0.75, 1.00, 0.75, 0.50, and 0.25, and the value of f_i is chosen as mentioned in the preceding section, taking into account the character (rigid or flexible) of the nearest neighbors. [These weights will be different when using the factors obtained by Vihinen et al. (1994), in which case a window of 9 residues is recommended.]

For some predictions, it may be worthwhile to construct also a hydrophilicity-accessibility (HA) profile, using for the individual residues the average of their h- and a-factors.

The set of values X_i define the corresponding profile for the protein under consideration. These profiles are usually presented in graphic representation, with a plot of X_i *versus* the sequence position i.

Interpretation of the profiles

As a rule, these profiles present a series of maxima, which can be correlated with a corresponding property of the peptide chain on the basis of intuitive considerations, as shown below (Thornton *et al.* 1991).

It seems evident that the maxima in the H- and A-profiles should appear at those sequence positions characterized by high hydrophilicity and accessibility, respectively. Hydrophilic regions, in general, are expected to be located on the surface of the protein, thus being highly accessible. This implies that there should be a correlation between the maxima in the H-profile and the maxima in the A-profile.

Similarly, one might expect that the maxima in the F-profile would identify the regions of the sequence with high flexibility. However, they may also correspond to bends in the peptide chain. The proper interpretation of the F-profile may be obtained from consideration of the R-profile.

Taking into account the definition of the r-factors, the maxima in the R-profile should correspond to those regions of the sequence where highly-interactive residues are found. Consequently, those maxima may identify the positions of bends in the peptide chain.

An analysis of the F- and R-profiles for 88 proteins (Thornton *et al.* 1991) showed a 81% correspondence (within ±4 sequence positions) between the maxima of the F- and R-profiles. This correspondence suggests the possibility of defining three types of regions, depending on the characteristics of these two profiles: rigid turns (with high recognition and low flexibility), easily distorted bends (with intermediate recognition and high flexibility), and flexible regions without turns (with low recognition and high flexibility).

8.3 Summary

The Brookhaven Protein Data Bank and the Cambridge Structural Database, as well as the complementary databases, readily available in the literature, constitute a very important source of information, to be consulted prior to any simulation procedure.

Regarding the profiles, it is appropriate to indicate that they are easily constructed and intuitively interpreted and that their usefulness in the prediction of surface regions, structural fragments, and antigenic determinants should not be underestimated.

Modeling of Isolated Systems and Associations

The points raised in the preceding sections would suggest that we should now proceed directly with the analysis of the results obtained in Monte Carlo (MC) and Molecular Dynamics (MD) treatments. Such an approach, correct as it would be, poses the danger of ignoring a wealth of information, obtained by other methods.

Let us, however, disregard that fact and simply consider that our goal is to perform highly-sophisticated T-dependent simulations. We already know that tertiary structures, modelled in T-independent treatments, could constitute appropriate starting points for the T-dependent simulations. Therefore, it seems logical that we should first analyze the results obtained in T-independent simulations.

We are aware, however, of the risks in such an endeavor. Risks which may be considerably reduced through the use of as much auxiliary information as possible: for example, as in the construction by homology or from fragments. It is in this connection that the sources of experimental information, reviewed in the preceding section, are of paramount importance. But it is also possible to proceed from scratch, which explains the great efforts dedicated to the prediction of secondary structures.

Consequently, in this section, we will start with the prediction of secondary structures, examine then the diverse ways in which the T-independent simulations may be carried out, and conclude with a review of the results that finally may be obtained in proper T-dependent simulations.

This section is then completed with an examination of the problems to be faced when dealing with peptides/proteins in solution or interacting with other molecules, a subject of extreme relevance in conjunction with the practical applications in molecular design. The main difficulties in the modeling of solvated proteins and of complexes are of a computational nature, due to the increased size of the system under study, which explain the efforts in the development of new techniques (such as the use of Fast Fourier Transforms) and approximations (as in the complementarity schemes).

9 Prediction of Secondary Structures

9.1 Introduction

Prediction of secondary structure should be viewed with caution, in spite of all the efforts described in the literature. At present only the network procedures offer the best chances for success.

The reason for this statement is to be found in the discussion (Section 5.1) on the forces responsible for the protein folding. The dilemma mentioned at that point concerned whether secondary structure units were formed first, followed by a hydrophobic collapse, or *vice versa*. In fact, as it is usually the case when trying to model reality, one had to admit that most probably both steps take place simultaneously and that *all interactions* (short, intermediate, and long range forces) play a role.

Therefore, if intermediate- and especially long-range forces determine to a certain point the secondary structure to be adopted, the prediction of the latter should proceed hand in hand with the simulation of the tertiary structure. The result is that we are back at square one: we have to simulate the tertiary structure in order to be able to predict the secondary structure.

This pessimism is well supported by the numbers obtained: the (percent) prediction success ranges from 50 to 80%, and there is doubt that it will be possible to progress much farther, unless more and more experimental information becomes available. For example, a considerable improvement would be possible if the secondary structure of all the possible triplets of amino acids were known. A knowledge of the secondary structure of all the possible quartets or, even better, all the possible quintets, would certainly provide a firmer basis for all the prediction algorithms. That is, it would help if we were to know the secondary structure of all the 8,000, 160,000, or 3,200,000 combinations of three, four, and five amino acids. And that information will not be available from the scant number of tertiary structures known at present (see Chapter 8).

A bright note, however, arises from the fact that a 100% success rate in the prediction of the secondary structure may not be needed after all (Rost *et al.* 1994), taking into account that *considerable variations in the position and length of secondary structure units can be accommodated within a given 3-dimensional structure.*

In any case, awareness of the risks will stimulate the care exercised in both the prediction procedure and the use of its results. It is in this connection that additional experimental information may be very useful. Other techniques [nuclear magnetic resonance (*NMR*), circular dichroism (*CD*), and Raman and Fourier transform infrared (*FTIR*) spectroscopy], in addition to *X*-ray diffraction, are being increasingly used for the determination of protein structures (Gronenborn and Clore 1990, Dyson and Wright 1991, van Gunsteren *et al.* 1991, Wüthrich 1991, Wishart *et al.* 1991, 1992, Perczel *et al.* 1993]. In this regard, it is appropriate to mention that, for peptides in solution, Raman and *FTIR* spectroscopy may give information on the population of the different (possible) conformers but that the *NMR* data may correspond to an ensemble of rapidly interconverting conformers; this is an important point to be taken into consideration whenever it is being contemplated to use that information as a restraint in a simulation procedure (see Section 10.6). In addition, it should be noted that the solvent may be a very important factor, which could override the propensity for a secondary structure due to the sequence (Waterhous and Johnson 1994).

The designations *probabilistic, physicochemical, pattern recognition, neural network,* and *knowledge-based* systems are those most commonly assigned to the methods available for the prediction of secondary structures. Those designations reflect the main characteristics of the corresponding algorithms but it must be emphasized that, ultimately, every method is based on a set of rules which attempt to reproduce the patterns observed in proteins. As such they are all based on knowledge of experimental information, with the exception (to some extent) of the *a priori* work of Ptitsyn and Finkelstein (1983,1989).

The most popular methods, whether because of simplicity, originality, or success, are those of Chou-Fasman and Garnier-Osguthorpe-Robson (probabilistic), Lim and Cohen-Abarbanel-Kuntz-Fletterick (physicochemical and pattern recognition) and Lambert-Scheraga (pattern recognition). Their basic characteristics will be reviewed here but no attempt will be made to repeat (because of their length) the sets of rules devised in each case. As an example, it will suffice to examine briefly those put forward originally by Chou and Fasman, in order to provide the reader with a general idea of what is involved in this kind of work. A comparison of the patterns invoked in the various procedures may, however, be more interesting and useful, even in connection with the neural network and the knowledge-based methods. It is our belief that such methods will become the common tool of future work and for that reason they will be examined in detail in this Chapter.

In addition to the original references, where specific details will be found, the reader is also referred to the works of Richardson (1981), Kabsch and Sander (1983), Rose *et al.* (1985), Thornton (1988), Yada *et al.* (1988), Pascarella *et al.* (1990), Huang *et al.* (1990), and Rost *et al.* (1993,1994). [The work of Pascarella *et al.* (1990), in particular, provides information on corresponding software packages for microcomputers.]

9.2 Non-Learning Algorithms

The essence of the non-learning algorithms may be summarized as follows: The available, experimental information on secondary structures is analyzed and interpreted in terms of residue probabilities or pattern existence and a corresponding set of rules. As such, these methods are knowledge-based (i.e., based on experimental information) but they are non-learning, in the sense that the rules are established once and for all and used thereafter without any modification [see Section 9.3 for additional comments].

9.2.1 Ptitsyn-Finkelstein

The method of Ptitsyn and Finkelstein (1983ab) is based on a Statistical Mechanics treatment of chain molecules, whereby the short-, middle-, and long-range interactions, which decide the secondary structure, are taken into account. The model adopted, with the protein chain floating on a hydrophobic surface, may be described within the framework of the Ising model for a linear cooperative system. The Ising model (Ising 1925), although developed for the simulation of the structure of a ferromagnetic substance, may also be used to simulate other systems (Huang 1963).

The conformational formalism of Zimm and Bragg (1958), in which the state of each residue is determined by its conformation as well as the conformations of several preceding residues, is not of use in this case, because of the large number of residues that should be taken into account. This difficulty is bypassed by the use of the positional formalism (Finkelstein 1975, 1977), in which the state of each residue is described exclusively by its position in the given type of secondary structure.

The parameters used are based on experimental and stereochemical estimations of secondary structure, stability of synthetic peptides, and estimates of the free energy of interaction with the hydrophobic surface (which simulates the long-range interactions).

Table 9.1 presents the α-helix and β-chain character assigned to each individual residue in the work of Ptitsyn and Finkelstein (1983).

Table 9.1 α-helix and β-chain character of individual residues (Ptitsyn and Finkelstein 1983)

	α-helix	β-chain
former	leu, met, lys, glu, ala, arg, ile, phe, trp	val, ile, phe, tyr, trp, leu, cys, met, thr, his, lys
indifferent	tyr, gln, lys+, glu-, his, arg+, val, cys	glu, arg+, ala, gln, ser
breaker	his+, asn, asp, thr, ser, asp-, gly, pro	his+, lys+, arg+, glu-, asp, asn, asp-, glu, pro

9.2.2 Chou-Fasman

The statistical analysis of experimental information, performed by Chou and Fasman (Chou and Fasman 1974ab, 1978ab, 1979; Fasman 1989) resulted in the assignment, regarding α-helices and β-chains, of strong former (F), former (f), weak former (I), indifferent former (i), breaker (b), and strong breaker (B) character to the individual residues (see Table 9.2), with corresponding potentials P.

Table 9.2 α-helix and β-chain character of individual residues (Chou and Fasman 1974)[a]

	α-helix	β-chain
strong former (F)	glu-, ala, leu	met, val, ile
former (f)	his+, met, gln, trp, val, phe	cys, tyr, phe, gln, leu, thr, trp
weak former (I)	lys+, ile	ala
indifferent former (i)	asp-, thr, ser, arg+, cys	arg+, gly, asp-
breaker (b)	asn, tyr	lys+, ser, his+, asn, pro
strong breaker (B)	pro, gly	glu-

[a]The symbols F, f, I, i, b, and B are to be ascribed a subscript α or β, respectively, depending on whether they refer to α-helices or β-chains.

The prediction rules are based on a preliminary search for α-helix and β-chain segments, according to specified conditions, and the values of their corresponding potentials, $<P_\alpha>$ and $<P_\beta>$, obtained by averaging the P_i values of their residues.

The basic conditions are:

(a) α-helices

1. Nucleation. A cluster of 4 residues (F_α or f_α) out of 6 initiates a helix. I_α residues count in this respect as $(1/2)f_\alpha$.

2. Propagation. The helix is extended on both sides as long as adjacent tetrapeptides are not helix breakers. If overlapping segments satisfy the nucleation rule, they are lined together. At least half of the residues in the extended helix must be helix formers and less than one third of them should be helix breakers.

3. Termination. The tetrapeptides b_4, b_3i, b_3f, b_2i_2, b_2if, b_2f_2, bi_3, bi_2f, bif_2, and i_4 (as well as those with B, I, and F) are helix breakers. Some of the residues (f, i) in the tetrapeptides may be incorporated at the helix ends. Adjacent β-regions, with $<P_\beta> > <P_\alpha>$ terminate the helix.

4. Proline. It cannot occur inside a helix or at its C-end but it can occupy the first turn at is N-end.

5. Boundaries. Pro, asp$^-$, and glu$^-$ are incorporated at the N-end of the helix and his$^+$, lys$^+$, and arg$^+$ at its C-end. Near the N-end, pro and asp are accepted as I_α and glu as F_α; near the C-end, arg and lys are considered as I_α and his as f_α.

(b) β-chain

1. Nucleation. Three f_β or F_β residues or a cluster of three β-formers out of four or five residues will initiate a β-chain.

2. Propagation. The β-chain is extended on both sides as long as adjacent tetrapeptides are not β-chain-breakers. At least half of the residues in the extended β-chain must be chain formers and less than one third of them should be chain breakers.

3. Termination. Adjacent α-regions, with $<P_\alpha> > <P_\beta>$, terminate the chain.

4. Breakers. Glu and pro should not be incorporated into β-chains unless they occur in tetrapeptides with $<P_\alpha> < <P_\beta> > 1$.

5. Boundaries. Charged residues and pro should not be incorporated into β-chains unless they occur in tetrapeptides with $<P_\alpha> < <P_\beta> > 1$.

The prediction rules are then:

Rule 1: A segment of 6 or more residues, with $\langle P_\beta \rangle < \langle P_\alpha \rangle > 1.03$, satisfying the above conditions, is predicted as α-helical.

Rule 2: A segment of 3 or more residues, with $\langle P_\alpha \rangle < \langle P_\beta \rangle > 1.03$, satisfying the above conditions, is predicted as β-chain.

A third rule was also given in order to get rid of possible ambiguities but it is not presented here because the purpose of the above summary is to bring to the attention of the reader the empirical character of the prediction.

9.2.3 Garnier-Osguthorpe-Robson (*GOR*)

The *GOR* algorithm (Garnier *et al.* 1978) is based on the use of parameters $I(s_i;$ sequence), which define the state (α-helix, extended chain, turn, coil) of the residue at the i-th position, with consideration of all the residues in the sequence.

This ideal definition of the parameters I cannot be translated into practice on the basis of a statistical analysis of a finite sample of protein structures and recourse must be made to approximations, reducing the number of residues to be taken into account as influencing the state of the one at the i-th position. The simplest approximation consists of using a single-residue dependence but using an appropriate window. Thus, the approximation used in the *GOR* algorithm is

$$I(s_i; \text{sequence}) = \sum_m I(s_i; r_{i+m})$$

with $-8 \le m \le 8$ (i.e., a 17-residue window). That is, the prediction requires a set of (20 amino acids) $\times$ (4 states) $\times$ (17 sequence positions) = 1,360 parameters.

In this case, the existence of patterns is masked through the definition and use of the corresponding numerical parameters.

9.2.4 Lim

The original algorithm of Lim (1974) [modified by Thornton *et al.* (1991); see also the work of Parker and Hodges (1991b)] is based on a set of rules, with chosen patterns, further characterized by auxiliary conditions specifying relationships for pairs and triplets of residues. The general rules are extremely simple but the auxiliary conditions for the specification of the patterns are extremely complex.

The residues are classified as hydrophobic (h), hydrophilic (g), or passageway (s), according to the values of the corresponding hydrophilicity-accessibility factors (see Sections 8.2.1 and 8.2.2). The basic classes of residues are

hydrophobic residues
h_1: ile, leu, phe
h_2: cys, trp, val, met
h_3: ala, tyr, his, (pro)

hydrophilic residues
g_1: asp, glu, lys
g_2: asn, gln, arg
g_3: ser, thr, (pro)

passageway residues
s: gly, ser

and the groups used to define the patterns are

hydrophobic residues
$H_{LL} = h_1$: ile, leu, phe
$H_L = h_1 + h_2$: ile, leu, phe, cys, trp, val, met
$\bar{H}_L = H_L$-residues as well as tyr, gly
$\tilde{H}_L = \bar{H}_L$-residues as well as ala
$H_S = h_3$: ala, tyr, his, (pro)

hydrophilic residues
$G_{LL} = g_1$: asp, glu, lys
$G_L = g_1 + g_2$: asp, glu, lys, asn, gln, arg
$\bar{G}_L = G_L$-residues as well as gly
$G_S = g_3$: ser, thr, (pro)

passageway residues
$S = s$: gly, ser

The list of patterns is presented in Table 9.3.

9.2.5 Cohen-Abarbanel-Kuntz-Fletterick (*CAKF*)

The basic concepts of this algorithm (Cohen *et al.* 1983, 1986), which has been applied to α/β proteins, are as follows:

(a) proteins consist of one or more sequentially contiguous domains;

(b) each domain is made from α-helical and/or β-chain units connected by turns;

Table 9.3   Patterns used in the Lim algorithm (Lim 1974, Thornton *et al*. 1991)[a]

α-helices

$H_L \text{Gly} GHG\text{Gly} H$ $G_L(/R/RR'/RR'R'')H_L R_1 R_2 R_3 H_L$

$H_L R_1 R_2 R_3 H_L$ $G_L(/R/RR'/RR'R'')H_L(R/RR')H_L R_1 R_2 R_3 H_L$

$H_L(R/RR')H_L R_1 R_2 R_3 H_L$ $G_L(/R/RR'/RR'R'')H_L R_1 R_2 R_3 H_L(R/RR')H_L$

$H_L R_1 R_2 R_3 H_L(R/RR')H_L$ $H_L R_2 R_2 R_3 H_L(/R/RR'/RR'R'')G_L$

$G_L(G)_\ell H(G)_m G_L{}^b$ $H_L(R/RR')H_L R_1 R_2 R_3 H_L(/R/RR'/RR'R'')G_L$

$G_L(G)_\ell HH(G)_m G_L{}^b$ $H_L R_1 R_2 R_3 H_L(R/RR')H_L(/R/RR'/RR'R'')G_L$

β-chains

$(R)_\ell (\tilde{H}_L)_p (R)_m{}^c$ $\bar{H}_L \bar{G}_L (\tilde{H}_L \bar{G}_L)_p \bar{H}_L$

---

[a]See the text for the definitions of H, H_L, $\bar{H}_L$, $\tilde{H}_L$, G, G_L, and $\bar{G}$. R_1, R_2, and R_3 denote any amino acid (subject to certain specifications). The notations (R/RR') and $(\cdot/R/RR'/RR'R'')$ are used to indicate variable-length insertions of 1 or 2 and 0, 1, 2, or 3 residues, respectively.

[b]$\ell = 0$, $m = 2,3,4$; $\ell = 1$, $m = 2,3$; $\ell = 2$, $m = 0,1,2$; $\ell = 3$, $m = 0,1$; $\ell = 4$, $m = 0$.

[c]$p > 1$, with ℓ and m ranging from 0 to 2.

[d]$p \geq 1$

(c) the core of each domain is that subset of helices or extended units whose primary packing interactions are within the domain;

(d) the major interactions that position the secondary units with respect to each other derive from the packing of hydrophobic residues.

In order to bring out the similarities between this algorithm and that of Lim, the residue grouping of Cohen *et al*. (1983) has been adapted as follows. The basic groups of residues are

hydrophobic residues
h_1: ile, leu, val
h_2: cys, met
h_3: phe, trp, tyr,

hydrophilic residues
g_1: asp, glu, lys
g_2: asn, gln, arg, his
g_3: ser, thr

passageway residues
s: ser, gly, pro

charged residues
c_+: lys$^+$, arg$^+$
c_-: asp$^-$, glu$^-$

The groups of residues, used to define the patterns, are: s_1, s_2, s_3, h, a_1, a_2, p_1, +, -, n_1, and n_2 for α-helices; b_1, o, I, f, c, and n_1 for β-strands; e_1, e_2, n_1, and n_3 for edges of β-sheets; and t_1, t_2, and t_3 for turns. They are defined, in terms of the basic groups of residues, as follows:

a_1: $h_1 + h_2 + h_3 + \text{lys} + \text{pro} = h + \text{pro}$
a_2: $a_1 + g_3 + \text{gly} = h + g_3 + \text{gly} + \text{pro}$
b_1: $h_1 + h_2 + h_3 + \text{ala} + \text{pro}$
c: h_2
e_1: $g_1 + g_2 + g_3 + s = p_1$
e_2: b_1
f: $h_3 + \text{ala}$
h: $h_1 + h_2 + h_3 + \text{lys}$
I: $h_1 = s_1$
n_1: s
n_2: $h_1 + h_2 + h_3$
n_3: $h_1 + h_2 + h_3 + \text{ala}$
o: $g_1 + g_2 + g_3 + \text{gly}$
p_1: $g_1 + g_2 + g_3 + s = e_1$
s_1: $h_1 = I$
s_2: $h_1 + h_2$
s_3: $h_1 + h_2 + \text{ala}$
t_1: $g_1 + g_2 + g_3 + \text{gly} = o$
t_2: $t_1 + \text{tyr}$
t_3: $t_2 + \text{ala}$
+: c_+
-: c_-

Using the notation $H = h_1 + h_2 + h_3$ and $G = g_1 + g_2 + g_3$, the above list of groups may be rewritten, in a more concise manner, as h_1, h_2, $h_1 + h_2$, $h_1 + h_2 +$ ala, $h_3 + \text{ala}$, H, $H + \text{ala}$, $H + \text{ala} + \text{pro}$, $H + \text{lys}$, $H + \text{lys} + \text{pro}$, $G + \text{gly}$, $G + \text{gly} +$ tyr, $G + \text{ala} + \text{gly} + \text{tyr}$, $G + s$, and s, which brings out more clearly the similarities between the Lim and *CAKF* approaches.

The corresponding list of patterns is presented in Table 9.4.

Table 9.4 Patterns used in the Cohen-Abarbanal-Kuntz-Fletterick algorithm (Cohen *et al.* 1983)[a]

α-helices

$s_2{**}s_1s_1{**}s_1$	$h{**}hh{**}h$	$a_2{**}a_1a_1{**}a_1$
$s_1{**}s_3s_1{**}s_1$	$h{***}hh{**}h$	$a_1{**}a_2a_1{**}a_1$
$s_1{**}s_1s_3{**}s_1$	$hh{***}h{**}h$	$a_1{**}a_1a_2{**}a_1$
$s_1{**}s_1s_1{**}s_2$	$hh{**}h{***}h$	$a_1{**}a_1a_1{**}a_2$
	$hh{**}hh$	
$p_1{***}p_1{***}p_1$	$+{***}-$	$(3n_1,3*)$
$p_1{**}p_1{***}p_1$	$-{***}+$	$n_2n_2n_2n_2$
$p_1{***}p_1{**}p_1$	$+{**}-$	
	$-{**}+$	

β-strands

$b_1b_1b_1$	$ooii{**}oi$	$(3\ell,2f)$	$(2n_1,2*)$
$b_1{*}b_1b_1$	$ooi{*}i{*}oi$	$(3\ell,1f)$	
$b_1b_1{*}b_1$	$oo{*}ii{*}oi$	$(4\ell,1*)$	
	$ooi{**}ioi$	$(3\ell,1f,1c)$	
	$oo{*}i{*}ioi$		
	$oo{**}iioi$		

edges of β sheets

$e_2e_1e_1e_2$	$(2n_1,2*)$
	$n_3n_3n_3n_3n_3n_3$

turns

$(3t_1,1t_2)$	$(3t_1,1*)$	$(4t_1,4*)$
$(4t_1,1*)$	$(4t_1,2t_3,1*)$	$(5t_1,4*)$
$(5t_1,2*)$	$(3t_1,4t_3)$	$(4t_1,2t_3,3*)$

[a]Expressions of the form (mx,ny) mean that in the sequence of $(m+n)$ residues, m of them are of type x and n of them are of type y. The asterisks stand for other residues.

9.2.6 Lambert-Scheraga

The approach of Lambert and Scheraga (1989abc), based on the analysis of 6,733 tripeptides, consists of a pattern recognition-based importance-sampling minimization (*PRISM*).

The tripeptides are classified into 64 different conformational types (from four different ϕ, ψ angle combinations) and their conformational probabilities are evaluated from the available sequences using pattern recognition techniques.

Representing the conformation of the entire chain by a sequence of single-residue conformational states and denoting as chain-states the distinct conformations in this representation, one can then proceed to the evaluation of the probabilities of the chain-states from the tripeptide probabilities. The chain-state probability estimator is the product of conditional and marginal probabilities, obtained from the tripeptide probabilities, with a penalty factor which will eliminate conformations containing helices and strands of excessive length. The probability estimator considers short-range conformational information, medium-range sequence information, and some long-range information.

The procedure is then completed with an energy minimization (see Chapter 10).

9.2.7 Combination Methods

A selective combination of (some of) the above algorithms, choosing the best components of each one, will undoubtedly lead to improved predictions.

Thus, Biou *et al.* (1988) use the *GOR* algorithm, complemented with a homology component, with a bit pattern method added whenever no agreement is found by the other two methods. The procedure proposed by Fraga and co-workers (Fraga *et al.* 1990, Thornton *et al.* 1991ab) represents essentially an extension of the Lim algorithm, complemented with a *GOR* component and consideration of the recognition factors (see Section 8.2.1). Similarly, the procedure of Parker and Hodges (Parker *et al.* 1986, Parker and Hodges 1991ab) uses a combination of *HPLC* hydrophilicity as well as accessibility and flexibility parameters to predict surface/loop/turn regions and a combination of Chou-Fasman and Lim algorithms to predict the structural character of the units bounded by the surface (break) regions. In another example, the secondary structure prediction method of Wishart *et al.* (1994) combines the Chou-Fasman and the Garnier-Gibrat algorithms with homology- and motif-based predictions.

Extension of the above argument suggests the use of most of/all the existing methods in order to reach a consensus, which could be complemented by consideration of additional information. The most complete proposal, in this respect, is embodied in the *flowchart* of Taylor (1987), presented in Fig. 9.1.

9.3 Learning Systems

Although, of necessity, the complete details of the preceding non-learning procedures have been omitted, their essential components have become evident. Essentially, all those algorithms represent variations of a same approach, differing only in the conditions considered and the weight attached to them. The predictions are based on the consideration of the nature of the individual residues (former, indifferent, or breaker/hydrophobic, hydrophilic, or passageway), short-range

interactions (doublets and triplets), medium-range interactions (patterns), and long-range interactions.

The rules embodied in the respective algorithms are derived through *training* on the basis of a chosen set of sequences of known secondary structure. When the algorithm is then used for the predictive study of a new protein, those rules are followed strictly. In that sense, it could be said that those algorithms represent flowcharts.

In order to change those rules (that is, to upgrade them), a new training session must be carried out, using as input an extended set of sequences with known structure. Therefore one could say that those algorithms are capable of being upgraded. The deficiency in the upgrading process arises from the rigidity in the corresponding flowchart, so that its change, when implemented in a software package, may require a laborious effort. That is the reason for having denoted them as non-learning procedures, even though such a designation does not do them justice (especially in the case of the *GOR* algorithm, with its use of decision constants).

These difficulties may be overcome through the use of learning systems, such as networks, instead of flowcharts, as discussed below. But first, some additional comments are needed, because of the some-times confusing terminology in use. The designations in mind are *training, learning, machine-learning, logic-based machine-learning*, regarding the development of the operating rules, as well as *knowledge-based architecture* or *knowledge-based systems*, with terms such as *flowcharts, networks* (in general), *neural networks, knowledge-based networks*, and *symbolic* and *non-symbolic learning methods*, when describing the complete system.

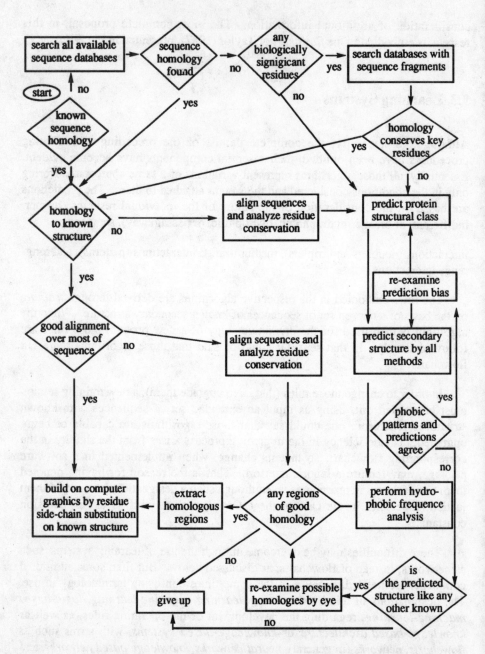

Figure 9.1 Flowchart of possible paths in the prediction of protein structures [from W.R. Taylor, *Protein Structure Prediction*, in *Nucleic Acid and Protein Sequence Analysis. A Practical Approach*, edited by M.J. Bishop and C.J. Rawlings, pp.285-322 (1987). Copyright © 1987 IRL Press, Oxford, U.K. Reprinted by permission of Oxford Univesity Press.]

All the training/learning, when performed through the use of computer software, constitutes machine-learning and, hopefully, all the machine-learning should be based on logic, thus being logic-based machine-learning. For simplicity, the terms *training* and *learning* will be used, interchangeably, to denote logic-based machine-learning.

Both flowcharts and networks are based on descriptive and strategic knowledge as well as logical inference (Clark *et al*. 1990) and therefore represent knowledge-based systems (architecture). The distinction is made in the literature between knowledge-based and neural networks, depending on whether they use symbolic or non-symbolic learning methods, respectively. Taking into account that both are knowledge-based, the above designations may lead to confusion and, consequently, hereafter the terms non-symbolic networks and symbolic networks will be used.

Non-symbolic networks are closely related to statistical learning methods whereas symbolic networks manipulate symbols or patterns. Non-symbolic networks have numerical outputs that cannot be interpreted as chemical properties while symbolic networks can lead to undiscovered patterns in chemical properties (King and Sternberg 1990).

9.3.1 Non-symbolic Networks

The applicability of non-symbolic networks (hereafter denoted as *NS*-networks) is quite wide, but here we will restrict the discussion to their use in the prediction of secondary structures. [The reader is referred to the work of Hopfield (1982) for a more general discussion and to the works of Bohr *et al*. (1988, 1990), Qian and Sejnowski (1988), McGregor *et al*. (1989, 1990), Holley and Karplus (1989), and Hayward and Collins (1992) for specific details.]

A *NS*-network consists of a number of layers, each of which is in turn constituted by a number of nodes. The number of layers and of nodes depends on the problem under consideration. The first layer is denoted as input layer while the last one is the output layer; the layers between the input and the output layers are the hidden layers, which are not needed when the problem under study is linearly separable. There is no communication between the nodes of the same layer and transmission takes place only from one layer to the one immediately above it, each node in the lower layer sending signals to all the nodes in the upper layer (see Fig. 9.2).

Learning in a *NS*-network is achieved through the so-called back-propagation of errors, that is, by minimization of the difference between the actual output of the network and the desired output, as known for the training set of data (which, in the present case, is a set of sequences with known secondary structure).

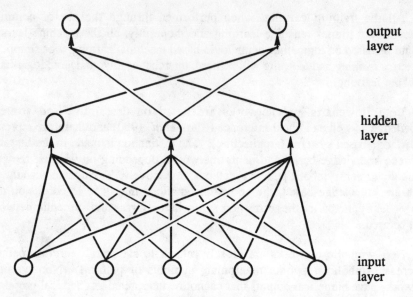

Figure 9.2   Schematic representation of the topology of a feed-forward layered neural network with 5 input nodes, 3 hidden nodes, and 2 output nodes.

The mathematical formulation for the working of a NS-network is straightforward and simple. Denoting as $x_{n(i)}$ the input of the n-th node in the i-th layer, which is generated from the outputs $y_{m(i-1)}$ of the m nodes in the $(i-1)$-th layer, one can write

$$x_{ni} = \sum_m w_{m(i-1)} y_{m(i-1)} + b_{ni}$$

where $w_{m(i-1)}$ denotes the appropriate weights (positive or negative) and b_{ni} are appropriate biases. The output and input of a given node are related by a sigmoid curve

$$y_{ni} = 1/(1 + e^{-x_{ni}/k})$$

where k is a constant.

Initially, the weights are assigned random values. The input (from the training data) sets the state (i.e., a number between 0 and 1) of each node in the input layer. This information constitutes the output of the input layer, which is sent to the first hidden layer, and so on, until the final information reaches the output layer. At the end of the cycle the weights are adjusted and a new cycle is started. The adjustment of the weights results (Holley and Karplus 1989) in a gradient descent in the total output error

$$\varepsilon = \sum_c \sum_j (O_{jc} - D_{jc})^2$$

where O_{jc} and D_{jc} denote the observed and desired output in unit j in the cycle c; the training continues until ε becomes asymptotic or, in practice, smaller than a given threshold.

The specific characteristics of a NS-network for the prediction of secondary structures are easily determined. In each cycle, the complete set of training sequences is presented, one at a time, to the input layer and, therefore, the first point to be settled is the size of the window, N_w, which will transfer the sequence information to the input layer. For example, for a full prediction in terms of α-helices and β-strands, a 17-residue window (see Section 9.2.3) may be appropriate, while for the exclusive prediction of β-turns (which involve 4 residues) a 4-residue window will suffice. Taking into account that there are 20 natural amino acids, it is necessary to assign 20 input nodes to each position in the window. This number, to be denoted by N_r, may be increased to 21 if the appearance of the N- or C-terminus within the window is to be taken into account; naturally, when studying peptides with non-natural as well as natural amino acids, that number should be increased appropriately. Therefore, the input layer will consist of $N_w \cdot N_r$ nodes, which will be labelled by the indices $i = 1, ..., N_w$ and $j = 1, ..., N_r$.

The input information is then transmitted as follows: the state of the j-th node for the i-th position will be one if the j-th residue is found at that i-th position in the window and 0 otherwise. That is: when the input has been fully loaded, N_r of the $N_w \cdot N_r$ input nodes will have the value 1 while all the others are set to 0.

The input nodes, therefore, contain the information corresponding to the character of the individual amino acids. This is the information which is sent as input to the first hidden layer. The number of nodes in this layer depends on how the information is to be processed, taking into account that each node in that layer receives information from all the nodes in the input layer. The processing of the information in this layer implies consideration of the pattern presented through the window.

The number of nodes in the output layer depend on the output desired. Thus, for the prediction of α-helices and β-strands one needs 2 output nodes while in the prediction of β-turns one might use 4 nodes (for type-I, type-II, non-specific turn, and non-turn) or increase that number appropriately if more types of β-turns are to be studied.

Thus, Holley and Karplus (1989) used a NS-network consisting of 3 layers, with 357 nodes in the input layer, 2 nodes in the hidden layer, and 2 nodes in the output layer, while McGregor *et al.* (1989) used also a 3-layer network, but consisting of 80, 8, and 4 nodes, respectively.

The output is the value found in the output node with the highest value and it may correspond to, e.g., the central residue in the window or the complete sequence in the window, depending on how the network has been trained. The output numbers are then translated into an actual prediction, with the use of an appropriate threshold value.

The performance of the *NS*-networks may be improved by a modification of the selection procedure of the data set in the training/testing sessions, increasing the number of units in the hidden layer, taking into account the correlation of consecutive patterns, averaging the results from independent networks, and using multiple sequence alignments instead of single sequences as input. These improvements, as incorporated into the *PHD* method of Rost and Schneider (1993), are summarized below.

The training/testing was performed in 7 sessions, using a set of 130 proteins. In each of the first 6 sessions, the testing was performed on 19 different proteins, with the remaining 111 proteins being used for the training. In the last session, only 16 chains were used for testing and the remaining 114 proteins for the training. In this way all the proteins were tested once.

The correlation of patterns was taken into account by using a network system with 2 network levels. The networks in the first level use a window of 13 residues, with 21 basic units per window position. The value of the output unit i for pattern v is calculated as

$$s_{i(2,v)} = f \left\{ \sum_{j=1}^{N_{i}+1} J_{ij(2)} \, f \left[\sum_{k=1}^{N_0+1} J_{jk(1)} \, s_{k(0,v)} \right] \right\}$$

where N_0 is the number of input units, N_1 is the number of units in the hidden layer, $s_{k(0,v)}$ is the input of pattern v to the k-th input unit, f is the sigmoid function (see above), and $J_{ij(\lambda)}$ is the junction between unit j in layer λ-1 and the unit i in layer λ (with $\lambda = 0, 1, 2$ for the input, hidden, and output layers, respectively). At each training cycle t, the values of the junctions are changed

$$\Delta J_{t+1} = J_t - \varepsilon \left(\frac{\partial E}{\partial J} \right)_t + \eta \Delta J_{t-1}$$

so that the error decreases (see above). The initial values of the junctions are chosen at random in the interval -0.1 through 0.1; ε and η are two constants. The second level, with a window of 17, takes as input the output from the preceding level; its output is the secondary structure classification for the central input cell.

The results from independent networks are then combined by means of an arithmetic average.

The main criticism of *NS*-networks is centered on the numerical nature of the output, which is considered as less comprehensible to the user (King and Sternberg 1990).

9.3.2 Symbolic Networks

A symbolic network (hereafter abbreviated as *S*-network) resembles a flowchart, in the sense that it consists of nodes and links, but there are important differences regarding representation, modularity, initiation, knowledge, and functionality (Clark *et al.* 1990).

The basic idea in a *S*-network is the same as for the non-learning methods: given a set of proteins of known sequence and whatever additional information is available, develop general relationships/rules that describe well that information and may, therefore, be used for prediction purposes. The goal may be not only the prediction of secondary structures but also the prediction of tertiary structures as well as other pertinent information, with integration of all the available knowledge and existing auxiliary tools. What distinguishes them from the non-learning methods is the symbolic approach, with *if-then* rules, which allows for updating as the knowledge available increases, as well as the possibility of detecting relationships which otherwise might have escaped observation. Their deficiencies arise from those in the databases used as well as in the auxiliary tools (see below).

Before proceeding with the inspection of specific examples of *S*-networks it may be appropriate to expand on the subject of databases, which constitute the foundation for all the predictive work (see Section 8.1). Within the context of this Chapter, mention must be made, in particular, of the relational (*RDB*) and object-oriented (*OODB*) databases of Islam and Sternberg (1989) and Gray *et al.* (1990), respectively, which might be considered as incipient *S*-networks. The difference between *RDB* and *OODB* databases, as discussed below, is how the database is set up and how the information is accessed.

The *RDB* of Islam and Sternberg (1989), with the *ORACLE* management system and the *SQL* (structured query language), is based on the data for 294 proteins in the Brookhaven Data Bank (see Section 8.1) and may be used in order to develop semiempirical rules for protein conformations. The information is stored in tables (structure, crystal, chain, site, residue, neighbor, *H*-bond, salt bridge, disulfide bridge, atom), with one row per entry and one column per term. For example, the table for residues has columns for the terms unique identifier, *BDB* code, type of residue, residue name, 1-letter code, unique entry number, sequence number of residue from *N*-terminus, ϕ dihedral angle, ψ dihedral angle, secondary structure assignment, etc., and there is one row per residue (as well as ligands) in the protein chain. The search is fast when scanning by columns but it is slow when scanning by rows for cross-correlation of tables (Gray *et al.* 1990).

The *OODB* of Gray *et al.* (1990), with routines in both the logic programming language *PROLOG* and the query language *DAPLEX*, can integrate calculations with the retrieval of data from the database. The *OODB* consists of objects (such as protein, chain, structure, helix, strand, sheet, loop, *H*-bond, salt-bridge, residue, atom, ...) and relationships. The queries are processed fast by following relationships from one object to another but, as the objects are not stored sequentially, the equivalent of term scanning is slow. The fact that the data are partitioned into modules adds flexibility to this database, as the user has the possibility of incorporating new modules for supplementary information or for intermediate working objects and relationships.

Although the Protein Machine Induction System (*PROMIS*) of King and Sternberg (1990) has been superseded by more recent work, as discussed below, it contains the main elements of a *S*-network. Its input consists of a set of proteins of known sequence and secondary structure and the chemical and physical properties of the residues. The learning (see Fig. 9.3), based on the idea of searching for powerful generalizations (Mitchell 1982, Dietterich and Michalski 1983, Carbonell and Langley 1987) in the database, ends when more powerful rules cannot be found.

PROMIS only specified 3 residues and, in order to overcome this deficiency, the residues are grouped into classes (Taylor 1986), which in turn are used to specify patterns. The basic classes of residues are:

size
    tiny: a, g, s
    small: p, v, c, a, g, t, s, n, d
    large: q, e, r, k, h, w, y, f, m, l, i

hydrophobicity/hydrophilicity
    very hydrophobic: f, m, l, i, v, a, g
    hydrophobic: h, w, y, f, m, l, i, v, c, a g, t, k
    hydrophilic: s, n, d, e, q, r

charge
    neutral: a, q, n, l, g, s, v, t, p, i, m, f, y, c, w
    polar: t, s, n, d, e, q, r, k, h, w, y
    positive: r, k, h
    negative: d, e

chemical character
    aliphatic: l, i, v
    aromatic: h, w, y, f

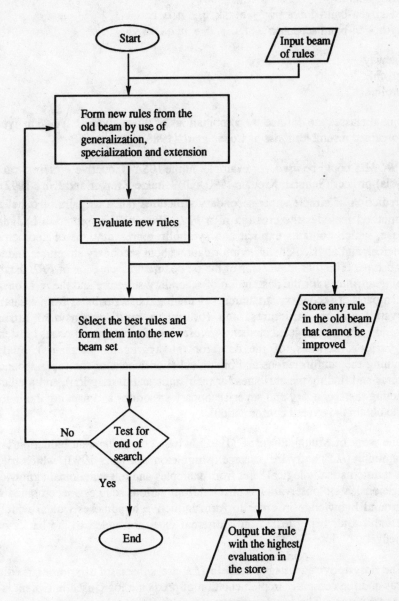

Figure 9.3 Flow of the *PROMIS* algorithm [From R.D. King and M.J.E. Sternberg, *Machine Learning Approach for the Prediction of Protein Secondary Structure*, Journal of Molecular Biology *216*, 441-457 (1990). Copyright © 1990 Academic Press Ltd., London, U.K. Reprinted by permission of Academic Press Ltd.]

hydrogen-bonding
    hydrogen-bond donor: w, y, h, t, k, c, s, n, q, r
    hydrogen-bond acceptor: y, t, c, s, d, e, n, q

passageway
    glycine
    proline

Additional classes are defined by combinations of those given above. The type of rules obtained resembles those of Cohen *et al.* (1983).

PROMIS could be used, for example, in the *IPSA* (Inductive Protein Structure Analysis) project (Schulze-Kremer 1990 and Schulze-Kremer and King 1992) for the prediction of simple super-secondary structures (such as pairs of α-helices). This method includes the creation of a *PRL* (Protein Representation Language) database, which contains explicit and symbolic representations of geometrical, topological, and physico-chemical information about secondary structures and their relationships. The four main steps of the procedure are: creation of *PRL* database (defining attributes for the description of secondary structure and the relationships between pairs of secondary structures, calculating those attributes from a database of crystallographic structures, and forming a multi-dimensional attribute description of the aggregated elements), clustering (forming a consensus clustering, from various methods, of the data for the aggregated elements), analysis (examining the conformations and biological significance of the super-secondary structures and finding the attributes for their statistical description), and prediction (predicting the secondary and super-secondary structures and packing the latter in order to obtain the overall conformation).

The work of Muggleton *et al.* (1992) is based on *Golem*, an Induction Logic Programming (*ILP*) software package (Muggleton and Feng 1990), which tries to discover automatically logical rules from examples and relevant domain knowledge (Muggleton 1991): observations collected from various sources are combined with background knowledge in order to form inductive hypotheses (rules) which, if found valid after being tested on additional data, are added to the background knowledge.

The descriptive languages used in *ILP* are subsets of first-order predicate calculus and the computer implementation of predicate logic used in *Golem* is the language *Prolog*. The input consists of positive examples, negative facts, and background knowledge and the output are rules based on the logical idea of Relative Least General Generalization: that is, they are the least general rules which, given the background knowledge, can reproduce the positive examples but none of the negative examples.

The various predicates used are the sequential relationships (which index the protein sequence relative to the residue being predicted, with a 9-residue window), sequences important in the formation of α-helices (Lim 1974), chemical and physical properties (see below), relative sizes, and relative hydrophobicities. The chemical and physical properties are described by predicates corresponding to the basic residue classes used by *PROMIS* as well as the following ones: not aromatic, small or polar, not *p*, not *k*, aromatic or very hydrophobic, either aromatic or aliphatic or *m*.

The *S*-network of Clark *et al.* (1990), written in *PROPS2* (a *Prolog* production system interpreter and knowledge programming language), is intended for a purpose far more general than the prediction of secondary structures. Entities (data) and relations (processing methods and constraints) have been built up in a network of 29 nodes and about 100 links, which represents the architecture for a prototype knowledge-based system that strives to simulate logically the processes used in the prediction of protein structures.

The entities used in this work are: biological substance (biological source, *DNA*, protein), structural description (*DNA* sequence, protein sequence, secondary structure, topology, tertiary structure, quaternary structure), classifications and identifiers (*DNA ID*, protein *ID*, functional classification, tertiary structural class, quaternary structural class), results of biophysical and biochemical assays (*2D-NMR* distance constraints, secondary structure composition, results of proteolysis, disulfide linkage, domain positions, results of *X*-ray studies, sequence composition, results of mutagenesis experiments, results of other experiments), and results of database queries, sequence analyses and other software (internal repeats, similar sequences, alignment, important patterns/regions, sequence profiles, similar structures). The relations or links correspond to either constraints (i.e., consistency requirements between entities) and minimal preconditions (i.e., processes that relate entities).

The prototype system (Fig. 9.4) includes an input/output module, a network description (set of propositions), a transaction manager (which contains a transaction database and decides which transactions are permissible at each stage), a browser (which explains the features of the network and of the transaction database), and an advice manager (which responds to the queries from the user).

The formal transactions are knowledge entry, node derivation (derivation of values for one entity from another), consistency checking (for the permissibility of values for two entities), node updating (by changing the value of an entity according to new constraints or in order to make it consistent with another node), and retraction (eliminating user-supplied information or withdrawing system processes). The system can operate in either a data-driven mode (in which data are supplied and possible actions are requested) and goal-directed mode (when the system is informed of the goal to be attained and the system is requested to provide the actions required).

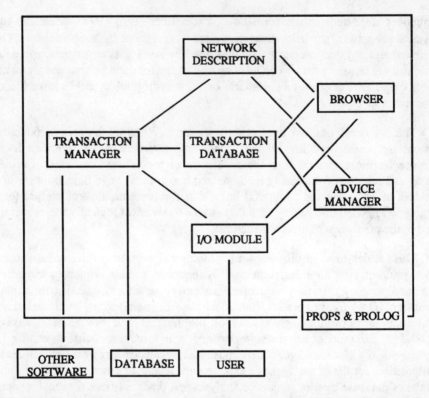

Figure 9.4   Architecture of the prototype system for *S*-network [From D.A. Clark, G.J. Barton, and C.J. Rawlings, *A Knowledge-Based Architecture for Protein Sequence Analysis and Structure Predictions*, Journal of Molecular Graphics *8*, 94-107 (1990). Copyright © 1990 Elsevier Science Inc., New York, U.S.A. Reprinted by permission of Elsevier Science Inc.]

In this system, the prediction of the secondary structure is based on a variety of procedures, such as Chou-Fasman and *GOR* (see Sections 9.2.2 and 9.2.3, respectively), and may make use of additional constraints, knowledge of the structural class, the *majority rule* of Cohen *et al.* (1986), etc.

9.4 Summary

A satisfactory prediction of the secondary structure of proteins is yet to be attained, generally speaking, although there are particular cases where the various methods have succeeded. What is undeniable is the staggering efforts made in the development of the algorithms.

We feel that the future belongs to the S-networks, but clearly some work is needed in the improvement of algorithms used in them for the secondary structure predictions. As conceived at this moment, that translates into an upgrading, from new experimental information, of the patterns to be used. That is the reason for the importance attached to patterns in this Chapter.

But taking into account the power of the NS-networks, one wonders whether in the future a NS-network for the prediction of secondary structures could not be incorporated into a general-purpose S-network.

Discussion of those methods based preferentially on the use of homologous sequences has been postponed to Chapter 10, where they will be examined in conjunction with the subsequent prediction of the tertiary structure. The reader is referred to the work of Benner et al. (1994) for a summary of the successes achieved with them.

10 Modeling of Tertiary Structures

10.1 Introduction

Preceding chapters, but particularly Chapters 6-9 (covering methods, software packages, data, and secondary-structure predictive procedures), provide the basis for the discussion of specific details and practical applications to be presented in this and following chapters. But before proceeding with that discussion some comments may help the reader in understanding the reasons for the approach adopted.

These comments refer to the goal of this work, which is to assist the readers (especially those who venture into this field for the first time) in obtaining an overall perspective of the type of work which is possible, the means available for its execution, and the possible results that one might obtain. All this while keeping in mind that, even though there is a very important component of research *per se*, ultimately one is striving for the development of new enzymes and catalysts, switches, vaccines, and drugs, just to mention a few of the practical rewards which are within reach.

Therefore, we have not attempted to present a complete summary of all the practical contributions to date. The number of such contributions is staggering and keeps on growing constantly, with the result that any such summary is immediately outdated. In addition, with presently available literature-search systems, the reader should have no difficulty in locating those contributions on a specific subject.

Consequently, our attention will be focused on the approaches that have been tried and the problems to be faced. Paramount among them is the existence of multiple minima and the desired accuracy of the results. In this regard it suffices to mention, for example, that enzyme mechanisms are sensitive to displacements of as little as 0.1 Å in the positions of the atoms involved in the enzymatic process (Levitt 1976).

Therefore, we will discuss first the problem of multiple minima before proceeding with an examination of the energy minimization procedures and the Monte Carlo (*MC*) and Molecular Dynamics (*MD*) simulations.

10.2 The Problem of Multiple Minima

A polypeptide chain (corresponding, say, to an intermediate-size peptide but particularly in the case of a globular protein) will have a considerable number of degrees of freedom, with the result that its energy hypersurface may present a large number of local minima, in addition to a global minimum (see Section 6.2.1).

This statement is still valid, but more general, if one substitutes *system* for *peptide* chain, in which case the designation system may stand for a peptide chain by itself or in interaction with other molecules. This is an important point as it relates to the different conditions in real life situations, experiments, and theoretical simulations.

In a real life situation (such as for a protein in a biological environment or an enzyme in an industrial process), the protein may interact with a variety of systems, such as, for example, other cellular components, the aqueous/non-aqueous medium in which it is immersed, receptors, substrates, etc. In addition, the influence of the thermal environment as well as the fact that the protein may be involved in a dynamic process will determine that its conformation may undergo fluctuations, most noticeable for short peptides in solution. But even the crystallographic structure of a protein is the result of other interactions besides the intramolecular ones. Finally, the results of theoretical simulations may correspond to a protein in vacuum, unless the solvent, substrates, etc., have been considered purposefully.

Therefore, for a given protein, one may distinguish between native conformation (with the ambiguity that such a designation implies, taking into account the above comments), its crystallographic structure, its solution structure and the lowest-energy conformation obtained in a theoretical simulation. The question then arises of whether and how such structures are related, which translates into whether one should be looking for the global minimum. The answer is a qualified yes, because, as discussed in Section 6.1, the corresponding structure may be used as starting point for more sophisticated calculations. A more appropriate answer, however, is that one should try and identify not only the global minimum but also that subset of local minima in its vicinity (i.e., within a given energy interval above the global minimum). In either case, the problem of ascertaining whether the global minimum has been identified will subsist and it is in this connection that it is of interest to examine the behavior of the distribution of potential energy minima [Perez *et al.* 1994; see also the work of Straub and Thirumalai (1993) on the distribution of potential energy barriers in proteins].

In the work of Perez *et al.* (1994), a comparison is made between the results obtained in an actual conformational search and those derived from a theoretical model. The calculations were performed for the hexanoic, octanoic, nonanoic, and dodecanoic acids, using the QUANTA/CHARMm software package, and the search was carried out by cumulative rotation of all the proper dihedral angles at random. Initially, 500 conformations were generated for each system. These conformations were then minimized, until a gradient smaller than 10^{-5} kcal/Å mol was reached. Of all the resulting structures, only those differing by at least 15° in any dihedral angle were retained. The procedure was then continued by minimizing 500 additional conformations, derived randomly from the last conformer obtained in the preceding step. For the hexanoic, octanoic, and nonanoic acids the search was continued until no new conformational minima were found; for dodecanoic acid, however, the search was stopped after 50 cycles. The number of identified minima was 183, 1,912, 5,196, and 13,170, respectively, the last figure being probably too small. [The results for hexanoic and octanoic acids were confirmed through a search by systematic nested rotations, whereby the conformations were generated sequentially by scanning each dihedral angle from 0° to 360°, in steps of 30°, excluding those conformations with contacts of the van der Waals spheres. The number of identified minima were 180 and 1,404, respectively, which indicates that the sequential search is inefficient, missing most of the low- and high-energy conformations.]

The results of these calculations may be very easily visualized in graphic form. Using a chosen value for the energy window (i.e., energy interval), one can construct a histogram of the number of minima *vs.* their energy (referred to the energy of the global minimum, taken as zero). Such a histogram shows a qualitatively Gaussian behavior, first with a fast rise towards a maximum value and then a fast decrease for the high-lying minima.

The theoretical derivation, for a model chain, is based on the rotational isomeric-state approach of Volkenstein (1963) and Flory (1969). Considering f torsional degrees of freedom, with $(m_i + 1)$ rotational minima associated with the i-th degree of freedom (with equal spacing ε_0), the total number of minima is given by

$$\Omega_{tot} = \sum_{E_0=0}^{\max E_0} \Omega(E_0) = \prod_{i=1}^{f} (m_i + 1)$$

where

$$E_0 = U_0/\varepsilon_0 \qquad \max E_0 = \sum_{i=0}^{f} m_i$$

and U_0 denotes the potential energy and $\Omega(E_0)$ the number of conformational minima with the (dimensionless) energy E_0. In the case of identical torsions one has

$$\Omega_{tot} = (m + 1)^f \qquad \max E_0 = mf$$

(with a single value m). The solution of the corresponding combinatorial problem yields

$$\Omega(E_0) = \sum_{s=0}^{int[E_0/(m+1)]} (-1)^s \binom{f}{s} \binom{E_0 + f - 1 - s(m+1)}{f-1}$$

which shows that the distribution of minima is symmetric around a maximum value Ω^* (corresponding to $E_0^* = mf/2$), which represents the most probable number of conformational minima (per energy window) and which may be approximated as

$$\Omega^* \simeq e^{\sqrt{m}(af-1)-bf}$$

with $a = 0.913\pm0.022$, $b = 0.243\pm0.035$, and a square correlation coefficient of 0.9988. This equation, together with the dependence of the total number of minima Ω and the position of the maximum E_0, may be used to compare this model with the results from the conformational search. This comparison suggests an average value $m \simeq 1.10$ (using an energy window $\Delta U_0 = 0.5$ kcal/mol), indicating the existence of 2 to 3 minima per rotational degree of freedom.

Two main points arise from this work. On one hand, we can easily see the staggering number of conformational minima for a protein with a large number of degrees of freedom. More important, however, is the fact that, being able to estimate the distribution of minima (for a given energy window), it will allow us to decide whether, in a conformational search, one is zeroing on the global minimum.

The implications of this work with respect to conformational search strategies have been analyzed by Centeno and Perez (1995). For met-enkephalin and bradykinin, a comparison of the distribution of minima, obtained in a thorough conformational search, with the density of states obtained within the rotational isomeric approximation confirms the above description. The conclusion is that *the genetic algorithms* (see Section 10.3.4), *simulated annealing* (see Section 10.6), *or techniques biased to obtain lower-energy conformations* (Berg 1993) *are the most efficient in finding the global minimum*. To this list, one should add the *taboo search technique* (Cvijovic and Klinowski 1995), which is a learning procedure that makes use of information obtained during preceding iterations.

10.3 Energy Minimization

The simplest application of energy minimization, though without predictive value, consists of the refinement of crystallographic structures. But, ultimately, energy minimization simulations are performed with a predictive goal in mind: that is, with the purpose of determining the structure of the peptide/protein under the appropriate conditions to be considered.

The problem of the multiple minima renders such a task extremely difficult and it is for that reason that a variety of procedures have been developed in order to overcome the problem. Those procedures may involve either an approximation (through the use of a simplified *PEF*) or an appropriate variation in the build-up process of the peptide chain.

Some comments on practical details (regarding Cartesian *vs.* internal coordinates, sequential *vs.* parallel calculations, and user control and visualization), which may have relevance for all the methods, are better examined first.

The structures of the individual amino acids (see Appendix 2) as well as those of peptide chains obtained in the simulations are given in terms of the Cartesian coordinates of their constituent atoms. The search through the conformational space implies changes in those coordinates, which are best introduced by changing the internal coordinates, i.e., the bond lengths and angles and/or the torsion angles; only the latter are changed in rigid-geometry calculations (see Section 6.4.3 for comments in the case of molecular dynamics simulations).

In actual energy calculations, however, some of the contributions are usually evaluated from the Cartesian coordinates. Thus, for example, in the evaluation of the non-bonded interactions, Eq. (5.32), the interatomic separations R_{ab} are evaluated from the Cartesian coordinates of the two atoms. The calculation of this interaction is expensive because it involves a double summation over the atoms in the two regions A and B. Such a double summation is usually coded, in a FORTRAN program, as a double '*do-loop*', and the calculation is performed in a *sequential* fashion: first, the interaction of atom a_1 of region A with all the atoms in region B is evaluated; then the interaction of atom a_2 of region A with all the atoms in region B is evaluated; and so on.

An increase in the speed of the calculation may be achieved very easily with a multi-processor computing facility. Let us assume, for example, that a 4-processor facility is to be used for the calculations. The task would be distributed so that each processor performs the evaluation of the contribution of one of the terms (in R_{ab}^{-1}, R_{ab}^{-4}, R_{ab}^{-6}, or R_{ab}^{-12}, respectively) of Eq. (5.32). The calculations in each processor are nearly equivalent and therefore the computing time would be reduced by almost a factor 4. A parallelization developed by Totrov and Abagyan (1994), operating in the space of the internal coordinates has, in fact, achieved a speed-up by a factor of 3 using 4 processors. One can imagine the savings in computing time that would be attained with a computing facility consisting of, say, 1,024 processors. [See the work of Totrov and Abagyan (1994) for additional references on parallelization.]

Visualization is an extremely important requirement in computer simulations, whose progress could not be followed so easily from the resulting numerical values. It is even more important whenever user participation is required, say, when the course of the simulation is to be altered (e.g., in order to escape from a local minimum, as mentioned in Section 6.2.1). A system combining visualization with user control will help in the modeling of large proteins and in understanding the role of the various interactions. The modeling system *SCULPT*, developed by Surles *et al.* (1994), with a *PEF* consisting of non-bonded interactions, *H*-bonds and torsion angles contributions, has been applied to the reversing of the direction of bundle packing in a 4-helix bundle protein, the folding-up of a 2-stranded β-ribbon into an appropriate β-barrel, and the design of the sequence and conformation of a 30-residue peptide that mimics one of the partners of a protein subunit interaction.

10.3.1 Refinement of Crystallographic Structures

The crystallographic structure of a protein corresponds to the global minimum in the energy hypersurface for the conditions existing in the crystal. Therefore, an energy minimization of such a structure (isolated, in vacuum) will result in its relaxation. The extent of that relaxation will depend on the strains present in the crystal as well as on the quality of the *PEF* used and care must be exercised in order to ensure that the distortion observed is not an artifact due to a poor *PEF*.

These refinements serve a variety of purposes. On one hand, of course, they may help in judging the quality of the *PEF* but, more important, they will yield a structure which may be used for a discussion of the possible function of the protein or as a starting point for a *MC* or a *MD* simulation.

As an example, we can mention the refinement of the *X*-ray crystallographic structure of hen egg-white lysozyme performed by Levitt (1974), with a *PEF* including R^{-6}- and R^{-12}- terms for the non-bonded interactions as well as terms for the changes in bond lengths, bond angles, and dihedral angles (see Sections 5.3.3-5.3.5), but ignoring the *H*-atoms. The energy minimization, through a steepest-descent procedure, was continued until the stereochemistry was satisfactory, that is, with relaxed bond lengths, bond angles, and dihedral angles, without contacts between non-bonded atoms, and an increased number of *H*-bonds.

10.3.2 Low-resolution Predictions

A preliminary, low-resolution exploration of the energy hypersurface may be performed with a simplified *PEF*.

Crippen and Viswanadhan (1984) used a residue-residue potential function, with one point (centered at the C_α) per residue. This function consists of a sum of pairwise, isotropic interaction terms, which should present a single minimum. The potential energy function was parameterized, from a set of 35 known crystal structures, as follows. First, considering groups of points (i.e., residues) that lie close together in the crystal structures and assuming a Boltzmann distribution of the unknown potential energy, that energy was deduced in order to produce such a distribution. Then, considering slightly larger groups of residues, the function derived so far was tested in order to see whether it yielded a Boltzmann distribution for the new clusters, adding new interaction terms if needed. The long-range interaction terms were adjusted so that the crystal structure of bovine pancreatic trypsin inhibitor (*BPTI*) would have a lower energy than other (distorted) conformations. The resulting *PEF* does not attempt to compress or to expand the crystal structures and maintains the features of the secondary structure.

Wallqvist and Ullner (1994), similarly, reduce each residue to two or three interaction points; one point represents the backbone while the side chain is represented by one or two points, depending on its size and complexity. The *PEF*, representing an effective potential, is parameterized according to a hydrophobic criterion, with the additional conditions that it will reproduce the typical *H*-bond patterns of polypeptides and yield conformations with allowed ϕ- and ψ-angles. Within the framework of the hydrophobic criterion, the assignment of sites for the side chains is as follows: one hydrophobic site for ala, val, pro, and thr; two hydrophobic sites for leu, ile, phe, tyr, trp, met, and cys; one hydrophilic site for ser, asn, and asp (as well as for the protonated *N*-terminus and the ionized *C*-terminus); one hydrophobic site and one hydrophilic site for arg, lys, glu, gln, and his. The *PEF* terms for the interaction between side chains are of the Lennard-Jones type, attractive for the hydrophobic chains and repulsive for the hydrophilic chains, with two parameters related, respectively, to the free energy of transfer of the residue from water to octanol and the size of the chain (which, in turn, is related to the solvent-accessible area). The interaction terms for non-connected backbone sites are of a modified Lennard-Jones type, with R^{-4} (instead of R^{-6}) and R^{-12} terms, with chosen parameters. In addition, the *PEF* includes terms for *H*-bonding, orientation of the side chains, bond angles, and disulfide bridges. The resulting *PEF* was tested for avian pancreatic polypeptide and a parathyroid hormone-related protein, with a satisfactory reproduction of the experimentally determined native structures.

A widely used procedure is that one developed by Sippl (1990,1993). The energy protentials for the atomic interactions for all the amino acid pairs were obtained from an analysis of known protein structures. The potential of mean force of the interactions, as a function of distance, was obtained using Boltzmann law from the distribution of distances between the interacting atoms (see also Section 11.3.1). In this procedure, the prediction of the conformation of an amino acid sequence of given length is performed as follows: take the conformations of all the segments with that same length in the database; mount (*thread*) the given sequence over each of those segments; evaluate the corresponding conformational energies; sort the conformations according to their energies; cluster the low-energy conformations accordig to conformational similarity; the results will predict a stable conformation (with only one cluster of very similar conformations), a flip-flop conformation (with two or more large clusters of different conformations), a metastable conformation (with one predominant cluster and a number of small clusters), or an unstable conformation (with many clusters of different conformations, with one or few conformations in each cluster).

10.3.3 Build-up Procedures

Starting from the residue-by-residue procedure (already described in a general fashion in Section 6.2.1), a number of variations have been proposed for the building and optimization of peptide structures. Some of those variations were developed as stand-alone procedures but they are included in this section because it may be argued that the resulting structure should perhaps be optimized, just as it is done for crystallographic structures. This statement applies in particular to those methods grouped below under the general designation of build-up from models. Figure 10.1 presents a schematic diagram of the set-up for the modeling of a protein conformation by energy minimization. The discussion below will examine the details of the four different build-up procedures: residue-by-residue, step-by-step, from structural fragments with flexible joints, and by comparative modeling (in increasing order of the size of the units used in the construction of the peptide chain).

(a) residue-by-residue

The brute-force optimization of a peptide chain, constructed residue-by-residue, does not have much chance of success except for short peptides. However, it is of interest to examine it once more, with consideration of the discussion in Section 5.1 on the forces in protein folding. After all, the build-up and subsequent optimization represents an attempt at simulating reality.

For simplicity in the discussion we will consider a hypothetical peptide, assumed to be composed of three α-helices (α_1, α_2, α_3), joined by flexible loops of appropriate length. Figure 10.2 presents three possible structures, with the understanding that none of them represents the native conformation. This simple example suffices to highlight the dilemma when tackling the simulation. Two possible ways come to mind, paralleling the discussion in Section 5.1:

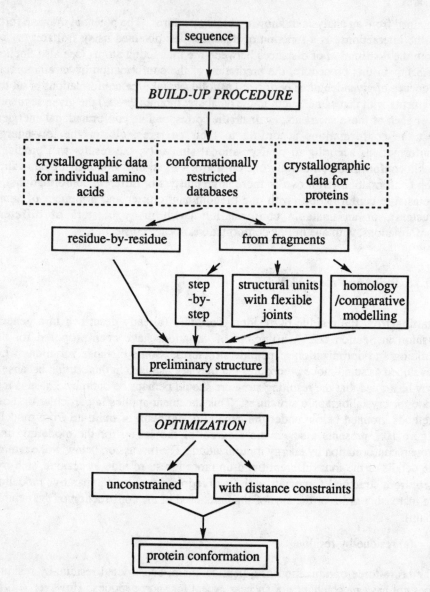

Figure 10.1 Energy minimization for the modeling of a protein structure, proceeding from the sequence, through a preliminary structure, to the final protein conformation. The three main types of input data are presented in the dashed-line box.

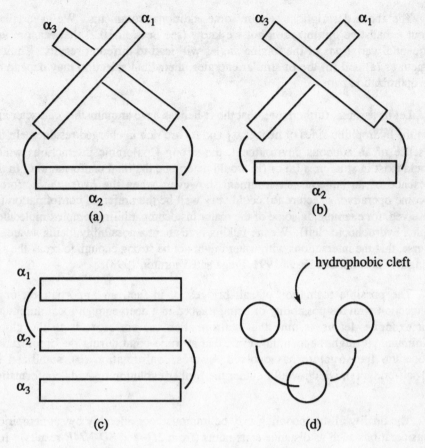

Figure 10.2 Schematic representation of three possible structures (a, b, c) of a hypothetical peptide chain with three α-helices; (d) presents a transversal perspective of structure (c), showing the existence of a cleft formed by the three helices.

(i) If we assume that the secondary structure is the result of short-range interactions, then one should optimize exhaustively the structure after the addition of each new residue. As the build-up progresses, intermediate- and long-range interactions will come into play, dictating how the tertiary structure will evolve. It is in the later stages of the process that hydrophobic interactions would be of importance (see below).

(ii) It might be, however, that the formation of the secondary structure takes place simultaneously with the adoption of the final tertiary structure, with all the interactions (from short- through long-range) playing a role. In such a case it would be advisable to construct the peptide chain at a low level of optimization, letting it adopt a somewhat extended conformation, which would, during the final optimization, collapse into the appropriate tertiary structure with its corresponding structural units.

The above descriptions require some additional comments. We have talked about exhaustive optimization but we know (see Section 10.2) that random and sequential variations of the torsion angles will lead to different results. Thus, if structures (a) and (b) are of similar energies, their final selection may depend on the optimization path.

Let us assume, furthermore, that the α-helices have amphipathic character and that the hydrophilic sides of helices α_1 and α_3 are rich in charged residues. In the absence of an aqueous environment, the strong Coulombic interactions which would exist in structures (a) or (b) would determine the final conformation. In the presence of an aqueous environment, however, when the *hydrophobic* forces become operative, structure (c) could very well be the preferred conformation (as observed, for example, in some of the major histocompatibility complex molecules, with a hydrophobic cleft). We are talking here about a possibility, being aware, of course, that the interactions with water might not be strong enough to break the salt bridges (Thornton and Fraga 1991, Fraga and Thornton 1993).

The possible formation of salt bridges is, in fact, an important factor in connection with the possibility of being trapped in a deep-enough local minimum. For example, let us assume that conformation (*a*) corresponds to the global minimum. It might happen, however, that at some point during the optimization procedure the structure has evolved towards conformation (*b*), stabilized by (a) salt-bridge(s), which would hinder the further evolution towards conformation (*a*).

The quality of the modeling may be improved considerably by semiempirical considerations such as distance constraints (from *2D*- and *3D-NMR* results). It is expected that an energy-minimized structure, which at the same time satisfies experimentally-known distance constraints, will be more satisfactory (Crippen and Havel 1988, Havel and Snow 1991) but both efficient algorithms and sophisticated coding are required in order to reduce the computing time (Glunt *et al.* 1993). [See also Section 10.6.]

A second, simpler possibility of improving the simulation may be found in the use of a conformationally-restricted database (Wu and Kabat 1973, Lambert and Scheraga 1989*a*, Unger *et al.* 1990), which may be viewed as being equivalent to the use of an empirical *PEF*. The computing time is considerably reduced (Palmer and Scheraga 1991, Rooman *et al.* 1991) as the search is restricted to a more probable conformational space: the search is performed directly through the generation of a large number of conformations, but the evaluation of their energies is more economic than the normal brute-force energy minimization.

An example of a conformationally-restricted database is the one obtained by Parker (1994) for tripeptides. This database was derived from the occurrence of dipeptides and tripeptides in the crystallographic structures of 150 non-homologous proteins. The data for 6,900 of the possible tripeptides were found in the 33,172 tripeptides in the sequences of those proteins. The information for the remaining 1,100 tripeptides, not found in those proteins, was generated from the data obtained for dipeptides; that is, for the triplet $R_1R_2R_3$, the information was obtained from the data for the sequences R_1R_2X and YR_2R_3, where X and Y stand for residues with the same polar/non-polar character as R_3 and R_1, respectively.

(b) step-by-step

This procedure, which has a better chance of locating the global minimum as well as the local minima in its vicinity, was used by Vasquez and Scheraga (1985) for the study of Met-enkephalin, Tyr-Gly-Gly-Phe-Met (using the software package ECEPP/2).

First, the minimum energy conformations of the terminally-blocked single residues (Vasquez *et al.* 1983) were determined, ordered according to increasing energies, labelled according to the codes of Zimmerman *et al.* (1977), and sorted; a conformation was kept if there was no other one with lower energy and the same label for the important χ angles and any conformation lying more than 5 kcal/mole above the global minimum was rejected. All the resulting conformations were then used to construct the appropriate dipeptides (Tyr-Gly, Gly-Gly, Gly-Phe, and Phe-Met). These conformations were minimized, labelled, and sorted and the conformations which were kept were minimized again, with the ω- and χ-angles fixed at the values obtained by Isogai *et al.* (1977). Finally, the tripeptides (Tyr-Gly-Gly and Gly-Phe-Met) were constructed and the resulting conformations used for the build-up of the pentapeptide.

The number of minima, obtained at each step and presented in Table 10.1, show how laborious this procedure may be. In fact, it may not be appropriate for larger chains because the number of smaller peptides to be constructed will increase significantly.

(c) from structural units and flexible joints

An improved variation, in terms of effort, of the above procedure consists of building up the peptide chain from the appropriate structural units, with flexible joints, with subsequent optimization only of the latter. The reduction in the work arises from the fact that the number of such units (which are very easily constructed) is usually much smaller than the number of shorter peptides to be constructed in the above procedure.

There is, in addition, a justification for this method: it has been observed (Kuntz 1972) that the turns appearing on the surface of proteins contain a high percentage of gly residues, are highly flexible, and may be considered to play a passive role (Wetlaufer and Ristow 1973). Ptitsyn and Rashin (1975) were able to rationalize the structure of myoglobin in terms of the most favorable packing of

Table 10.1. Low-energy conformations in the step-by-step procedure for Met-enkephalin (Vasquez and Scheraga 1985).[a]

Residues	Tyr	Gly	Phe	Met
	16	12	18	37
Dipeptides	Tyr-Gly	Gly-Gly	Gly-Phe	Phe-Met
initial	192	144	216	666
optimized	88	77	91	157
restricted	45	73	44	31
Tripeptides	Tyr-Gly-Gly	Gly-Phe-Met		
initial	316	233		
optimized	147	157		
Pentapeptide	Tyr-Gly-Gly-Phe-Met			
initial	2792			
optimized	288			
selected[b]	14			

[a]See the text for details. All the chains are terminally blocked.

[b]Conformations lying within 5 kcal/mole of global minimum.

helices with amphipathic character and Levitt and Warshel (1975) represented *PTI* as a series of relatively rigid fragments connected by flexible joints.

The implications are that the secondary structure is fundamentally determined by short-range interactions (between residues near each other in the sequence) and that the interaction of those units, probably modified to some extent during the process, results in larger structural fragments, which in turn may undergo further variation (Honig *et al.* 1976).

Formally, the procedure is very simple, equivalent to the residue-by-residue method, with the difference that the build-up proceeds from both fragments (the structural units) as well as residues (for the joining loops). Thus, for the hypothetical protein of Fig. 10.1, one would proceed first with the construction of the three α-helices, join them with the residues corresponding to the loops, and proceed with the optimization of the latter. Taking into account the discussion given above for the residue-by-residue procedure, there is no need of further comments here, being far more instructive to consider the inverse problem: that is, the freedom of movement of structural fragments in an experimentally determined structure.

Honig *et al.* (1976) applied a separation procedure, without bad contacts, to the experimental structure of subtilisin *BPN* (Wright *et al.* 1969), using a *PEF* with only non-bonded interactions (Coulomb and van der Waals) and *H*-bond terms (see Section 5.3.4); the CH-, CH_2-, and CH_3-groups were treated as single atoms and the energy minimization was performed by the Davidon method [see Section 6.2.1 for the general problem of energy minimization and the work of Fraga *et al.* (1978) for specific details].

Subtilisin *BPN* contains three distinct (1-100, 100-175, and 175-275) regions. The calculations showed that the joints at 50-54 and 76-80 are extremely rigid, due to short-range interactions, while the joint at 193-197 is also rigid due to the fact that the region 200-275 seems to be threaded in a cleft formed by the rest of the molecule. On the other hand, the joints at 98-103, 118-122, 128-132, 143-147, and 168-172 are extremely flexible, indicating a dependence on only long-range interactions.

Two main conclusions emerge from these observations. On one hand there is the fact that, in a conformational study, *it will be possible to vary only a small number of torsion angles*, maintaining large fragments with a fixed geometry. Furthermore, *breathing vibrations, fluctuations upon substrate binding, and low frequency modes may involve motion of rigid structural units about the connecting joint(s)*.

This study suggests that the regions 1-100 and 100-150 were formed independently, interacting as entities with one another at a relatively late stage in the folding, and that the region 200-275 must be in place before this process takes place. These facts, if not known when undertaking a *de novo* simulation, would affect the quality of the modeling; that is, it appears that it may be difficult to predict the tertiary structure unless information on the folding pathway is available. In other words: *a bias may be introduced depending on the optimization path*.

The influence of the side chains on the close packing of the rigid structures is minimized, within the context of a simulation, by the fact (Gelin and Karplus 1975) that *they appear in the conformations corresponding to the energy minima of the free residues*.

(d) homology/comparative modeling

In these methods, at the higher end of the scale of build-up procedures (in term of fragment sizes), it is attempted to construct the structure of the protein under study from the structure(s) of protein(s) with sequence identities/similarities, on the basis of the argument (Osguthorpe 1989) that tertiary structures are less variable than the sequences. This modeling may be performed with or without energy considerations (see, e.g., the work of Warme *et al*. 1974, Greer 1981, Furie *et al*. 1982, Jones and Thirup 1986, and Blundell *et al*. 1988).

This procedure has been consistently denoted in the literature as modeling by homology, whereby the structure of the protein is constructed from the structures of homologous proteins, on the basis that homologous proteins (related by divergent evolution) are more or less similar. On the other hand, non-homologous structures (a consequence of convergent evolution) may show a high degree of similarity and, consequently one may decide to proceed with the modeling on the basis of similarity with non-homologous proteins. Because the view has been expressed (Sternberg and Islam 1990) that structure predictions based on local sequence similarities might not be appropriate in general, except in those cases where an evolutionary or functional correspondence exists, it is necessary to qualify appropriately the designations in order to avoid confusion. Thus, one should distinguish between *homology modeling* and *comparative modeling* (Bajorath *et al*. 1993), with the added distinctions of *structural homology* and *structural similarity* as well as *sequence identity* and *sequence similarity*.

Whether one is proceeding with a homology modeling or a comparative modeling, first one must obtain the necessary information on sequence identity/similarity. The importance of this subject has motivated considerable attention, resulting in a wealth of procedures. For the purpose of this work it is not necessary to go into the details of these methods, sufficing to assume that those identities/similarities have been found. Consequently, the reader is referred for details to the work of Bacon and Anderson (1986), Bryant (1989), Taylor and Orengo (1989), Sali and Blundell (1990), Argos *et al*. 1991, Sander and Schneider (1991), Schuler *et al*. (1991), Tyler *et al*. (1991), Vriend and Sander (1991), Orengo *et al*. (1992), Russell and Barton (1992), Saqi *et al*. (1992), Zhu *et al*. (1992), and Henikoff and Henikoff (1993). There are, however, two important points that must be mentioned: sequence relatedness may be pinpointed unambiguously whenever an identity of over 30% exists, but problems may arise otherwise especially if parts of one sequence are missing in the other (Taylor 1988); *solvent-inaccessible cores, more conserved within a family of homologous proteins, constitute a better starting point for the modeling* (Hubbard and Blundell 1987). It should be noted, however, that it may be possible to obtain useful models, say, for the study of enzymatic activity, on the basis of a weak homology if the essential residues for that activity are well conserved (Nakamura *et al*. 1991).

The alignment of the sequences will identify the structurally-conserved and variable regions of the protein under study. The aligned regions are then ascribed the structure in the homologous protein, in the case of sequence identity, or they are *mutated*, if not completely identical, to the new sequence, retaining the original structure. Those regions are then joined by the variable regions, using the sequence of the protein under consideration, and their structure is determined in a variety of ways (Osguthorpe 1989). For example, one can impose distance constraints, such as between the *N*- and *C*-termini of the two sections to be joined (Jones and Thirup 1986, Greer 1990) or for disulfide bridges (Sali and Overington 1994), or proceed with a systematic search based on packing and/or energy considerations (Moult and James 1986, Dudek and Scheraga 1990, Hart and Read 1992, Kono and Doi 1994). In particular, the database of Sali and Overington (1994) contains 105 alignments of similar proteins or protein fragments, with 416 entries, 78,495 residues, 1,233 equivalent entry pairs, and 230,396 pairs of equivalent alignment positions.

Dudek and Scheraga (1989) determine the structure of the surface loops by means of a global energy minimization procedure, which generates a trajectory of local minima: First, a collection of low-energy backbone structures is obtained by a local energy minimization procedure (including the interactions within the surface loop and between the surface loop and the rest of the protein) applied to a large collection (200-2,000) of backbone structures generated by deforming a given initial structure; one low-energy side-chain structure is then generated for each of those backbone structures and a new local energy minimization procedure is performed for that collection of structures, and the structure with the lowest energy is then used as the next point in the trajectory. Kono and Doi (1994), on the other hand, use an automata network (Hopfield and Tank 1985; see also Section 9.3.1) to predict sets of sequences and their side-chain conformations from a given backbone geometry; the extent of side-chain packing is given by the network energy, which is lowered by the interaction of the connected automata, corresponding to every side-chain conformation in the rotamer space, at each residue position.

The fact that there are about 80,000 known amino acid sequences and that only 120 distinct folding patterns have been identified so far (Bowie and Eisenberg 1993, Orengo *et al.* 1994) suggests that the same fold might be adopted by many different sequences, in which connection it is of interest to note that the range of folds that may be adopted by a sequence depends significantly on its overall pattern of hydrophobicity (Hinds and Levitt 1994). Consequently, one could attempt to obtain the structure of the protein under study by fitting its sequence onto the backbone of known structures, followed by an energy minimization: that is, one could perform an *inverse folding* of the sequence with (a) given structure(s), guiding the spatial placement of the amino acids by *threading* the sequence through the structure (Sippl 1990,1993; Lathrop 1994) and determining the mutual compatibility of the sequence and the structure (Godzik *et al.* 1993). The procedure may be automated through the use of a library of protein folds or motifs (Fig. 10.3), all of which are tested for the given sequence, with the final selection made on the basis of a minimum-energy criterion using, e.g., empirical potentials derived from known crystal structures (Jones *et al.* 1992, Bryant and Lawrence 1993). Godzik *et al.* (1992) use a lattice Monte Carlo algorithm, based on a structural-fingerprint library of contact maps and buried/exposed patterns of residues. [The method of Taylor (1993), similar to threading, uses a correlation of hydrophobic conservation and inter-residue distances.]

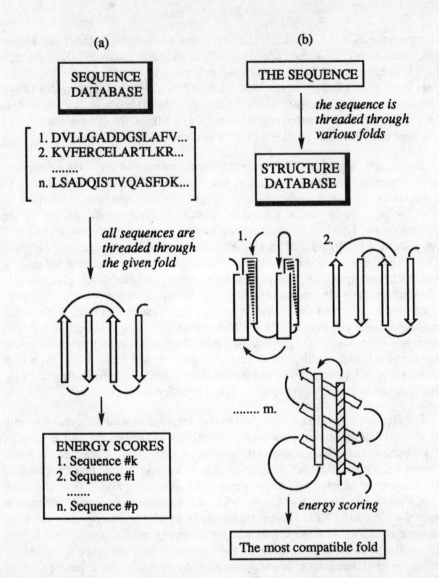

Figure 10.3 Schematic representation of an inverse folding algorithm: (a) the algorithm finds all the sequences compatible with a given structure; (b) the algorithm finds, from a library of known structures, that structure that fits best the given sequence (Godzik *et al.* 1993) [From A. Godzik, A. Kolinski, and J. Skolnick, *De Novo and Inverse Folding Predictions of Protein Structure and Dynamics*, Journal of Computer-Aided Molecular Design 7, 397-438 (1993). Copyright © 1993 ESCOM Science Publishers B.V., Leiden, The Netherlands. Reprinted by permission of ESCOM Science Publishers B.V.]

One can also proceed by *tinkering*, which involves both folding and inverse folding, using a *constrained hydrophobic core construction method*, with the following steps: design a tentative sequence for a target structure; fold the sequence to all its native conformations; modify the target structure and the sequence in order to reduce the degeneracy; repeat until obtaining a structure encoded in a sequence of the lowest possible degeneracy (Yue and Dill 1995, Yue *et al.* 1995).

10.3.4 The Genetic Algorithm

The genetic algorithms, which have been described as *intelligent stochastic methods* (Brodmeier and Pretsch 1994), try to overcome the problem of the multiple minima through an evolutionary process, starting from an initial generation, which will result in the survival of the fittest individuals in the final generation (Holland 1975, Goldberg 1989, Brodmeier and Pretsch 1994).

A *generation* consists of a large number of individuals (denoted as *chromosomes*), each chromosome containing the parameters (denoted as *genes*) that describe the individual. Thus, when studying the problem of protein folding, a chromosome denotes a conformation, whose genes may be the torsion angles (when operating within the context of the rigid-geometry approximation). The fitness of each individual of a generation may be estimated according to various criteria (see below).

The next generation may be constructed by either *copying* selected individuals into the new generation, *mating* (with single or multiple crossovers), or *mutating*, with the possibility of external control of the relative weights of the three procedures. The parent selection may be performed by the *roulette-wheel* technique (resulting in selection proportional to the fitness) and a *dynamic mutation* technique may be incorporated for fine-tuning of the population, by switching from full mutations to incremental mutations after a given number of generations. Because the algorithm has no memory, a *generation-gap* technique may be used, whereby the fittest individuals are copied into the new generation. The problem of a premature convergence into a local minimum may be avoided by a *sharing* technique, which results in a *forced mutation*. Each new individual is tested against all the individuals already existing in the generation being produced. The criterion for this test may be, for example, that the sum of the differences (in absolute value) of all the torsion angles of the new individual and each existing ndividual should be greater than a given threshold; if such is not the case, random increment values are given to the torsion angles of the individual being tested (Brodmeier and Pretsch 1994).

Brodmeier and Pretsch (1994) define the fitness factor of the i-th individual of a generation, in terms of energies evaluated with a $MM2$ potential, as

$$f_i = \exp\{(-s \times \text{value}_i - \text{best value})/(\text{average value} - \text{best value})\}$$

where the s factor (with values between 0 and 1) controls the walk strategy through the energy hypersurface; a value $s = 0$ corresponds to a random walk while a discrimination towards the best individuals is obtained with values close to 1.

Dandekar and Argos (1994) have used the genetic algorithm of Goldberg (1989) for the study of the backbone structures of the four-helix bundle proteins cytochrome$_{562}$, cytochrome c', and hemerythrin, within the rigid-geometry approximation and restricting the values of the angles ϕ, ψ to those in seven standard conformations; the side chains were not considered explicitly, but they were mimicked by means of parameters attached to the C_α atoms. The fitness of the individuals was estimated as

$$f = C + \text{clashes} + \text{secondary structure} + \text{tertiary structure}$$

The results suggest that *hydrophobic forces, as well as the existence of stable secondary structure units, are important, while local forces and H-bonding play a much less significant role.*

McGarrah and Judson (1993) have performed a detailed analysis of the GA in a test application to cyclic hexaglycine, examining the selection of the GA parameters (number of generations, population size, and rate of interaction between subpopulations) and the use of a hybrid method, combining a global GA search with a local gradient minimization. In particular, it was found that large numbers of generations are not useful and that, given a fixed product (number of generations) times (populations), it is better to have a large population with a smaller number of generations. Diversity is maintained with a high selection rate, with equal probability for all the eligible parent chromosomes.

The GA may be extended to consideration of a continuum solvent and the use of a distance-dependent dielectric constant but it is not clear (McGarrah and Judson 1993) how it could be applied to the case of explicit-solvent simulations (see Section 10.4.3).

10.4 Molecular Dynamics Simulations

As already emphasized throughout this work (see, in particular, Chapter 6), energy minimization procedures, which are time- and temperature-independent, should be viewed only as a first step towards the description of the structure and the function of a protein.

The proper description, providing a prediction of the structural, statistical thermodynamics, and spectral properties of the protein, as well as of its dynamic conformational history, must be obtained through a Molecular Dynamics (*MD*) simulation. Those properties are evaluated as time-averaged quantities over the complete trajectory, while the structural fluctuations may be monitored either graphically or by an analysis of the trajectory (Hagler *et al.* 1985). The graphic representation, in particular, may help in understanding the (biological) function of the protein and even in identifying its active site (see the work of Nilsson 1990 and references therein). *MD* simulations have also a potential use as a computational technique for a fast sweep of the conformational energy hypersurface (Gronbech-Jensen and Doniach 1994), in order to identify conformations of interest, but the reader is referred to the comments given below on timescales, which are of particular relevance in this context.

Since the pioneer work of McCammon *et al.* (1979) (see Section 6.4.3), *MD* simulations have proliferated and will continue to do so in the future. The reader should have no difficulty in finding whether *MD* simulations have been performed for a particular system and therefore we will focus our attention on a few illustrative examples, with emphasis on the practical problems and the corresponding attempts to solve them.

10.4.1 Illustrative Applications

MD simulations are useful in the refinement of crystallographic structures, determining protein conformations by homology/comparative modeling as well as from *NMR* data, evaluating the free-energy changes resulting from mutations, confirming the stability of *H*-bonds, investigating the significance of the secondary structure and of hydrophobic forces in the protein folding, etc. (Levitt 1981, Brünger *et al.* 1986, Karplus and Petsko 1990, Hermans *et al.* 1992, Zhang and Hermans 1993, van Gelder *et al.* 1994).

Van Gelder *et al.* (1994) have proposed the use of *MD* simulations for the build-up of protein structures in those cases where only the backbone or just the C_α coordinates of a template structure are available. The protein backbone is built up in a crude fashion by homology/comparative modeling and refined by energy minimization. The side chains are then incorporated and positioned by means of a *MD* simulation, and the final structure is obtained by energy minimization (after having cooled it to 0 K).

Brünger *et al.* (1986) have performed a *MD* simulation on crambin (with 46 residues), with constraints for 240 interproton distances (of less than 4 Å), obtained from *NMR* data (see Section 10.6). The potential function used was CHARMm (with a modified Coulombic term in order to simulate the solvent effect; see below), incremented with constraint terms of the form $c(R_{ij} - R_{ij}^0)^2$, where R_{ij} is the calculated interproton separation and R_{ij}^0 is the corresponding target distance between the i-th and j-protons; the coefficient c is expressed as $(1/2)sk_BT(\Delta R_{ij})^{-2}$, where s is a scale factor, k_B is Boltzmann constant, and ΔR_{ij} is the positive (for $R_{ij} > R_{ij}^0$) or negative (for $R_{ij} < R_{ij}^0$) error estimate of R_{ij}. The calculations were performed in two stages, with timesteps of 1 femtosecond (1 $fs = 10^{-15}\,s$). First, the secondary and tertiary structure folding was accomplished in several phases (with variable scale factor), with energy minimization after each phase in order to reduce the distortions provoked by the increase in temperature. The second stage consisted of a 5 picosecond (1 $ps = 10^{-12}\,s$) heating and equilibration period, followed by a 12 ps period at 300 K, and a final trajectory of 10 ps for the evaluation of average structures and properties. An interesting observation is that, within the framework of the constraints introduced, *the correctly folded tertiary structure is achieved only when the secondary structure units are formed, at least in part, before the tertiary structure folding takes place* (see Section 5.1 for comments on the secondary structure *versus* hydrophobic collapse dilemma).

The significance of the hydrophobic interactions has been studied by Zhang and Hermans (1993), with *MD* simulations on a model system consisting of two chains with the identical sequence Ac-AVAA-(LAAAVA)$_3$-NMe. The *MD* calculations were carried out with timesteps of 2 fs, at 300 K and 1 atm (both maintained constant by small adjustments at each step of the simulation), with cut-off threshold for non-bonded interactions (see below), using the program CEDAR (with the *PEF* parameters of GROMOS) and the SHAKE algorithm, in order to maintain the bond lengths constant. The free-energy changes arising from mutations at given sequence positions (by substitution of Ala with Val or Leu) were obtained from the thermodynamic cycle involving the four structures random coil (X_n), random coil (Ala_n), coiled coil (Ala_n), and coiled coil (X_n), where n indicates the sequence position at which substitution has been made. For the coiled coil model, the system was aligned in a box and solvated with almost 3,000 water molecules, with a minimum distance of 3.8 Å between water and peptide atoms; the water structure was first optimized by energy minimization (with the model system maintained fixed) and then the *MD* simulation for the whole system was performed in three stages (for 10 ps at 100 K, 10 ps at 200 K, and 10 ps at 300 K), with a cut-off threshold of 8 Å for the non-bonded interactions. The values for the random coils in the above thermodynamic cycle were estimated from the values obtained for the peptides Ac-Ala-Ala-Ala-NMe and Ac-Ala-X-Ala-NMe, where X stands for either Val or Leu; the *MD* simulations for these peptides were run as for the coiled coil, but with a cut-off threshold value of 6 Å. The results indicate that the contribution to the stability of the coiled coil decrease in the order Leu > Val > Ala.

10.4.2 Practical Difficulties

The coding of a computer program for *MD* simulations (see Section 6.4) does not offer great difficulties. The problems arise, however, when the simulations are performed. One could say that all the *MD* simulations to date have been performed at a *state-of-the-art* level but that the present *state-of-the-art* is not adequate in terms of computer capabilities.

Let us delve into this statement in more detail. The purpose of a *MD* simulation, as mentioned above, is to predict the dynamic behavior of a protein through the solution of the corresponding equations of motion, given a specific *PEF* (which, for the sake of the argument, will be assumed to be satisfactory). The timescales of the transformations occurring in proteins are determined by the rates of overcoming free-energy barriers, which may range from a few to many tens of kcal/mol (Brooks III and Case 1993); the timescales may therefore range from picoseconds to milliseconds ($1 \ ms = 10^{-3} \ s$) and even larger (Sneddon and Brooks III 1991): thus, vibrational times are of the order of *ps* or smaller, rotational isomerization times are of the order of nanoseconds ($1 \ ns = 10^{-9} \ s$), nucleation times for the formation of helices from random coils are of the order of hundreds of *ns*, and the times for other structural fluctuations (of possible biological interest), such as unfolding, may be of the order of *ms* (Nilsson and Tapia 1992, Gronbech-Jensen and Doniach 1994). On the other hand, the accurate solution of the equations of motion requires the use of rather small timesteps, say, of the order of *fs*, with the result that a simulation of some hundreds of *ps* or a few *ns* represents already a major undertaking: for example, the *MD* simulation of human lysozyme, performed by Saito (1992), required 400 hrs. of computing time on a *FUJITSU VP-400E*. The possibility exists of using different timesteps for different interactions (Streett *et al.* 1978, Grubmüller *et al.* 1991) on the basis that only a small percentage of interactions require very small timesteps for an accurate numerical integration but this approach offers some dangers, as shown by Biesiadecki and Skeel (1993).

Therefore, until unrestricted computing times on fast, large-capacity computers will be available to every researcher in this field, drastic measures are needed or the results from the simulations may fail to provide information on the interesting modes of motion of the protein. Some of the possible approaches adopted in order to reduce the computing times are examined briefly below, with the understanding that they are not necessarily exclusive of each other.

(a) The simplest solution, naturally, consists of restricting the calculation to only a subset of atoms of interest such as, *e.g.*, those involved in the active site and its surroundings (Karplus and Petsko 1990).

(b) The evaluation of the non-bonded interactions, as already mentioned in Section 10.3, constitutes the most expensive component of the calculations and, therefore, any simplification in such a calculation may result in a drastic reduction of the computational time. Such a reduction could be achieved at the simplest level, by consideration of a cut-off threshold, usually in the range of 6 to 10 Å. That is, the evaluation of the non-bonded interaction between two atoms is omitted if their separation is greater than the adopted cut-off threshold. This simplification still involves the costly evaluation of the separations between all the non-bonded atoms and therefore one may think of a further simplification: at a given step in the simulation, the separations between all the non-bonded atoms are calculated and a *list of non-interacting* pairs of atoms (with separations greater than the threshold) is prepared; then the interactions between those pairs of atoms is disregarded for a number of following cycles. Because the structural changes that take place during those cycles will result in a change of the separations between the non-bonded atoms, it is necessary to update the list of non-interacting atoms every so many cycles, at the discretion of the user, with a compromise between a drastic reduction in the computing time (with updating of the list after a large number of cycles) and the danger of a biased simulation (Allen and Tildesley 1987), due to the errors introduced; thus, for example, Saito (1992) found that the Coulomb interaction energy was evaluated with an error of 18 kcal/mol per atom with a truncation cut-off of 10 Å.

The above description has been presented in order to illustrate the use of the so-called *atom-based truncated potential* method. The problem, however, is that the energy will not be conserved if the forces change excessively during one timestep and the only solution would be to reduce the timestep, with the corresponding increase in the computation time. Because of its relevance, the truncation problem has attracted considerable attention (see the work of Steinbach and Brooks 1994 and references therein) and a number of truncation procedures have been proposed: *atom-based force truncated potential, group-based truncation, potential switching, shifted potential, force shifted potential, atom-based force switch, switched group-shift potential,* and *generalized force shifted potential* (where the designation *group* stands for a cluster of atoms). The potential switch, with a switching region bounded by two thresholds, results in very large forces in the switching region; the shifted potential and the force shifted potential damage the short-range interaction of charged groups; the atom-based force switch is very expensive because of the need for the evaluation of logarithms or, in a modified formulation appropriate for the Lennard-Jones interactions, may yield artificial minima when a narrow switching region is used; the switch group-shift-potential is equivalent or preferable to the group-based potential switch depending on whether either group is neutral or charged groups/ions are involved.

(c) Significant savings in the computational times may be attained with a rigid-geometry approach (i.e., with fixed bond lengths and bond angles), which may be formulated in either Cartesian or internal coordinates. Two comments are required, however, in this connection. On one hand, the reader is referred to Section 6.4.3 regarding criticism of this approach. On the other hand, it may be emphasized that the approach to be discussed here is conceptually different from the case whereby the geometry is constrained to essentially remain fixed through the use of large force constants in the corresponding terms of the *PEF* (see Section 5.3.3); such an approach would be subjected to the same criticism, with an added disadvantage: the timesteps required in order to describe the motion would be smaller, as a result of the increase in the oscillation frequencies.

While still operating with Cartesian coordinates, the rigid-geometry approximation may be introduced by projecting out all the components of the non-bonded interactions, which would alter the bond lengths and angles. With this mathematical restraint of the geometry, the long-time dynamics of proteins might be simulated with timesteps of the order of 0.5-10 *ps*, but there may be need for a parameterization procedure (Gronbech-Jensen and Doniach 1994) (see Section 10.5).

(d) A very promising approach is the one proposed by Amadei *et al.* (1993), in which the simulation would be restricted to an *essential subspace* of limited dimensionality, corresponding to those few degrees of freedom associated with the motions relevant to the function of the protein; the remaining subspace, with a narrow Gaussian distribution, may be considered to be physically constrained, corresponding to irrelevant local fluctuations.

Elimination of the overall translational and rotational motions yields a molecule-fixed Cartesian coordinate system, in which the trajectory is given by the $3N$-dimensional column vector $\mathbf{x}(t)$ of all the atomic coordinates in the system. The correlation between the atomic motions is expressed by the symmetric, covariance matrix $\mathbf{C}$ of the positional deviations,

$$\mathbf{C} = <(\mathbf{x} - <\mathbf{x}>)(\mathbf{x} - <\mathbf{x}>)^{\dagger}>$$

where the bra-ket notation denotes average over time. Diagonalization of this matrix with an orthogonal coordinate transformation matrix $\mathbf{T}$, $\Lambda = \mathbf{T}^{\dagger}\mathbf{C}\mathbf{T}$, yields the corresponding eigenvectors and eigenvalues (λ_i), which may be ordered according to decreasing values of the λ_i, so that the first, say, n eigenvectors (with n much smaller than N) will represent the largest positional deviations. The equations of motion in the essential subspace may then be obtained by applying the transformation $\mathbf{T}$ to the equations of motion of x.

The study of the essential subspace of lysozyme (with 1,258 atoms) has been performed from the results of two *MD* simulations, in vacuum (Amadei *et al.* 1993) and in solution (Smith *et al.* 1993). The simulation in vacuum, for 1 *ns*, was carried out using the *PEF GROMOS*, with timesteps of 2 *fs*, at 298 K, using the SHAKE algorithm to maintain fixed bond lengths, with the translational/rotational motions removed every 0.5 *ps*, and saving the conformations every 0.5 *ps*. The simulation in solution, with 5,345 water molecules, was carried out for 900 *ps*, at 300 K, using the *PEF GROMOS*, and saving the conformations every 0.05 *ps*. The results indicate that 35 eigenvectors (out of 3,792) contribute to 90% of the overall motions and confirm that *simulations in vacuum appear not to be appropriate for the study of biologically relevant motions.*

(e) The ultimate solution, as already pointed out, lies in the availability of fast, large-capacity computers, whether with a single *CPU* or, for example, with a system of parallel processors (see Section 10.3). An equally acceptable alternative is to be found in the use of special-purpose computers, such as the one designed by Ito *et al.* (1994), which can calculate the interaction (Coulombic as well as van der Waals) of one particle with all the others.

Another possible difficulty in *MD* simulations arises from the possible dependence on the starting structure. Such a dependence may be reduced through the weighting of the atomic masses (Mao and Friedman 1990) with the same numerical factor (>1.0) for all the atoms. This weighting determines that the potential barriers, say, for dihedral rotations, will be overcome more easily. The results of Mao *et al.* (1991) for FMRF amide and CYFQNC (both linear and in a pseudo-cyclic conformation) suggest that the method may be useful for the conformational analysis of constrained molecules as well as for the modeling of the loops on protein surfaces.

At a formal level, a variety of techniques (such as umbrella sampling, thermodynamic perturbation theory, chemical perturbation through thermodynamic cycles, etc.) have been proposed in order to obtain statistically-valid samples of the biologically-relevant conformations and the reader is referred to the reviews of Straatsma and McCammon (1992), Hermans (1993), and Brooks III and Case (1993).

10.4.3 The Solvent Effect

The solvent effect should be properly taken into account through consideration of an appropriate number of solvent molecules (*explicit-solvent method*). The increased computational effort, particularly for a realistic number of solvent molecules, makes it advisable to resort to some approximation, such as the use of an appropriate dielectric constant ε in the Coulombic term of the non-bonded interactions (*implicit-solvent method*), with either ε = constant or ε = kR (which could be used in the explicit-solvent method). The analysis of Steinbach and Brooks (1994), however, rejects the use of a distance-dependent dielectric constant. The explicit-solvent treatments will be discussed in Chapter 11 and here two representative examples of implicit-solvent calculations will be examined.

McKelvey *et al.* (1991) performed a *MD* simulation for the peptide *N*-acetyl-Tyr-Ile-Gly-Ser-Arg-NHCH$_3$. Each *MD* cycle consisted of three stages of 10 *ps* each (at a high temperature of 1500, 800, or 600 K; cooling to 300 K; and a final trajectory at 300 K, with saving of the conformations at intervals of 1 *ps*).

The last structure was then recycled through the same procedure, yielding altogether 1,200 conformations. These structures were then optimized by a process of 10,000 steps of an Adopted Basis Newton-Raphson (*ABNR*) energy minimization. The *MD* cycles were carried out with a dielectric constant $\varepsilon = 10$, while the energy minimizations were performed for three different values ($\varepsilon = 1, 4, 10$). Among other results, it was concluded that *the relative stabilities of the various conformations of peptides with highly polar side chains will be highly dependent on the dielectric constant of the medium.*

Nilsson and Tapia (1992) studied a (polyglycine) homopolymer, with 438 atoms, of the bacteriophage T4 glutaredoxin in four $100\,ps$ simulations at 293 K with timesteps of $2\,fs$, using the *GROMOS* force field (with the Coulomb interactions scaled to 0, 81, 100, and 225% of the normal strength), with analysis of the trajectory by interactive animation with the *mdFRODO* package (Nilsson 1990). The bond lengths were constrained by *SHAKE*, non-bonded interactions were estimated by the twin-range charge-group technique (with cut-offs of 8 and 13 Å), the list was updated every 10 steps for all non-bonded interactions below 8 Å, and the Coulombic interactions were reevaluated also every 10 steps. The results indicate that *with increased electrostatic interactions, the global relative arrangement of the secondary structure elements is retained,* but with some local deformations, while *the secondary structure melts in a cooperative fashion when the Coulomb interaction is reduced,* yielding a collapsed, molten globular conformation.

10.5 Long-time Dynamics

The contrast between the biological timescales *vs* the simulation timescales in traditional *MD* simulations has highlighted the need for long-time simulations. Among the approaches suggested, a purposefully brief mention of the rigid-geometry treatments has been made, with the idea in mind of reexamining it here.

The problem with constraining degrees of freedom arises from the vanishing momenta conjugate to the constrained degrees (Fixman 1978). The Jacobian of the metric tensor, which characterizes the potential energy hypersurface of the unconstrained degrees, must be taken into account for a correct evaluation of ensemble averages (Ryckaert *et al.* 1977, van Gunsteren and Berendsen 1977, Fixman 1978, van Gunsteren 1980); otherwise, a non-uniform distribution of dihedral angles will be obtained (Knapp and Irgens-Defregger 1993).

The study of slowscale motions (with times greater than $1\,ps$) may be carried out, for example, *via* a Langevin Dynamics (*LD*) algorithm (see also Section 10.7) (Gronbech-Jensen and Doniach 1994) or through a *MC* simulation of a *MD* run (Knapp and Irgens-Defregger 1993).

The fast bond oscillations average about the equilibrium geometry (i.e., the equilibrium bond lengths and bond angles). Therefore, if the dynamics simulation were to be based only on the non-bonded interactions, in which case on has to deal only with the rotational isomerizations, it would be possible to use larger timesteps (ensuring, of course, that the discrete integration algorithm converges). This can be achieved through the use of mathematical constraints on the chemical bonds, thus eliminating the corresponding degrees of freedom, the associated fast modes of oscillation, and the thermal heat bath (in which case the system is not described any longer as having its total energy conserved; see also Section 10.7). In the LD approach, a local noise source acting on each atom is introduced, balanced through a friction term via the fluctuation-dissipation theorem [see the work of Gronbech-Jensen and Doniach (1994) for details and a comparison with the treatments of Ermack and McCammon (1978), McCammon et al. (1980), and Pear et al. (1981)].

The MC technique has been applied to the simulation of MD of polymers (Baumgartner 1984, 1987, Kremer and Binder 1988, Fichthorn and Weinberg 1991), in the on-lattice modality; that is, with the monomers (as point particles) positioned on the grid points of a lattice with a geometry consistent with the bond lengths and angles of the system under study. Application of this approach to the case of proteins faces the difficulty of accounting for the individuality of the residues (Kolinski et al. 1987, Skolnick and Kolinski 1991). The off-lattice modality, on the other hand, with unconstrained bond lengths and angles, allows a considerable amount of stretching/bending, with a corresponding high rejection rate in the MC steps due to the high energies encountered.

In the approach of Knapp and Irgens-Defregger (1993), the bond lengths, bond angles, and amide planes are held fixed. The eliminated degrees of freedom serve as heat bath and impose a damping mechanism, resulting in a diffusion-like motion on the potential energy hypersurface, with coherent motions of molecular groups due to the remaining degrees of freedom. A window algorithm (for 3 consecutive amide planes) is used for the study of local conformational changes, without affecting other parts of the system (Go and Scheraga 1970). In the MC steps, the window position is first chosen at random and then arbitrary values are given to the dihedral angles. A complete scan along the backbone, consisting of N amide planes, is supposed to consist of (N-2) randomly-chosen windows.

The interpretation of the results of such a treatment, in terms of a MD simulation, is established through the definition of the timestep: the timestep is defined as a complete scan of the window algorithm along the backbone of the protein. The magnitude of this timestep, which could be obtained by comparison with benchmark calculations, has been estimated at about 15 ps. [The model calculations of Knapp and Irgens-Defregger (1993) were performed without non-bonded interactions, whose consideration will increase the execution time of the MC steps.]

With this technique, both time-independent equilibrium properties and correlation functions (see Section 6.4.4) may be obtained, so that the model could be used for the simulation of the long-time evolution of proteins.

10.6 Simulated Annealing

The conceptual foundation of simulated annealing is rather simple: starting from a high-energy system (at high T, hence the designation of *annealing*), a controlled cooling will lead to a conformation close to the global minimum (Kirkpatrick *et al*. 1983, Aarts and Korst 1989). This idea is already present in *MD* simulations (see Section 10.4), where the simulation is first performed at high T, which is then decreased towards the desired value, with the expectation of avoiding the trapping in local minima (see below). Simulated annealing (*SA*) represents a very powerful technique for the prediction of protein structures but it must be emphasized that the success of *SA* simulations depends on the quality of the *PEF* used (Laughton 1994) and to a large extent on the temperature protocol.

Within the context of the *MC* technique, with a random walk on the potential energy hypersurface, the high T decreases the rejection rate (see Section 6.3); that is, the hills on the hypersurface are more easily overcome and the walk samples large regions of the hypersurface. As the T is lowered slowly, the hills start to become higher and the conformation evolves towards a minimum energy. At each T, the MC calculation is performed in the usual way: random change of the conformation, evaluation of the potential energy, and Metropolis test. The T is lowered according to a given protocol (e.g., on the basis of conformational changes accepted or rejected since the last T change).

Most of the calculations have been performed with *on-lattice* models (Wilson and Doniach 1989, Sikorski and Skolnick 1990, Snow 1992, Covell 1994, Kolinski and Skolnick 1994, Sali *et al*. 1994) and only some of their more distinctive features will be mentioned below. Wilson and Doniach (1989) evaluate the energy as the sum of the interactions between adjacent or near-neighbor lattice sites, using semiempirical potentials derived from the C_α distances in a database [see also the work of Tanaka and Scheraga (1976) and Crippen and Viswanadhan (1985), the latter discussed in Section 10.3.2.], and conclude that the *formation of the secondary unit is sequence-specific and is influenced by long-range interactions*. Snow (1992) has proposed a gradual lowering of both T and the stepsize (*i.e.*, the angle changes) over the course of the run. At the end of each iteration (consisting of several sweeps through all the angles), the acceptance rate is evaluated and the T and stepsize are adjusted or not depending on the value of the acceptance rate: if that rate is within the accepted range of 40-60%, both T and the sizestep are decreased if the best energy obtained during the iteration is not lower than in the preceding iteration. Kolinski and Skolnick (1994) use a coarse lattice for folding from random coil and a finer lattice for the later stages; their potential energy function consists of a sequence-independent statistical potential for the main-chain C_α conformation, a cooperative potential simulating H-bonding in real proteins, a rotamer energy term (in a single ball representation of the side chains), a local angular correlation term for the side-chains orientations, an amino acid centrosymmetric interaction, a pairwise interaction for side groups, and a four-body, side group contact map template interaction. The potential of Sali *et al.* (1994) depends only on nearest neighbor contacts, independent of other aspects of the chain conformation, with parameters obtained from a Gaussian distribution; the calculations for 200 sequences (using a *bead* model), with the purpose of investigating the Levinthal paradox (see Section 2.3.3), show that *the necessary and sufficient condition* (within that model) *for a sequence to fold rapidly is that the native state corresponds to a pronounced-energy minimum.*

More sophisticated treatments have been carried out by Nayeem *et al.* (1991), Holm and Sander (1992), Nakazawa *et al.* (1992), and Collura *et al.* (1993). Nayeem *et al.* (1991) have studied Met-enkephalin (in the absence/presence of water), using the *ECEPP/2* potential energy function, with random or arbitrarily generated starting conformations; the run is started at a high T (in the range 1000-5000 K), the *MC* search is performed for a finite number of steps, the T is lowered according to the protocol of Kirkpatrick *et al.* (1983), and the run is terminated when the *MC* acceptance rate drops below a given threshold (say, 10%). Holm and Sander (1992) have studied the side-chain optimization (in homology modeling, with a given backbone model; see Section 10.3.3), using random initial conformations and a modified Metropolis test (which gives an initial probability of 50% for the acceptance of uphill steps); the energy is calculated from truncated, pairwise 6-9 potentials (with a 6 Å cut-off and with the interactions precalculated and stored). Nakazawa *et al.* (1992) carried out a simulated annealing for the 16-36 fragment of *BPTI*, using *ECEPP/2* (with $\epsilon = 2$), starting at 1000 K and ending at 250 K, with T decreasing exponentially through 10^4 *MC* steps; the relevance of these calculations arises from the fact that *a simulation for a fragment* (more manageable than the modeling of the complete protein) *seems to help in understanding the folding pathway.* Collura *et al.* (1993) performed calculations for segments of immunoglobulin, *BPTI*, and bovine trypsin, using the program ESAP (Higo *et al.* 1992); the calculations started from extended conformations and the energy was calculated from a potential (Robson and Platt 1986) with a united atom representation for the non-bonded interactions (i.e., with the H-atoms of nonpolar groups treated implicitly) and a harmonic, distance-constraint term for the backbone atoms of the terminal residues (for the appropriate simulation of loops), with $\epsilon = 5$, a cut-off threshold of 10 Å, and renewal of the pairlist every 10 *MC* steps; within the rigid-geometry approach, the scaled collective variables of Noguti and Go (1985), i.e., the eigenvectors of the second derivative matrix of the potential energy with respect to the torsion angles, were used, with the second derivative matrix evaluated every 2,000 steps (with cut-off at 5 Å for the non-bonded interactions), and the vector step constructed by linear combination of its eigenvectors.

In the *PEACS* (Potential Energy Annealed Conformational Search) algorithm of van Schaik *et al.* (1992), without conservation of the total energy, the simulation is biased towards low-energy states by coupling the potential energy of the system with an external potential bath. The annealing schedule is defined by a potential energy relaxation time (which controls the exchange rate between the potential energy of the system and the external bath), a correction factor (in the evaluation of the velocities), and the reduction of the energy level of the potential bath (according to the lowest energy values found in the simulation process). Depending on the schedule used, entrapment may occur (in the case of fast cooling) or a wider conformational space is sampled (in the case of slow cooling) (Byrne *et al.* 1994).

The essential difference between the usual *SA* and the *Threshold Accepting (TA)* algorithms (Duek and Scheuer 1990) is to be found in the acceptance rule: *SA* accepts, with small probabilities, those conformations that are worse while *TA* accepts every new conformation which is not much worse. For small peptides, the results have indicated that *TA* is a better technique than *SA* for the determination of lowest-energy conformations (Morales *et al.* 1992).

There exists also the possibility a hybrid simulation, with consideration in the annealing procedure of given (distance and/or angle) constraints obtained, say, from experimental data from different sources (such as *NMR*); those constraints are introduced as additional terms in whatever *PEF* is being used. Thus, Nilges *et al.* (1988) performed such a simulation, but *within the framework of a MD procedure*, using a *PEF* consisting of terms for the changes in bond lengths, bond angles, and torsion angles and repulsive non-bonded interactions [in order to prevent bad contacts, incremented with terms for interprotein distance constraints (from *NMR* data), with adjustable parameters for the last two types of terms]. The procedure involved the following steps: generation of an approximate fold through *embedding* (which yields a set of substructures for a small subset of atoms); fitting of the residues, one at a time, to those substructures, with extended conformation; unrestrained minimization; *MD* simulation at 1000K with adjustment of the variable parameters; *MD* simulation at 300K; restrained minimization. The results for crambin and the globular domain of histone H5 indicate that the method allows for a *sampling of the conformational space consistent with the experimental data*. Variants and improved treatments for this type of simulation have been proposed (Borgias and James 1988, Nerdal *et al.* 1989, Yip and Case 1989, Baleja *et al.* 1990, Lane 1990, Bonvin *et al.* 1991, Nilges *et al.* 1991, Edmondson 1992, Mirau 1992) but the real stumbling block is the nature of the *NMR* data (particularly for peptides in solution), which correspond to averages, a fact that has lead to the development of the *time-average restraints* procedure (Torda *et al.* 1989, 1990, Pearlman and Kollman 1991). The reader is referred to the detailed analysis of this problem by Pearlman (1994), with the conclusion of *the need for a revision of the idea of what constitutes an experimental structure and how to simulate it*, a recommendation which has general validity for all types of simulations.

10.7 Hybrid Dynamics/Monte Carlo Simulations

The need for a hybrid method arises from two considerations (Guarnieri and Still 1994), namely, the problems associated with the existence of (large) potential barriers and the complementary characteristics of the *MD* simulations and the *MC* technique. The existence of large potential barriers may result in the traditional dynamics simulation spending all the time sampling the local space of the starting conformation and even 3-5 kcal/mol potential barriers may pose serious problems. That is, the efficient sampling ability of the dynamics simulation may be wasted. On the other hand, the *MC* technique has the ability, with large stepsizes, of crossing large potential barriers. Therefore, in a hybrid method, the *MC* component will drive the system from one conformation to another and the dynamics simulation will sample the local space of that new conformation.

The Langevin equation (used for the analysis of Brownian motion)

$$F = m\dot{v} = R - m\gamma v$$

(where R denotes a random force, γ is the friction coefficient, and all other symbols have the usual meaning) constitutes the basis of stochastic dynamics (*SD*) simulations (Chandler 1987). For example, the stochastic equation of motion of van Gunsteren and Berendsen (1988) is

$$m\dot{v} = f[x(t)] + R(t) - m\gamma v$$

with

$$<R(t)R(t')> = 2\,m\gamma\,k_B T\delta(t-t')$$

where the bra-ket notation denotes an equilibrium ensemble averaging.

Guarnieri and Still (1994) have developed a hybrid method by combining the above *SD* procedure with a *MC* sampling, alternating the *SD* and *MC* steps. A requirement for these calculations is the availability of an algorithm [developed by Guarnieri and Still (1994)] for the updating of the dynamics simulations, as the usual algorithms (such as the leap-frog, Verlet, and Beeman algorithms; see Section 6.4) require information from the previous timestep, which will not be available after a successful *MC* step.

The same idea of introducing *MC* sampling steps between Langevin iterations is present in the work of Chen *et al.* (1993), based on the development of Duane *et al.* (1987). In this method, the total (instead of the potential) energy is used and the time integration scheme is time reversible, so that it retraces its steps on reversal of the sign of the integration step. Numerical results show a poor global sampling of the conformational space but a good local sampling about the starting point (Byrne *et al*. 1994), with equilibration about an order of magnitude faster than temperature-rescaled *MD* simulations (Chen *et al*. 1993).

11 Molecular Associations

11.1 Introduction

It can be easily imagined that the computational difficulties in protein studies increase dramatically when their environment is taken into consideration, as compared with their study as isolated systems. Proteins may be in a (biological/industrial) solvent environment, which may affect their structure (Brooks III and Case 1993, Daggett and Levitt 1993), and in interaction with other systems. Foremost among such interactions are those involved in the intercellular communication and the various responses (such as the immune response) of biologically active cells, which are controlled by interactions with membrane receptors (Barinaga 1992 and Sezerman *et al.* 1993), and the antigen-antibody (Ag-Ab) binding. The common characteristic of all these cases is the existence of molecular associations, which are the result of the molecular recognition process (Jaenicke 1987). The study of these interactions, already of interest *per se*, has the added significance of extremely important practical applications, such as the rational design of vaccines and drugs, as discussed in Chapter 12.

What could be the way of tackling this problem? Let us consider the general problem of a peptidic/non-peptidic, flexible ligand binding to a protein in a given solvent. The theoretical study should, at least, be able to pinpoint the active site of the protein, determine the structure of the ligand-protein complex, and evaluate the corresponding binding energy, with consideration of the solvent effect. One could proceed directly with a Molecular Dynamics *(MD)* simulation of the complete system, i.e., protein, ligand, and solvent, but it would be computationally more effective to precede it, if possible, with a series of preliminary calculations, such as a statistical-mechanics Monte Carlo *(MC)* treatment of the solvent, a *MD* calculation of the protein by itself, an energy-minimization treatment of the ligand-protein system, a *MD* simulation of the ligand-protein system, without and with solvent, the latter perhaps preceded by a statistical-mechanics *MC* treatment of the solvent around the complex. Such calculations will yield useful information, of interest by itself (such as the thermodynamic quantities for the solvent) or for a simplification of the final treatment, such as the possible identification of the active site of the protein, if not known from experimental information.

This is a tall order and for that reason different approximations have been tried, as described below, where solvation and molecular recognition are discussed separately, for simplicity. Because both problems share the same computational difficulties (i.e., high computing times), it is appropriate to start with a discussion on the possible fast evaluation of interaction energies.

11.2 Fast Evaluation of Interaction Energies

The real bottleneck in the evaluation of interaction energies, as already mentioned in Section 10.3, is caused by the non-bonded interactions (see Eq. 5.32). A possible solution may be found in the application of the (discrete) fast Fourier transform (*FFT*) method, as suggested by Harrison *et al.* (1994). [See also the work of Press *et al.* (1992) for practical details and available software for the Fourier transform methods.] The main requirement for this procedure is the factorization of the terms in the non-bonded interactions: that is, the terms $c_{ab}^{(n)}/R_{ab}^n$ should be reexpressed as $c_a^{(n)}c_b^{(n)}/R_{ab}^n$ (Minicozzi and Bradley 1969, Jorgensen 1981, Hermans *et al.* 1984). Such a factorization has been discussed in detail in Section 5.3.2 for the term $c_{ab}^{(12)}/R_{ab}^{12}$.

The *FT* of two functions, F(u) and f(v), may be viewed as expressing the back-and-forth transformation of a same function (Press *et al.* 1992). Thus, for the two functions above, the direct and inverse integral *FT* are

$$F(u) = \int_{-\infty}^{\infty} f(v)\, e^{2\pi i uv}\, dv \quad f(v) = \int_{-\infty}^{\infty} F(u)\, e^{-2\pi i uv}\, du \tag{11.1}$$

Given two functions, f(v) and g(v), and their Fourier transforms, F(u) and G(u), their *convolution* and *correlation* are defined as

$$C_n(f,g) = \int_{-\infty}^{\infty} f(w)g(v-w)dw \quad C_r(f,g) = \int_{-\infty}^{\infty} f(w+v)g(w)dw \tag{11.2}$$

respectively, with the properties that the *FT* of the convolution is equal to F(u)G(u), i.e., the product of the individual *FT*, and that the *FT* of the correlation is equal to F(u)G*(u), where G*(u) denotes the complex conjugate. If the function f(v) has been sampled, $f_k = f(u_k)$, at a finite number of points, $k = 0, 1, 2, ..., N-1$, with equal spacing Δ of the variable, $u_k = k\Delta$, then the discrete FT of f(u) is given by

$$F(u_\ell) = \int_{-\infty}^{\infty} f(u) \, e^{-2\pi i u_\ell v} \, dv \simeq \Delta \sum_{k=0}^{N-1} f_k \, e^{2\pi i u_\ell v_k} = \Delta \sum_{k=0}^{N-1} f_k \, e^{2\pi i k \ell / N} = \Delta F_\ell$$

where

$$u_\ell = \frac{\ell}{N\Delta} \qquad \ell = -\frac{N}{2}, \, ..., \, \frac{N}{2}$$

and

$$F_\ell = \sum_{k=0}^{N-1} f_k \, e^{2\pi i k \ell / N}$$

is denoted as the discrete *FT* of the N points f_k. The discrete *FFT* is based on the fact that the discrete *FT* of length N can be written as the sum of two discrete *FT*, each of length $N/2$, with the corresponding savings in computing time.

Let us consider the evaluation of the non-bonded interactions between a receptor molecule and a ligand. We can define a grid such that the receptor molecule is completely immersed in and surrounded by it, being sufficiently large also to accommodate the ligand in different relative positions (i.e., separations and orientations) with respect to the receptor molecule. If the non-bonded interactions have been factorized, one can rewrite the interaction energy, corresponding to any of the non-bonded interaction terms, as

$$\Delta E = \sum_a \sum_b \frac{c_a^{(n)} c_b^{(n)}}{R_{ab}^n} = \sum_b c_b^{(n)} \sum_a c_a^{(n)} (1/R_{ab}^n) = \sum_b c_b^{(n)} V_b^{(n)}$$

with

$$V_b^{(n)} = \sum_a c_a^{(n)} (1/R_{ab}^n)$$

where the summation over a extends to the atoms in the receptor molecule and the summation over b extends to the atoms in the ligand; $V_b^{(n)}$ denotes the potential, due to all the atoms in the receptor molecule, at the position of atom b of the ligand. In such a case, in order to obtain a mapping of the potential energy hypersurface, corresponding to the non-bonded interactions, one could proceed as follows. First one would evaluate the potential, due to all the atoms of the receptor molecule, at all the grid points and store the values. One would then position the ligand in interaction with the receptor molecule and evaluate the interaction energy using the stored values of the potential. For example: one positions the ligand so that its center of mass coincides with one of the grid points, with a given orientation. As the atoms of the ligand will usually not be found at grid points, an interpolation of the potential values will be necessary. At each grid point, the calculation should be repeated for different orientations, say, chosen randomly.

The complete map of the energy hypersurface will be obtained by performing similar calculations for all the grid points.

These calculations involve, in addition to the evaluation of the potential at all the grid points, a transformation of the coordinates of the ligand for its translation to each grid point, the transformations needed for the various orientations, and the evaluation of the interaction energy (with interpolation of the potential) for each relative position. The computing cost is, therefore, proportional to the product of the (number of grid points) times the (number of orientations). As the grid must be large, when considering a large receptor molecule, the cost of the calculation will be considerable.

It is in this connection that the FFT method is useful, bringing about a considerable reduction of that cost. In the usual application of the FT method, the variables are time and frequency; in the present application, they are (molecular and grid) coordinates. The actual calculations involve, in brief, the following steps (Harrison *et al.* 1994): define the molecular and grid coordinate systems and their transformations; construct the (real space) maps of the kernels ($1/R^n$); evaluate the discrete FT of those kernels; convolute the coefficients for each potential with the kernels; evaluate the interaction energy as a sum of weighted correlations.

This procedure might also be applied to other types of calculations, such as energy minimization for an isolated molecule, non-explicit consideration of the solvent effect, etc.

11.3 Solvation

Although a hydration shell is thought to be associated with proteins (about 0.3 g of water per g of protein), there is little water in the interior of proteins and whatever water is found is usually associated with large cavities (Williams *et al.* 1994). X-ray structures of the solvent surrounding several proteins have been obtained (Teeter 1984, Finer-Moore *et al.* 1992, Madhusudan *et al.* 1993, Lounnas and Pettitt 1994) and a survey of 16 high-resolution structures has revealed that 40-60% of the main chain atoms, 14% of the hydrophobic side chains, and 80% of the hydrophilic side chains were in contact with water (Thanki *et al.* 1988). The significance of water regarding protein stability (Edsall and McKenzie 1978, 1983) and of its entropy and free energy contributions to the folding process (Privalov and Makhatadze 1993) have been discussed, but there is still considerable debate regarding its role in that process (Levitt and Park 1993, Woolfson *et al.* 1993, Karplus and Faerman 1994). Additional details on the structure of water surrounding proteins, peptide hydration, heat capacities of amino acids and unfolded proteins, protein-water interactions, and methodologies for free-energy calculations may be found in the work of Hagler and co-workers (Hagler *et al.* 1974, Hagler and Moult 1978, Hagler and Osguthorpe 1980, Hagler *et al.* 1980),

Privalov (Makhatadze and Privalov 1990, Privalov and Makhatadze 1990), Teeter (1991), van Gunsteren and Mark (1992), and Kollman (1993).

The emphasis in this work, reflecting the approach adopted by most researchers, is placed on the Molecular Mechanics (*MM*) treatments. In the case of solvation, particularly when the solvent is water, more accurate treatments are possible, because of the availability of sophisticated potential energy functions (*PEF*) for the water-water interactions. Proper statistical-mechanics *MC* treatments have been performed (for the water itself) for bulk water and for ions and biomolecules in an aqueous environment, as illustrated in the early work of Barker and Watts (1969), Abraham (1974), Abraham *et al.* (1976), and particularly Clementi and co-workers (see, e.g., Clementi and Popkie 1972, Kistenmacher *et al.* 1973*abc*, 1974, Popkie *et al.* 1973, Clementi 1976, Ranghino and Clementi 1978, Romano and Clementi 1978,1980, etc.) For other solvents, a *MC* simulation could be preceded, in order to start it from an appropriate configuration, by a calculation performed using the algorithm of Iglesias *et al.* (1991) for the build-up of stable molecular associations.

At the *MM* level, mention has already been made (see Section 10.4.3) of non-explicit solvent treatments, where the effect of the solvent is approximated through the use of a dielectric constant ($\varepsilon \neq 1$) in the Coulombic term of the non-bonded interactions. Alternatively, that effect may be simulated with semiempirical *PEF* derived from the crystallographic database of proteins, on the basis of thermodynamic or electrostatic considerations (involving the Boltzmann and Poisson-Boltzmann equations, respectively), or with a *PEF* incremented with correction terms dependent on solvent accessible surface areas.

Explicit-solvent calculations are possible, but at an additional computational cost, if an appropriate *PEF*, which includes the necessary terms for the interaction with the solvent molecules, is available (see, e.g., Section 5.3 for the case of aqueous solutions).

11.3.1 Potential Energy Functions

The development of empirical *PEF* from known protein structures has been carried out by Gelin and Karplus (1979) and Sippl (1993), the latter through the use of the inverse Boltzmann law, as described below. The argument underlying this approach is that the protein structure is the result of all the existing interactions, including those with the solvent in which it has been immersed.

The probability, p_i, associated with the energy U_i, was expressed (see Section 4.2) as

$$p_i = \frac{1}{Q} e^{-U_i/k_B T}$$

where Q denotes the partition function. Therefore, the inverse Boltzmann law may be written as

$$U_i = -k_B T \ln Q - k_B T \ln p_i$$

At constant T, Q is a constant and it may be assigned the value 1 without affecting the energy difference between states (see below), so that

$$U_i = -k_B T \ln p_i$$

The subscript i, labeling the state, stands for a set of variables. In a more general fashion, we may subdivide that set into two subsets, say, j and k, and consider the case when the subset k is fixed at certain values. Correspondingly we will use the notation p_{jk}, p_j, and p_k to denote the probabilities, with the conditions

$$p_{j(k)} = p_{jk}/p_k \qquad p_j = \sum_k p_{jk} \qquad \sum_j p_j = 1$$

The associated energies are then

$$U_{j(k)} = -k_B T \ln p_{j(k)} \qquad U_j = -k_B T \ln p_j$$

and their difference

$$\Delta U_{j(k)} = U_{j(k)} - U_j = -k_B T \ln (p_{j(k)}/p_j)$$

from which one can obtain the corresponding forces by differentiation with respect to the variables in the subsets. With the above formulation, all the forces common to all the subsystems have been removed. For actual calculations, the variables chosen are the residues, atom types, sequence separation between the residues, spatial separation between the atoms, and the number of protein atoms, within a sphere of chosen radius, around each protein atom (at the center of the sphere); the latter number, related to the solvent exposure of that particular atom, is used in order to extract the solvent effect.

The significance of the electrostatic interactions in solvation as well as in ligand binding (Kozack and Subramaniam 1993) suggests the possibility of proceeding with the study of both problems in terms of only such interactions (with added considerations; see Section 11.4.2). The electrostatic interactions in a solvated system (protein or ligand-protein) consist (Gilson *et al.* 1985) of the interactions between pairs of charged atoms in the system itself as well as the interaction of those charged atoms and the solvent, the latter being responsible for the fact that those atoms tend to be found on/near the surface of the protein (Paul 1982, Rashin and Honig 1984), defining its hydrophilic regions (see Sections 8.2.1 and 12.2.3). Those interactions may be obtained (Warwicker and Watson 1982, Rogers *et al.* 1985, Klapper *et al.* 1986, Warwicker 1986, Gilson and Honig 1987,

1988*ab*, Davis and McCammon 1989, Kozack and Subramaniam 1993) from finite difference solutions to the (linearized) Poisson-Boltzmann equation.

The Poisson equation (see, e.g., Menzel 1960)

$$\nabla^2\phi + \rho/\varepsilon = 0 \text{ (in } mks \text{ system)} \qquad \nabla^2\phi + 4\pi\rho/\varepsilon = 0 \text{ (in Gaussian system)}$$

relates the electrostatic potential ϕ at a point in space with the charge density ρ (per unit of volume); ε is the dielectric constant and $\nabla^2 = \nabla\cdot\nabla$ is the Laplacian operator, defined from the gradient vector operator ∇ (see Section 3.1.1). For solutions (e.g., a solution consisting of a protein, the solvent, and an ionic atmosphere), the so-called Poisson-Boltzman (*PB*) equation is obtained within the framework of the Debye-Hückel theory (Debye and Hückel 1923). In the Gaussian system of units, the *PB* equation (McQuarrie 1976, Oberoi and Allewell 1993, Holst *et al.* 1994)

$$\nabla\cdot[\varepsilon(r)\nabla\phi(r)] - \varepsilon(r)\kappa^2(r) \sinh[\phi(r)] + 4\pi\rho(r) = 0$$

where $\phi(r)$ is the electrostatic potential (expressed in units of k_BT/e, where e is the elementary electric charge), $\rho(r)$ is the charge density within the protein, $\varepsilon(r)$ is the dielectric constant, and κ is the inverse of the Debye length or Debye screening distance (Oldham and Myland 1994); *sinh* denotes the hyperbolic sine. The linearization of the *PB* equation is based on the assumption that $\phi(r)$ is small, in which case one may use the approximation $\sinh[\phi(r)] \simeq \phi(r)$, obtaining

$$\nabla\cdot[\varepsilon(r)\nabla\phi(r)] - \varepsilon(r)\kappa^2(r)\phi(r) + 4\pi\rho(r) = 0$$

(see Gilson and Honig 1987, Davis and McCammon 1989, Kozack and Subramaniam 1993). The charge density $\rho(r)$ is to be evaluated for the protein configuration obtained from the crystal structure coordinates.

In the model of Gilson and Honig (1988) (see also Honig *et al.* 1993) the system is represented by a cavity (with uniform dielectric constant ε_m), defined by the spheres (with radii corresponding to the atomic radii and dielectric constants dependent on the atomic polarizabilities) centered at the positions of the atoms, considered as point charges, and it is surrounded by the solvent (which may contain an ionic atmosphere) with dielectric constant ε_s. One can then write

$$\Delta G^\circ = \Delta G_c^\circ + \Delta G_p^\circ + \Delta G_a^\circ = \Delta G_c^\circ + \Delta G_s^\circ = \sum_i (\Delta G_{c(i)}^\circ + \Delta G_{s(i)}^\circ)$$

where ΔG_c°, ΔG_p°, and ΔG_a° correspond to the charge-charge, charge-polarizable solvent, and charge-ion atmosphere interactions, respectively; ΔG_s° contains all the interactions with the solvent (ΔG_p° and ΔG_a°), and the summation extends to all the charges. For the (linearized) Poisson-Boltzmann equation one has

$$\Delta G^\circ = \frac{1}{2} \sum_i q_i(\phi_{c(i)} + \phi_{s(i)})$$

where ϕ denote the potentials, at the position of charge q_i, produced by the remaining charges and the solvent, respectively. For the computations, a two-step thermodynamic process is used, starting from the reference state, which corresponds to the uncharged system (with the charges turned off) immersed in a medium with dielectric constant equal to the adopted dielectric constant of the system. The first step of the process involves turning on the charges and the corresponding ΔG_c° may be evaluated directly by the Coulomb law. In the second step, the dielectric constant of the medium is changed from ε_m to the value adopted for it, ε_s, resulting in a change in the potential due to the medium, $\Delta\phi_{(i)}$, at the position of each charge, so that

$$\Delta G_c^\circ = \frac{1}{2} \sum_i q_i \Delta\phi_{(i)}$$

The finite-difference calculations for the (linearized) Poisson-Boltzmann equation may be performed using the software package DelPhi (Gilson and Honig 1988ab; see also Press $et\ al.$ 1992).

At a lower level of approximation, the effect of the solvent (in non-explicit solvent calculations) may be taken into account through the addition to the original PEF of a new term, $\sum_i g_i A_i$, where the summation extends to all the groups in the system under consideration, A_i denotes the solvent-accessible surface area of the i-th group, and g_i is the free-energy of hydration per unit of surface area of the i-th group (Wako 1989, Nayeem $et\ al.$ 1991, Vila $et\ al.$ 1991, Williams $et\ al.$ 1992).

11.3.2 Molecular Dynamics Simulations

Since the first MD simulations were performed (van Gunsteren $et\ al.$ 1983, van Gunsteren and Berendsen 1984), the number of such calculations has grown considerably, with studies at different temperatures (including high temperatures, for the unfolding of proteins in solution) and pressures (Kitchen $et\ al.$ 1992). As such calculations have become routine (given the appropriate computer power), we will limit the discussion here to some specific points, after considering the illustrative example of van Gunsteren $et\ al.$ (1983).

Van Gunsteren $et\ al.$ (1983) studied $BPTI$, with 454 heavy atoms, 113 H-atoms (using the united-atom model for the remaining H-atoms) and 560 water molecules, for a total of 11,844 degrees of freedom, with thermal bath (at 300 K), $\Delta t = 0.002$ ps and a run of 20 ps, obtaining an average structure deviating only about 1 Å from the X-ray structure and observing that *the mobility of the water molecules increases gradually with their distance from the protein.*

The practical points to be kept in mind, regarding possible errors in the calculations, concern the water model (which should reproduce the experimentally-observed structure), the boundary conditions and their adjustments (with the consequent need for the truncation of non-bonded potentials), the truncation of such non-bonded potential [with temperature increases due to spurious forces, to be corrected by rescaling of the atomic velocities or the use of a thermal bath (Berendsen *et al.* (1984)], the starting point and length of the simulation [with the corresponding effects on the random (noise) and systematic (histeresis) errors (Hermans *et al.* 1992)]. All of this, of course, is under the assumption that a correct *PEF* is being used.

The existence of ions may be taken into account if the *PEF* includes the appropriate terms. Thus, Kalko *et al.* (1992) carried out a *MD* run (of about 120 *ps*) for the study of ion channels of icosahedral viruses, using the *PEF AMYR* and the software package *PCAP* [developed for *MM, MD*, electrostatic, and free-energy calculations (Snow and Amzel 1986, Cachau *et al.* 1990)]. The preference binding of an ion *vs.* another at a binding site is evaluated

$$p = R_i/R_j = e^{-(\Delta G_i - \Delta G_j)/k_B T}$$

from the corresponding rate constants. The required free-energies are evaluated from a four-step thermodynamic cycle, involving the ion *i* in solution, the ion *j* in solution, the ion *j* at the binding site, and the ion *i* at the binding site. The free-energy change corresponding to the exchange of the ions in solution is evaluated from the difference in hydration energies while the free-energy change corresponding to the exchange at the binding site is evaluated by a free-energy perturbation (*FEP*) method (Beveridge and DiCapua 1989, Jorgensen 1989, Straatsman and McCammon 1992), whereby the parameter defining an ion is changed slowly into the parameter defining the other ion during the *MD* simulation. The difference $(\Delta G_i - \Delta G_j)$ is then calculated from the difference of those two free-energy changes above. [It should be noted that the *FEP* method works well whenever the exchange or mutational path connects two systems which are similar, such as the two ions considered in the calculation. See Section 12.2.4 for an alternative of use in those cases where the above condition is not met.]

11.3.3 Other Approaches

The normal mode analysis has been used by Yamato *et al.* (1993), for the study of the effect of pressure on the conformation and volume in sperm whale deoxymyoglobin, and Yoshioki (1994), for the study of the frequency distributions of the internal motions in *BPTI* in different environments, by changing the magnitude of the external force field. In that work, using three different water configurations (with the positions of the water molecules kept fixed during the calculation), the protein is allowed to move in a hole surrounded by water. The water molecules are positioned in a rectangular box using *GROMOS* and the interactions are evaluated with *UNICEPP* (with the united-atom approximation for the *H*-atoms not to be involved in *H*-bonding and containing Coulomb and 6-12 non-bonded interactions as well as *H*-bonding terms for the interaction protein-water). The equilibrium point is assumed to have been reached when the conformational energy is approximated by a multidimensional parabola, using all the positive eigenvalues from the force constant matrix, which is obtained from the second derivatives of the conformational energy function (including the contribution from the water) with respect to the 310 variables.

11.4 Molecular Recognition

As indicated in Section 11.1, the difficulties in the study of a molecular recognition process are of a practical nature, due to the computing requirements for a complete treatment. As it is always the case, recourse must then be made to approximations, based on the characteristics of the process.

A molecular recognition process consists of several steps (Peradejordi 1989). There is, first of all, a kinetic component, whereby the ligand L_i (i.e., the system being recognized) will approach the receptor R (i.e., the recognizing molecule). Next, the molecular complex RL_i

$$R + L_i \; \overset{K_i}{\rightleftharpoons} \; \{RL_i\} \implies \text{response}$$

is formed, with a corresponding association constant K_i. It is the formation of such a complex that triggers the biological response. Usually, this process is reversible, and the response will be dependent on the diffusion of the ligand away from the receptor, as its concentration changes, e.g., if it is metabolized by the organism. In some instances, however, the process is irreversible, with formation/breaking of covalent bonds. The usual Molecular Mechanics treatment cannot handle the latter case, which must be studied by appropriate quantum-chemical methods. Therefore, the discussion here will be restricted to the general aspects of the formation of a molecular association, with the two systems in interaction, without covalent binding. If appropriate, such a binding may be studied later for a simplified model.

The formation of the complex is the result of specific interactions between (a) given region(s) of the ligand (i.e., its *recognition site*) and a particular region of the recognizing molecule (i.e., its *receptor* or *active site*). A given molecule may contain both recognition and receptor sites; thus, e.g., in a class I major histocompatibility complex (*MHC*) molecule, the receptor site binds the processed antigen while its recognition site (*antigenic determinant* or *epitope*) is recognized as non-self whenever cells presenting it on their membranes are transplanted (which is the reason for those molecules to be denoted as *transplantation antigens*). This topic will be discussed in more detail in Chapter 12, in conjunction with the design of synthetic vaccines.

Those interactions imply a dynamic complementarity, both steric and chemical, of the recognition and receptor sites and define the *intrinsic activity* of the ligand for the receptor considered. The overall *affinity* of the ligand for the receptor, however, depends on the complete interaction of both systems; that is, it depends on the *binding energy*

$$\Delta E_i = E_{RL_i} - (E_R + E_{L_i})$$

where E_R, E_{L_i}, and E_{RL_i} denote the molecular energies of the receptor, the ligand, and the receptor-ligand complex, respectively (see Section 3.8.1). Consequently, families of related ligands will be characterized by recognition sites with similar intrinsic activities, each ligand possibly having a different affinity. The optimization of the receptor-ligand complex should be carried out through a minimization of the binding energy (Brooks *et al.* 1988). A more detailed discussion of this subject will be presented in Chapter 12 when considering the *structure-activity relationships* (*SAR*).

This simple discussion suggests that the first step towards the study of a molecular recognition process, because of its steric requirements, should be the analysis of the spatial characteristics of the binding sites and of the molecular surface of the ligand.

11.4.1 Molecular Surfaces and Binding Sites

The concept of molecular surface is rather elusive and therefore different definitions (Richards 1977) are possible, such as *van der Waals* (Lee and Richards 1971), *molecular or Connolly* (Connolly 1986), consisting of contact and reentrant components, and *accessible surfaces*. A variety of algorithms have been proposed for their representation: *bit lattices* (Pearl and Honegger 1983), *bit lattices and fractals* (Higo and Go 1989), *numerical* (Connolly 1981), *analytical* (Connolly 1983, 1985, Eisenhaber and Argos 1993), *cartographic 2D-projection* (Taylor *et al.* 1983), *negative sphere description* (Kuntz *et al.* 1982), β-*splines* (Colloc'h and Mornon 1990), and *Fourier-series description* (Leicester *et al.* 1988). Their characteristics may be analyzed using fractal dimensionality (Lewis and Rees 1985), solid angles, β-splines, and Voronoi tessellation (Voronoi 1908, Brostow *et al.* 1978). The details of all these definitions and methods are omitted here, as they can be found in the comprehensive review of Lewis (1991).

The binding sites are usually found in clefts, holes, or in the boundaries between protein domains, with an average size of 5-10 Å (Lewis 1991) (see below). Some of their features, as well as of the corresponding recognition sites, have been determined by Janin and Chothia (1990) from the experimental data for protease-inhibitor and antigen-antibody (*Ag-Ab*) complexes: number of residues making contacts across interfaces, 27-40; accessible surface areas buried in interfaces, 1,250-1,950 $Å^2$; number of intermolecular *H*-bonds, 6-14; average chemical character, without any particular amino acid composition; mobility, 6.2-28.8 $Å^2$ [with rigid preformed binding loops in inhibitors (Ruhlmann *et al.* 1973, Janin and Chothia 1976) and mobile epitopes (Westhoff *et al.* 1984, Tainer *et al.* 1984)]; small conformational change upon formation of complex; rate constant, 10^5-10^7 M^{-1} s^{-1}; free-energy of dissociation, 11-18 kcal/mol, with a dissociation constant of 10^{-8}-10^{-13} M. The main conclusions are that most collisions at the target surface lead to *stable associations, favored by electrostatic, H-bond, and hydrophobic forces, and opposed by the loss of translational, rotational, and internal degrees of freedom.*

The particular case of the complexes involved in the immune response, whether cell-mediated or not, merits special attention in connection with the assumption of *complementarity* between the receptor and the recognition sites. On one hand, there is the interesting fact (Barinaga 1992) that a reduced number of *MHC* molecules are able to bind, with high affinity, a large variety of peptides. The experimental information indicates that the termini of the peptides are tethered at two pockets in the groove of the *MHC* molecule, with the result that larger peptides will bulge out of the groove in the middle, and that the *T*-cell detects the surface of the peptide protruding from the groove; two questions that arise are whether the variable parts of the peptide must be exposed in order to be recognized and whether the binding peptide induces a change in the shape of the *MHC* molecule. On the other hand, the experimental information for the complexes of a series of steroids, conformationally different, with the Fab' fragment of the anti-progesterone antibody DB3, shows that the antibody is unable to complement completely the shape of the antigen; the binding, which occurs without any major rearrangement of the binding site, is realized through different orientations of parts of the antigen in the binding site (Arevalo *et al.* 1993). A conformational sensitivity analysis (Susnow *et al.* 1994) of the complexes of the cytosolic protein FKBP-12 (with two loops of significance, with a direct role in the ligand binding or in the subsequent binding to an effector protein, respectively) has shown that the conformations of the residues in the active site and in the two loops are sensitive to the ligand conformation and that the orientation of the amide and keto carbonyls of one of the ligands is sensitive to the aromatic side chains of the binding site. It is of interest that recent studies on the binding of peptides to the *Src* homology 3 (SH3) domain have shown that peptide ligands can bind to the same site but in opposite orientations (Feng *et al.* 1994).

The conclusions that emerge from these observations are: (a) *a few key residues anchor the ligand* (Marshall 1992), as confirmed by the fact that single substitutions within the binding site may change the affinity by a factor $1\text{-}10^4$ (Laskowski *et al.* 1987); (b) *ligand flexibility*, especially in the long side-chains, *plays a significant role in the binding* (Warwicker 1989). Evidently, more observations are needed before a general conclusion is reached, but it seems that a full complementarity of the receptor and recognition sites may not be required. A partial complementarity (involving the anchor residues) and ligand flexibility may be the essential factors. Complementarity schemes, therefore, should be used with caution, probably as a first step towards the complete solution of the problem.

11.4.2 Docking Procedures

As already mentioned in Section 6.2.2, one may perform a docking simulation by a direct approach, whereby the two systems (rigid or flexible) are allowed to interact and approach each other until a stable complex has been formed. A comprehensive set of calculations [e.g., using the improved *AMYR* program (Torrens *et al.* 1991)], starting from different relative positions of the two interacting systems, could yield a mapping of the corresponding energy hypersurface, with due attention paid to the proper characterization of its stationary points (Rubio *et al.* 1993).

Three points of significance in this suggested procedure are the rigidity/flexibility of the interacting systems, the procedure for the sampling of the energy hypersurface, and the features of the *PEF* used. Consideration/disregard of the flexibility, a restricted or extensive sampling, and the use of an appropriate *PEF* or its substitution by other considerations define the different levels of approximation at which the simulations may be performed, as discussed below.

The use of an appropriate *PEF*, with consideration of (all) the important interactions will result in an expensive calculation, which may be simplified through the exclusive use of considerations such as surface complementarity, surface area burial, steric hindrance, etc. The sampling of the energy hypersurface may be performed in a manual manner (see, e.g., the work of Busetta *et al.* 1983, Pattabiraman *et al.* 1985, and Tomioka *et al.* 1987), whereby the user positions the systems as desired, or automatically (see, e.g., the work of Wodak and Janin 1978, Kuntz *et al.* 1982, Goodsell and Dickerson 1986, Connolly 1986, Billeter *et al.* 1987, Lipkowitz and Zegarra 1989, Goodsell and Olson 1990, and Jiang and Kim 1991). The disregard/consideration of the flexibility of the systems (or, at least, of the ligand) characterizes the rigid-body (*RBD*) (see, e.g., the work of Shoichet and Kuntz 1991, Bacon and Moult 1992, Jiang and Kim 1992, and Kuntz 1992), semiflexible (*SFD*) (see, e.g., the work of Wodak and Janin 1978, Billeter *et al.* 1987, DesJarlais *et al.* 1988, and Shoichet and Kuntz 1991), and flexible (*FD*) (see, e.g., the work of Ghose and Crippen 1985, Goodsell and Olson 1990, Yue 1990, Caflisch *et al.* 1992, Hart and Read 1992, and Leach and Kuntz 1992) docking procedures, the latter being the preferable one; in a *RBD* simulation, the search is restricted to the translational and rotational degrees of freedom, while in a *SFD* simulation a local energy minimization (i.e., with consideration of the flexibility) is performed at each position sampled in a *RBD* approach.

These characteristics and approximations are not mutually exclusive. Thus, Monte Carlo, Simulated Annealing, and Molecular Dynamics simulations are usually performed with more complete *PEF*, in an automatic fashion, preferably, but not necessarily, with flexible ligands. Complementarity-based simulations, on the other hand, are usually carried out in the *RBD* approximation, with consideration of some *PEF* (often denoted as scoring function and consisting of the electrostatic interaction terms) when it is desired to improve the prediction by pruning of the results or moving up to the *SFD* level. The details of some representative simulations are presented below.

(a) Complementarity-based simulations

Since the early work of Blow *et al.* (1972), docking studies, based on some sort of complementarity (shape, charges, potential), have been performed by Salemme (1976), Greer and Bush (1978), Kuntz *et al.* (1982), Lee and Rose (1985), Janin and Wodak (1985), Connolly (1986), Warwicker (1989), Jiang and Kim (1992), Leach and Kuntz (1992), Meng *et al.* (1992), Novotny and Sharp (1992), Walls and Sternberg (1992), Meng *et al.* (1993), Sezerman *et al.* (1993), and Helmer-Citterich and Tramontano (1994).

The various complementarity-based procedures represent variations of a general method that can be summarized as follows: the ligand (or its anchor part) is positioned in the receptor site and orientational sampling is performed on the basis of (usually shape) complementarity considerations; those configurations are pruned and clustered through the use of (usually electrostatic) scoring functions; and the final configurations are selected on the basis of additional criteria. The relevant points, therefore, are: a knowledge of the receptor site or a procedure for its identification; the possible need for a very extensive orientational sampling; reduction of the computing cost by precomputation (in a $3D$-grid) of the scoring function. The bottleneck arising from the extensive sampling may be avoided (at least in part) by biochemical knowledge (Walls and Sternberg 1992) or through a compromise between extensive sampling and moderate sampling with energy minimization (Meng *et al.* 1993).

The method of Helmer-Citterich and Tramontano (1994) for the identification of receptor (as well as recognition) sites is based on a projection of the solvent-accessible area (Connolly 1983) into bidimensional matrices. The procedure consists of the following steps: choose an appropriate cylindrical reference system for the molecule; cut the surface in slices (perpendicular to the cylindrical axis); divide the slices into regions of approximately constant area; analyze the regions for distance from axis and select in each region the point with maximum radial coordinate; store the radial coordinates of those points in a row of a matrix; substitute each element in that row with the difference between its value and the value of the preceding point. The result, which is independent of the distance from the axis, will identify the *knobs* (as a series of positive values followed by a series of negative values) and the *grooves* (as a series of negative values followed by a series of positive values) on the surface of the molecule. [The resulting matrices for the two systems may then be used in a complementarity search.]

Some of the specific features of the variations of the general procedure, schematically described above, may be observed in some illustrative examples.

Jiang and Kim (1991) use a cube representation of the molecular surface and volume, with a set of surface and a set of internal grid points. The docking procedure is carried out in three steps: the initial collection of configurations is obtained on the basis of size, shape, close packing, and steric hindrance; then a screening procedure is used in order to identify that subpopulation of configurations with favorable interactions between the buried surface areas; and *the correct configuration for the complex is finally identified on the basis of additional theoretical, biochemical, and genetic considerations*, as well as by visual inspection.

Leach and Kuntz (1992) have studied the docking of flexible ligands using the programs *DOCK* (Kuntz *et al.* 1982, DesJarlais *et al.* 1986, 1988, Shoichet and Kuntz 1991), *WIZARD* (Dolata *et al.* 1987), and *COBRA* (Leach and Prout 1990). First, the possible orientations of the anchor fragment of the ligand are determined. Then a clustering of the configurations is performed by a conformational analysis of the flexible ligand within the receptor site and an orientation is selected for each family of configurations, yielding a set of configurations to be used in a subsequent exploration. The procedure involves the evaluation of the surface of the macromolecule and of its negative image, generation of ligand orientations, check of unfavorable steric interactions, evaluation of interactions (*H*-bonding and electrostatic, using *AMBER*, stored in grid points, and using interpolation), and clustering on the basis of similarity/dissimilarity, defined from the *rms* distance separation for all the atoms in the ligand or its fragment. The subsequent exploration is performed with fixed anchor, searching the conformational space for the remainder of the ligand. The method (may) allow(s) the incorporation of experimental information (*H*-bonding, salt-bridge, binding at a metal cation site, etc.).

Meng *et al.* (1992) also precompute (in a 3*D*-grid) the terms of the *PEF* which involve sums over atoms of the receptor. The characterization of the receptor site is carried out using the *MS* algorithm of Connolly (1983*ab*) and the docking procedure finds the sets of ligand atoms that match sets of sphere centers. The scoring is done on the basis of electrostatic interactions [using potentials calculated with the *Del Phi* program (Klapper *et al.* 1986, Gilson *et al.* 1987); see Section 11.2.1] and *MM* interaction energies (with Coulombic and 6-12 van der Waals terms, using geometric mean values for the coefficient of the latter; see Section 11.2).

Walls and Sternberg (1992) generate large number of orientations and use steric scoring (with a soft Lennard-Jones-type potential) to assess complementarity and a simplified electrostatic model (Levitt 1976) in order to remove unacceptable configurations, with an algorithm developed specifically for a computing facility consisting of a distributed array of processors. [See also Sections 10.3 and 10.4.2 for the use of specialized computers.]

Sezerman *et al.* (1993) first determine the anchor residue locations and then find the conformation of the intervening chain using various algorithms (Bruccoleri and Karplus 1985, 1987, Vasmatzis 1992, Qiang *et al.* 1993).

(b) Monte Carlo simulations and Simulated Annealing

Most of the features of these procedures have been discussed elsewhere in this work (see Sections 4.3.1, 6.3, 10.5, and 10.6) and therefore only specific details, regarding approximations and variations, will be discussed here, with a detailed inspection of only the most distinctive procedures.

Thus, for example, Goodsell and Olson (1990) perform a docking with *SA*, accelerating the calculations through the use of grid energy evaluations (Goodford 1985). Cherfils *et al.* (1991) use rigid systems, with the protein modelled by one sphere per residue, with a *PEF* of attractive terms proportional to the interface area, and perform a *SA* to obtain clusters of orientations with steric fit, refining the complexes using the program X-PLOR (Brünger 1987), with *CHARMm* parameters and full atomic representation, and suggest that *some of the solutions found may represent alternate modes of association*, which might be observed when mutations or chemical modifications could prevent the formation of the native complex. Hart and Read (1992) use a *floating MC* procedure, with the Metropolis test replaced with a scoring function that measures the average distance of the ligand to the target surface, and a *SA* run with a pairwise *PEF*, with repetition of the complete procedure for a large number of different randomly generated initial configurations. Caflish *et al.* (1992) perform the *MC* docking in three steps: the positions and conformations are sampled; the molecular association energy is minimized for preselected docking conformations (using a *PEF* with intermolecular non-bonded terms for the elimination of bad contacts); the resulting configurations are subjected to *MC* minimization (using a *PEF* with *H*-bonding and Lennard-Jones-type terms for the heavy atoms and the polar *H*-atoms, in 400 cycles, with each cycle involving a random change of a random dihedral angle, 100 iterations of a conjugate gradient algorithm, and a Metropolis test at 310 K). Knegtel *et al.* (1994*b*), for the study of protein-*DNA* complexes, use the program *MONTY* (Knegtel *et al.* 1994a), with a *PEF* of square-well potentials for *H*-bonding and van der Waals interactions (for heavy atoms and polar *H*-atoms), with initial random selection of dihedral angles of selected protein surface side chains, rough minimization by rotation of those side chains (in order to eliminate bad contacts), random rotation/translation of the protein and variation (in random steps of ±10°) of the rotatable side chains, and energy minimization with acceptance of moves which lower the energy and relax the structure and conclude that *inclusion of biochemical/biophysical data as restraints in the simulations will increase the chances of obtaining the correct answers with a high degree of certainty* [see, e.g., Section 9.3.2].

Dauber-Osguthorpe *et al.* (1988) studied the binding of the antibacterial agent trimethoprim to *E. coli* dihydrofolate reductase (*DHFR*) in the presence of 259 water molecules, with *SA* and minimization with respect to all the Cartesian coordinates (with a total of 10,800 iterations). The conclusions reached were that salt-bridges and *H*-bonds provide directional interactions, favorable van der Waals interactions are important, a strain energy (with loss of vibrational, rotational, and translational energy) is induced in the ligand in exchange for the gain due to the interaction with the enzyme, and that *it is not sufficient to include only one layer of water molecules because water near, but not forming H-bonds with, the ligand are also important.*

Abagyan *et al.* (1994) perform an internal coordinate modeling (*ICM*) [previously used for peptides (Abagyan and Argos 1992), molecular design (Gibson *et al.* 1993, Borchert *et al.* 1993), and homology modeling (Eisenmenger *et al.* 1993)] in a double-energy Monte Carlo with Minimization (*MCM*) procedure. The internal coordinates (bond lengths and bond, torsion, and phase angles) are the variable parameters (although some may be kept fixed), a directed tree-like graph is imposed on all the atoms (real as well as virtual, but ignoring the ring-closing covalent bonds), with the atom positions determined by the bond lengths, the bond angles, and the torsion angles (for side chains) or the phase angles (for the main branch). Each molecule has two virtual atoms attached to it in a fixed configuration, and the molecular subtrees are connected by virtual bonds to form the final tree (with a default connection to a third virtual atom fixed at the origin of coordinates). The *PEF* contains term for the Coulombic, van der Waals, *H*-bonding, and torsion angle contributions (from *ECEPP*/2) as well as contributions for the solvent effect [in terms of solvent-accessible areas (Eisenberg and McLachan 1986, Wesson and Eisenberg 1992)] and the changes in bond lengths, bond angles, and phase angles, and restraints. The accessible surface is evaluated by the algorithm of Schrake and Rupley (1973) and the interaction lists (see Section 10.4.2) for non-bonded interactions and distance constraints are updated after any abrupt conformational change. The *MCM* is performed with random conformational changes [such as change of one angle at a time, Brownian-like step, or a biased-probability random step (Abagyan and Totrov 1994)], energy minimization of analytically-differentiable energy terms, evaluation of the non-differentiable energy terms, and application of the Metropolis test. Comparison of the results obtained for the docking of two rigid helices from a leucine zipper domain with torsion angle and Cartesian coordinate optimization, respectively, suggests that an appropriate approach to the problem of protein folding (and ligand binding) could consist of a *coarse global sampling followed by a local energy minimization procedure in the torsion angle space.*

The *fragment approach* of Friedman *et al.* (1994), in the study of *Ag-Ab* complexes has been used in order to identify the *lock and key* and *handshake* components in ligand binding. The results [obtained using the *MC* algorithm of Goodsell and Olson (1990), with a *PEF* incorporating Coulombic, van der Waals, and *H*-bonding terms, with fixed geometry] indicate that *the anchor interaction* (involving shape and electrostatic complementarity) *is essential and that the specificity and incremental binding energy are determined by additional interactions, involving the remainder of Ag.*

(c) Molecular Dynamics Simulations

The results of Mao (1991, 1992) in a mass-weighted *MD* treatment (see Section 10.4.2) for the binding of an octapeptide inhibitor to aspartyl proteinase from Rhizopus chinensis (rhizopuspersin), using the *CHARMm PEF* highlight the importance of ligand flexibility. That is, it is not sufficient to optimize the ligand conformation towards complementarity with the receptor site. *The ligand must have that flexibility that will allow it to enter the binding site and adopt then the optimal binding conformation as well as dissociate itself thereafter and find its way towards the opening of the receptor site.* This flexibility may also be simulated with a modified *MD* algorithm (DiNola *et al.* 1994), using different thermal baths for the center-of-mass motion of the substrate, its internal and rotational motions, and the motion of the receptor, respectively. Using a translational T of 1300-1700 K for the ligand will allow it to overcome high barriers.

The calculations of Nakagawa *et al.* (1993), on the other hand, show how it is possible to identify, e.g., the water molecule which is the potential nucleophile in a deacylation process. Those calculations were performed with the *CHARMm PEF* (in the united-atom approximation, with polar *H*-atoms, but without *H*-bonding terms), switching of non-bonded interactions (at 7.5-8.0 Å for the van der Waals interactions and 9.0 Å for the Coulombic interactions, with update of the interaction list every 20 *fs*; see Section 10.4.2), with a timestep of 1 *fs*. The procedure consisted of the following steps: minimization of the protein conformation; equilibration of the water at 300 K for 7 *ps*; addition of more water molecules in order to have the proper density; new equilibration of the water for another 10 *ps*; equilibration of the complete system, at 300 K, for 10 *ps* with velocity scaling and for another 10 *ps* without scaling; two production runs of 60 *ps* each (saving the coordinates every 40 *fs*). It is interesting to note the observation that *the side chains of the residues involved in the catalytic process are significantly less flexible* than other side chains in the protein, probably *the result of interconnected H-bonding networks* among the residues and between the residues and the solvent water in the catalytic site.

Applications

There is no doubt that a considerable effort has been directed towards the development of formulations and computational methods for the modeling of peptide and protein structures, as evidenced by the number of references included in this work, even though an exhaustive listing has not been attempted.

Successful treatments of chosen systems have also been reported, especially at the basic level. There is, however, a remarkable difference when considering the practical application of those methods to real problems. This difference may be due, in part, to the fact that perhaps not all the successful work is published. But there is, in addition, a factor which has not appeared in the preceding pages: economics.

It is quite different, on one hand, to propose a new method and apply it to a given system, and that is the end of the exercise, and, on the other hand, to perform a predictive study and propose a new series of compounds, which are then to be synthesized and tested for bioactivity. This is an expensive task, a factor which adds more responsibility to the work of the practitioner.

The question that arises then is whether all the necessary tools are available or whether more development work is still necessary. Our conclusion is that there is still much to be done, particularly regarding potential energy functions and structure-activity relationships.

12 Structure-Aided Molecular Design

12.1 Introduction

Ultimately, all the developments (formulations, numerical techniques, software packages, visualization procedures), described in the preceding chapters, will come to fruition with concrete results of practical value. Diverse, and very significant, possibilities exist, and more will be found. At this point, it will suffice to mention the development of conductors, molecular switches, synthetic vaccines, pepzymes, and drugs. All these projects may be grouped together under the general designation of *structure-aided molecular design* (Robertus 1994), although other designations, such as *structure-assisted, structure-based,* and *computer-aided* molecular design may also be found in the literature.

The common factor in all those endeavors is the design of either a peptide (or, in some instances, its non-peptide analog) or a protein, capable of a specific function. The differences arise simply from the difficulty of the task, related to the size of the system to be designed; that is, as a rule it will be more difficult to design a pepzyme than a small-size ligand. Computer simulations, based on Molecular Mechanics methods, may be used for the design work but the final steps in the *rational* design of peptidomimetics could undoubtedly profit from proper quantum-mechanical calculations.

For simplicity in their examination, the topics in this Chapter have been separated into two main sections, concerning the *de novo* design of peptides and proteins (in general) and the design of drugs, respectively, the latter subject having received far more attention.

12.2 *De Novo* Peptide and Protein Design

Most of the work has been carried out at the experimental level (Betz *et al.* 1993, Tuchscherer and Mutter 1995), especially for the so-called *template-assembled synthetic proteins, TASP* (Mutter and Tuchscherer 1988, Mutter and Vulleumies 1989, Anderson *et al.* 1993). The varied applications encompass the mimicking of discontinuous epitopes and of binding and catalytic sites (Kazmiersski *et al.* 1991, Hirschmann *et al.* 1992, Tuchscherer *et al.* 1992, 1993), channel formation (Ackerfeldt *et al.* 1992, Grove *et al.* 1993), multiple antigenic peptide (Tam 1988), synthetic vaccines (Kaumaya *et al.* 1992), and pepzymes (Sasaki and Kaiser 1989, Hahn *et al.* 1990; see also Section 12.2.4). Other approaches involve the stabilization of helical conformations through incorporation of C_α-alkylated (Altman *et al.* 1992, Toniolo *et al.* 1993) or trifunctional residues (Ghadiri and Choi 1992), initiation of helices or sheets (Feigel 1989, Diaz *et al.* 1992), generation of preselected secondary structure patterns (Kaiser and Kezdy 1983, Ackerfeldt *et al.* 1992, DeGrado 1993), design of peptides able to undergo conformational transitions (Mutter *et al.* 1991, Dado and Gellman 1993; see also Section 12.2.2), and the design of heterodimeric coiled-coil proteins (Zhu *et al.* 1992, Adamson *et al.* 1993, Zhou *et al.* 1994).

These possibilities are mentioned here, even though they fall outside the scope of this work because of their experimental character, as they might be preceded by, or supported with, appropriate computer simulations. Thus, for example, the generation of a structural unit with a desired pattern, using a limited set of amino acids and with rearrangement of the sequence, could be easily investigated with the methods outlined in Chapter 9, followed with an evaluation of the relative stabilities of the various predicted structures. In this connection, the *GRAFTER* software package (Hearst and Cohen 1994), which contains geometric search and evaluation functions, may be useful. Similarly, the metal-mediated assembly (Liberman and Sasaki 1991, Ghadiri *et al.* 1992, Ghadiri and Case 1993) may also be studied by theoretical means, with the corresponding savings in the necessary synthetic work (Yamamoto 1995); the difficulty in this case arises from the fact that very few of the potential energy functions (*PEF*) are able to handle metal cations.

We will consider now some other applications, where theoretical simulations have played (electron transport systems and synthetic vaccines) or will play (switch peptides and pepzymes) an important role.

12.2.1 Conductive Polymers

The objective is the design of a polymer, say, *poly*(Ala-Ala-X), where *X* stands for a non-natural amino acid with a suitable side chain, which will show electrical conductivity, probably through the overlapping of the π-type molecular orbitals of aromatic groups. Elegant work on such systems has been carried out by Oka and co-workers (Oka *et al.* 1990*abc*, 1991, 1992). Simulations have been performed, e.g., for *poly*(L-Phe-D-Ala-Gly), *poly*(L-Ala-D-Ala-L-Phe), and *poly*(L-Ala-D-Ala-L-NapAla), among others, where NapAla stands for 1-naphthylalanine. Thus, in the case of *poly*(L-Phe-D-Ala-Gly), the computer simulations were performed (using the *PEF ECEPP*) as follows. A total of 729 starting conformations for the tripeptide were prepared from all the possible combinations of the conformations corresponding to the single-residue minima. The results from the minimization of those conformations were used as starting points for the minimization of an octamer, denoted as *poly*(L-Phe-D-Ala-Gly). For this system, a new minimization procedure, after having varied some of the torsion angles, yielded a total of 246 minima, lying at less than 3 kcal/mol above the global minimum, which corresponds to a β^6-helix. The interesting fact is that *the planes of the aromatic rings of Phe were at a distance of 2.9 Å of each other, smaller than the separation of 3.4 Å between consecutive planes in graphite.*

12.2.2 Molecular Switches

The designation *switch peptides* is used to denote those peptides which are capable of undergoing conformational transitions, e.g., medium (pH) induced or redox-controlled. For example, Gallagher and Crocker (1994) have observed two types of crystals in *BPTI*, depending on the pH of the solution, with a marked difference in the number (19 *vs.* 8) of protein-to-protein *H*-bonds.

The predictive study of pH-induced or $\alpha \to \beta$ transitions in peptides does not offer any difficulty. An important component of these simulations will be the detailed evaluation of the energy hypersurface, in order to determine accurately the propensity of the peptide towards the interchanging conformations.

Insofar as the above definition ignores the time-component of the switching process, one could envisage the catalytic sites of (some of the) metalloproteins as representing redox-controlled molecular switches, with a different conformation depending on the oxidation state of the cation (see, e.g., the reviews of Sykes 1985, 1991 on plastocyanin). The first step in the study of an electron-transfer process involving two proteins should consist of a simulation of their interaction. Such a calculation could be approximated on the basis of electrostatic interactions, as discussed in Section 11.3.1, but with practical difficulties due to the size of the lattice required when dealing with two large proteins. An alternative exists in the *boundary element* technique, whereby the integral equations are transformed into linear algebraic equations (Zauhar and Morgan 1985, 1988, 1990, Yoon and Lenhoff 1990, 1992, Zhou 1993*ab*). In the formulation of Zhou (1993*b*), for a system of M proteins (labelled I) in an ionic solution, one has

$$4\pi\,\phi_{I(i)}(\mathbf{r}) = \frac{4\pi\,\phi_{I(i)}^{(v)}(\mathbf{r})}{\varepsilon_{(i)}} + \int_{S_I} dS \left\{ \frac{\mathbf{n}\cdot\nabla_S\phi(\mathbf{R})}{|\mathbf{r}-\mathbf{R}|} - \phi(\mathbf{R})\mathbf{n}\cdot\nabla_S\frac{1}{|\mathbf{r}-\mathbf{R}|} \right\}$$

$$-4\pi\phi(\rho) = \sum_{J=1}^{M} \int_{S_J} dS \left\{ \frac{\varepsilon_{(i)}\mathbf{n}\cdot\nabla_S\phi(\mathbf{R})}{\varepsilon_{(o)}|\rho-\mathbf{R}|}e^{-\kappa|\rho-\mathbf{R}|} - \phi(\mathbf{R})\mathbf{n}\cdot\nabla_S\frac{e^{-\kappa|\rho-\mathbf{R}|}}{|\rho-\mathbf{R}|} \right\}$$

where $\phi_{I(i)}(\mathbf{r})$ denotes the potential at the interior position $\mathbf{r}$ of protein I and $\phi(\rho)$ is the potential at position ρ in the solvent. The notation used is as follows: v, vacuum; i, interior; o, exterior; S, surface; $\mathbf{n}$, unit vector normal to surface at postion $\mathbf{R}$; ε, dielectric constant; κ, inverse of Debye length. When the positions $\mathbf{r}$ and ρ move to a position $\mathbf{r}_S$ on the I-th surface, the above expressions transform into

$$2\pi\,\phi_{I(S)}(\mathbf{r}_S) = \frac{4\pi\,\phi_{I(S)}^{(v)}(\mathbf{r}_S)}{\varepsilon_{(i)}} + \int_{S_I} dS \left\{ \frac{\mathbf{n}\cdot\nabla_S\phi(\mathbf{R})}{|\mathbf{r}_S-\mathbf{R}|} - \phi(\mathbf{R})\mathbf{n}\cdot\nabla_S\frac{1}{|\mathbf{r}_S-\mathbf{R}|} \right\}$$

$$-2\pi\phi_{I(S)}(\mathbf{r}_S) = \sum_{J=1}^{M} \int_{S_J} dS \left\{ \frac{\varepsilon_{(i)}\mathbf{n}\cdot\nabla_S\phi(\mathbf{R})}{\varepsilon_{(o)}|\mathbf{r}_S-\mathbf{R}|}e^{-\kappa|\mathbf{r}_S-\mathbf{R}|} - \phi(\mathbf{R})\mathbf{n}\cdot\nabla_S\frac{e^{-\kappa|\mathbf{r}_S-\mathbf{R}|}}{|\mathbf{r}_S-\mathbf{R}|} \right\}$$

These equations may be used for the interior, surface, and exterior potentials. For the interaction of two proteins in an ionic solution, an iterative procedure may be used (Zhou 1993*b*), with subsequent evaluation (see Section 11.3.1) and mapping of the interaction energy, for the study of the corresponding complex (such as cytochrome c and cytochrome c peroxidase). The lowest-energy complex conformation could then be used as a starting point for an approximate quantum-mechanical treatment (e.g., using a simplified model) of the electron-transfer process. [See Section 12.3.5 for software packages that could be used for these calculations.]

12.2.3 Synthetic Vaccines

The cell-mediated immune response of an organism to a foreign antigen (Ag) involves molecules of the immunoglobulin supergene family, expressed on the surface of antigen-presenting cells (APC), antibody-producing B cells, and $T4$ (helper T lymphocytes, T_h) and $T8$ (cytotoxic T lymphocytes, CTL) cells. The antigen, processed by APC, is displayed in the form of a peptide (Bjorkman *et al.* 1987*ab*, Brown *et al.* 1988), corresponding to an immunodominant region, in association with class I or class II molecules of the major histocompatibility complex (*MHC*) (Klein 1979, 1986, Dausset 1981, Kimball and Cooligan 1983, Kaufman *et al*. 1984). The *T*-cells, specific for either class of *MHC* molecules (class II for T_h and class I for the *CTL*) use similar receptors (Rupp *et al*. 1985, Marrack and Kappler 1986); in particular, *CD4* and *CD8* have been identified as recognition molecules on the surface of *T4* and *T8* cells, respectively (Engleman *et al*. 1981, Krensky *et al*. 1982, Meuer *et al*. 1982, Biddison *et al*. 1982, Wilde *et al*. 1983, Swain 1983, Karathas *et al*. 1984, Littman *et al*. 1985, Sukhatme *et al*. 1985, Maddon *et al*. 1985).

Antibody epitopes are believed to be non-sequential (i.e., conformational or discontinuous) (Barlow *et al*. 1986), being formed by non-adjacent regions brought together by the folding of the native structure. Predictive methods have focused on surface characteristics (see Section 8.2), such as hydrophilicity (Hopp and Woods 1981, Hopp 1985, 1989), accessibility (Janin 1979), mobility (Westhof *et al*. 1984, Tainer *et al*. 1984), and protrusion (Thornton *et al*. 1986). A knowledge of the tertiary structure of the protein is needed, however, to completely identify the conformational epitopes.

T-cell epitopes, on the other hand, are linear (i.e., sequential) (Barcinski and Rosenthal 1977, Berzofsky *et al*. 1979, Shimonkevitz *et al*. 1984, Manca *et al*. 1984, Schwartz 1985, Schwartz *et al*. 1985, Townsend *et al*. 1986, Livingstone and Fathman 1987) and may appear either on the surface or in interior regions of the protein. The *MHC* receptor site may accommodate an α-helical peptide of about 20 residues or an extended chain of about 8 residues; if the conformation of the bound peptide is bent, or only partly helical with unfolded ends, it may have an intermediate number of residues (Schwartz *et al*. 1985). [See also Section 11.4.1] Because in an individual, either class of *MHC* molecules must bind a very large number of foreign peptides, it is believed that they must share a common conformation. Thus, some of the criteria used for the prediction of *T*-cell epitopes are the amphipathic α-helical propensity (Watts *et al*. 1985, DeLisi and Berzofsky 1985, Berkover *et al*. 1986, Berzofsky *et al*. 1987, and Cease *et al*. 1987), the existence of general sequence patterns or motifs (Rothbard 1986, Lamb *et al*. 1987, Sette *et al*. 1988), or *MHC* allele-specific consensus sequences (Guillet *et al*. 1987, Rothbard *et al*. 1988).

Table 12.1 presents the correspondence between epitopes, secondary structure, and patterns for 14 proteins. The theoretical results (secondary structure and amphipathic patterns) have been obtained with *maPSI*, which predicts the secondary structure according to a modified Lim-*GOR*-recognition procedure (see Section 9.2.7); the general amphipathic patterns are *GHHGG, HGGHH, GGHHG,* and *HHGGH*, for α helices, while for β-chains the pattern is one of alternating hydrophobic and hydrophilic residues, and the special amphipathic α-helix patterns $GH_LH_LH_SS$, *GHHGS, HGGHS, GGHSG,* and *GSHHG* are accepted on the basis of the relative hydrophobicity/hydrophilicity of the two faces (e.g., the highly hydrophobic face H_LH_L *vs*. the face GH_SS) (see Section 9.2.4). The results show

Table 12.1 Correspondences between epitopes, secondary structure, and patterns.

Protein	Sequence positions[a]	Secondary structure[b]	Sequence[c]	Pattern[d]
Allergen (ragweed)	51-65	53-65(α)	EVWREEAY HAC(A)DIKD	$G_LH_LH_LG_LG_L(\alpha)$
Cytochrome (bovine)	13-25	10-18(α)	KCAQC(-)HT VEKGGKHK	$H_LG_LG_LSS(\alpha)$
Cytochrome (horse)	45-58	57-60(α)	GFTYTDANK NKGIT	$G_LG_LSH_LG_S(\alpha)$
Cytochrome (pigeon)	93-103	92-100(α)	DLIAYLKDA TA	$H_SH_LG_LG_LH_S(\alpha)$
Cytochrome (moth)	93-107 (88-102)	93-104(α)	NERADLIAY LKQATK	$H_SH_LG_LG_LH_S(\alpha)$
Foot & mouth virus (VP1)	145-160	151-156(α)	RGDLQVLAQ KVARTLP	$G_LH_LH_SG_LG_S(\alpha)$
Glycoprotein D (herpes)	26-48 (1-23)	26-37(α)	KYALADASL KMADPNRFR GKDLP	$G_LH_LH_SG_LG_S(\alpha)$
Insulin (B) (bovine)	5-16	1-20(α)	HLCGSHLVE ALY	$H_LH_LSG_SH_S(\alpha)$
Hemagglutinin (flu)	50-70 (48-68)	58-61(β) 64-66(β)	KICNNPHRIL DGIDCTLIDA L	$H_LG_LH_LG_SH_L(\beta)$
	110-145 (105-140)	109-111(β) 118-123(β)	SLVASSGTL EFIT(N)EGFT (N)WTGVTQN GGSN(Y)ACK	$G_SSG_SH_LG_LH_L(\beta)$
	305-328 (306-329)	311-315(α)	CPKYVKQNT LKLATGMR NVPEKQT	$G_LH_SH_LG_LG_L(\alpha)$ $G_LH_LH_LH_SS(\alpha)$ $H_LG_LG_LH_LH_S(\alpha)$
Lysozyme (egg white)	34-45	30-38(α)	FESNFNTQ(E) ATNR	none
	46-61	52-58(β)	NTDGSTDYGI LQINSR	$H_LG_LH_LG_L(\beta)$
	78-93	88-94(β)	IPCSALLSSDI TASVN	$G_LH_LG_SH_SG_SH_L(\beta)$

Myoglobin (sperm whale)	69-78	73-78(β)	LTALGAILKK	none
	106-121 (106-118, 110-121)	103-118(α)	**FISEAIIHVLH SRHPG**	$H_LH_LG_SG_LH_S(\alpha)$ $G_SG_LH_SH_SS(\alpha)$
	132-145	122-142(α)	**NKALELFRK DIAAK**	$G_LH_LH_LG_LG_L(\alpha)$
Nuclease (staphylococcal)	61-110 (61-80, 81-100, 91-110)	58-64(α) 65-68(β) 73-76(β)	**FTKKMVEN AKKIEVEFN (D)KGQ**	$G_LG_LH_LH_LG_L(\alpha)$ $G_LH_LG_LH_LG_LH_LG_L$ (β)
		90-94(α)	**RTDKYGRGL (K)AYIY(R)A DG(K)KMVN**	$G_LG_LH_LSG_L(\alpha)$ $SG_LH_LH_LG_L(\alpha)$
		99-113(α)	**YIY(R)ADG (K)KMVNEA LVRQGLAK**	$G_LH_LH_LG_LG_L(\alpha)$ $G_LSH_LH_SG_L(\alpha)$
Nucleoprotein (flu)	50-63	54-59(β)	**SDYEGRLIQ NSLTI**	$G_LH_SG_LSG_LH_L(\beta)$
	335-349	324-342(α)	**SAAFEDLRV LSFIRG**	$G_SH_LH_LG_LS(\alpha)$
	365-379	364-377(α)	**IASNEN*MD* (E)A(T)MESS TL**	$G_LH_SH_LG_LG_S(\alpha)$
Ovalbumin	324-340 (323-339)	323-336(α)	**ISQAVHAAH AEINEAGR**	$H_LG_SG_LH_SH_L(\alpha)$ $H_LG_LG_LH_SS(\alpha)$

[a]Corresponding to epitopes, with the sequence given in the fourth column. Sequence positions given in the literature, if different from the above, are in parenthesis. [See Rothbard 1986 for complete references.]
[b]Predicted using *maPSI*
[c]Differences between the sequence given in the literature and the sequence used are given in parentheses.
[d]Amphipathic patterns corresponding to the sequence regions given in boldface.

that, except in two cases, one or more amphipathic patterns are found within each predicted fragment. It is also interesting to test this hypothesis for human, *Mycobacterium bovis, Mycobacterium tuberculosis,* and *Mycobacterium leprae* heat shock proteins (*HSP*) (Kaufman 1990), for which detailed antibody and *T*-cell epitope identification is available (Lamb *et al.* 1987, Mehra *et al.* 1986, Young *et al.* 1987, van Eden *et al.* 1988, Munk *et al.* 1989, Hill Gaston *et al.* 1990), with the following consideration: (a) in spite of the fact that epitopes may be associated with β-chains in a minority of cases (see Table 12.1), the search has been carried out only for amphipathic α-helices; (b) taking into account the experimental observation (Schwartz *et al.* 1985) that *T*-cell epitopes generally consist of a minimum of 7, but often 8-15, contiguous amino acids, the predicted epitopes have been defined by extending the amphipathic α-helix pattern by 5 residues on each side (for a total of 15 residues); (c) overlapping or adjacent epitopes have been combined. Inspection of Table 12.2 shows that all the observed epitopes, including the antibody as well as the *T*-cell epitopes, are identified in the theoretical prediction: nine epitopes are overlapped and five are found adjacent to predicted epitopes. This is an interesting result, considering that the procedure was developed by correlation to only *T*-cell epitopes; it may be taken to confirm the postulate, based on recognition considerations, that antigenic determinants in general are predominantly helical (Fraga *et al.* 1988) and that *hydrophobic residues may be important in the genetic control of the immune response* (Singh *et al.* 1989).

Once the epitopes have been (tentatively) identified, as described above, confirmation of the prediction may be attempted through theoretical simulations. In all cases, of course, it would be appropriate to proceed with a modeling of the antigenic protein but simplified approaches are also possible. For example, in the case of discontinuous epitopes one could study (perhaps using a simplified model) the interaction antigen-antibody. On the other hand, one might simulate the peptide with the sequence corresponding to the sequential epitope. Some simple studies of this type are examined below.

The interaction antibody-antigen (*Ab-Ag*) may be computer simulated, as done by Kozack and Subramaniam (1993) for the complexation of anti-hen egg lysozyme monoclonal *Ab* HyHEL-5 with lysozyme. In this work, the linearized Poisson-Boltzmann equation is solved in order to determine the long-range electrostatic forces between *Ab* and *Ag* (Gilson and Honig 1987, Davis and McCammon 1989; see Section 11.3.1) and the relative motion, determined by those forces, is modelled within the context of Brownian motion theory (Northrup *et al.* 1984, 1988, Davis *et al.* 1990, Nambi *et al.* 1991). The diffusional motion is propagated in discrete timesteps Δt according to the Langevin equation algorithm of Ermak and McCammon (1978)

$$\Delta \mathbf{r} = \frac{D}{k_B T} \, F(t)\Delta t + \mathbf{R}(t)$$

where $\Delta \mathbf{r}$ is the change in relative position, $F(t)$ is the electrostatic force obtained

Table 12.2. Correspondence between theoretical and experimental epitopes of heat shock proteins.[a]

Human[b]	Experimental[c]	M. bovis[b]	Experimental[d]	M. Leprae[b]
		26-36	26-41	26-36
74-88		75-89	45-79	75-89
	109-200	94-108	90-110	94-108
114-128, 133-147		123-137	129-148	123-137
		148-162		148-162
179-193	206-214	176-190		176-196
		248-262		248-262
271-285	269-289	268-282		
	298-307	306-320		306-320
368-382				
388-402, 416-441	430-441	386-428	417-439	385-399, 401-429
485-499		468-482	450-467	468-482
501-515		501-515	508-521	503-517
520-534		520-534	525-539	520-534
543-557		537-551	553-566	540-553

[a]The sequence numbering corresponds to the human 65-kDa HSP (Jindal *et al.* 1989).
[b]Predicted using *maPSI*.
[c]In human, *M. bovis*, and *M. tuberculosis* HSP (Van Eden *et al.* 1988, Munk *et al.* 1989).
[d]In *M. leprae* HSP (Lamb *et al.* 1987, Mehra *et al.* 1986).

from the solution for the electrostatic potential ϕ, and $\mathbf{R}(t)$ is a random vector (which satisfies statistical constraints); $\mathbf{D}$ is a relative diffusion constant. In this procedure the *diffusing (lysozyme) particle does not follow a directed path to the binding site; it tends to random walk until it is captured by a region of attractive potential.*

Vega *et al*. (1992) have simulated a 9-residue peptide of the major antigenic site of the foot-and-mouth disease virus in a three-step procedure: generation of the initial conformations (using ECEPP, with ε = 2), optimization of the lowest conformers (using AMBER, with ε = 4R; see Section 10.4.3), and *MD* simulation. The calculations indicate the existence of a rather stable conformation for the peptide, exhibiting an initial α-helix followed by two γ-turns.

Fraga *et al*. (1993) studied the peptide conformation involved in the recognition of *(EYA)$_5$*, *EYK(EYA)$_4$*, and *EYAEAA*(EYA)$_3$ by *MHC* class II molecules and *T*-cell receptor. The procedure involved the simulation of the peptides as α-helices, β-chains, and with full optimization (see Section 6.2.1), with prediction of the corresponding $d_{NN}(i,i+1)$ *nOe* connectivities. The theoretical simulation was able to pinpoint the special requirements for recognition and confirm the experimental evidence (Boyer *et al*. 1990) that *the function of critical residues may extend beyond contacting the T-cell receptor or the MHC molecule.*

The synthetic-peptide vaccines must possess a high level of immunogenicity and induce antibodies that cross-react extensively with the pathogen (van Regenmortel 1989), including different *B*- and *T*-cell epitopes of the infectious agent (Eroshkin *et al*. 1993). These conditions could be met by using a four-α-helix bundle as template, which could permit the combination of *B*- and *T*-cell epitopes, with high immunogenicity and effectiveness (because of the *T*-cell epitopes), ability to induce immune response to various *B*-cell epitopes, and similarity of the *B*-cell epitope conformation in the vaccine and in the virus (Eroshkin *et al*. 1993). The proposed scheme for vaccine design could then consist of the following steps: prediction of the secondary structure of the viral antigen; theoretical analysis of the experimentally studied *T*- and *B*-cell epitopes; selection of epitopes with required functional and structural properties; exclusion of undesirable kinds of epitopes; combination of epitopes into a single chain; analysis of the resulting protein sequence; tertiary structure modeling; and redesign of the sequence, if necessary. This procedure, however, would not be appropriate for discontinuous epitopes or *T*-cell epitopes, with β-strand propensity and high hydrophobicity.

12.2.4 Pepzymes

The designation *pepzymes* is used to denote peptide enzymes. Their rational design must conform with a number of considerations, as existing in natural enzymes: they must have functional-group diversity, lower the energy of the substrate complex relative to the free systems, and be able to catalyze all the steps of the reaction, whose mechanism must be known (Jencks 1975, DeGrado 1993, Johnsson *et al.* 1993).

Leu/Lys peptides catalyze the hydrolysis of *RNA* (Barbier and Black 1988, 1992) and polymers of the types $(LK)_n$, as β-sheets, $(LKKL)_n$, as α-helices (Brack and Orgel 1975, Brack and Spach 1981), and $(LKKLLKL)_2$ and $(LKLKLKL)_2$ (DeGrado and Lear 1985) have been prepared; the last two, which exist in random conformation in solution, adopt an α-helix or β-sheet conformation, respectively, at the interface between water and a non-polar solvent. Atassi and Manshouri (1993) have designed two pepzymes to mimic the active sites of α-chymotrypsin and trypsin.

Evidently, this type of work is amenable to a preliminary study through computer simulations, if the active site of the enzyme under study is known. The significance of the simulations lies in the fact that it will be possible to try and design pepzymes with enhanced features, through manipulation of the sequence.

12.3 Drug Design

12.3.1 General Considerations

Biological processes involve a molecular recognition, in which a ligand (*L*) interacts with a receptor molecule (*R*), with formation of the corresponding complex (*LR*), which may be stabilized by non-covalent or covalent binding. The recognition may result in a response (as, for example, in the interaction with a membrane receptor) or constitute a step in one of the reactions taking place in the cell (e.g., the interaction of a specific substrate with an enzyme) or in a non-self organism.

A system, other than the ligand specific for the process under consideration, which binds to the receptor molecule, is denoted as either *agonist* or *antagonist*, depending on whether the receptor molecule is activated or not when the corresponding complex is formed, while the designation *inhibitor* is reserved for those systems that will block an enzyme. Drugs may be either agonists, antagonists, or inhibitors, depending on whether they are intended to potentiate or to lower the cellular response or to inhibit the activity of an enzyme, respectively. There are, however, drugs which do not fit into any of the above groups, either because they are the precursors of the corresponding bioactive systems or because their activity is due to a different type of interaction, as in the case of drugs that interfere with the membrane of a non-self cell. Drugs must be strong inhibitors (if such is their assumed role), with low toxicity, good (oral) bioavailability, and reasonable residence time (Robertus 1994), so that the last two conditions rule out the use of peptides as drugs (Zuckerman 1993); furthermore, their flexibility reduces their affinity (Dean 1994).

For simplicity and taking into account the general purpose of this work, we will restrict the discussion with the following considerations. We will assume that the target molecule is a protein and that the specific ligand is a peptide, with the goal of the drug design procedure being the identification of a non-peptide analog (denoted as *peptidomimetic*) with the appropriate bioactivity (as agonist, antagonist, or inhibitor). [See, e.g., the work of Dean (1994) for references on the mimicking of β-turn structures.]

Since the pioneer work of Goodford and co-workers (Bedell *et al.* 1976, Brown and Goodford 1977), the field of drug design has evolved considerably. At present, a systematic procedure for the structure-aided design of drugs involves several steps. First it is necessary to identify the pharmacologically-relevant target molecule, elucidate its structure, and pinpoint its active site. Then, once potential lead compounds with biological activity have been identified and modelled, an iterative procedure is carried out, whereby variants are proposed, synthesized, and tested for activity until a satisfactory compound has been found (Ealick and Armstrong 1993, Baun 1994). Figure 12.1 presents an illustrative view of such a general procedure.

The three points to be examined are then the identification of lead compounds and/or their modeling and the development of the appropriate analogs. The identification of the lead compounds is best achieved from experimental observation but it might also be attempted directly (on the basis of previous experience, chemical reasoning, and theoretical arguments), with modeling of the candidate(s) in interaction with the target molecule. This step is appropriate in any case, in order to obtain useful steric information. Finally, the development of analogs may be carried out by means of quantum-chemical calculations and of *quantitative structure-activity relationships (QSAR)*.

12.3.2 Libraries

The identification of lead compounds through the use of *libraries* has now become a standard procedure in drug design. There are libraries of phage displayed peptides, synthetic peptides, peptidominetics, biopolymers, ... (Hoess 1993, Zuckermann 1993, Kenan *et al.* 1994). [See also the work of Johnson and Chiswell (1993) regarding the use of phage antibody libraries for human antibody engineering.]

initial phase

- identification of the protein target followed by production and purification of the protein

- design and development of assays to monitor inhibition or activation of the target protein

- identification of a lead inhibitor molecule

- determination of structure of the native protein target

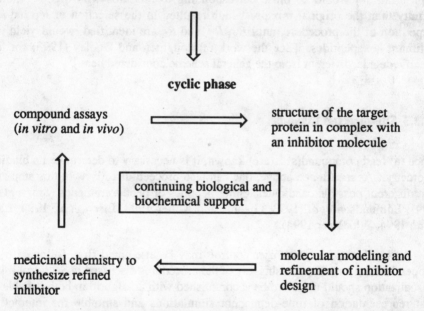

cyclic phase

compound assays
(*in vitro* and *in vivo*)

structure of the target protein in complex with an inhibitor molecule

continuing biological and biochemical support

medicinal chemistry to synthesize refined inhibitor

molecular modeling and refinement of inhibitor design

Figure 12.1. Schematic representation of a structure-aided drug design procedure. [From M.A. Navia and D.A. Peattie, *Structure-Based Drug Design: Applications in Immunopharmacology and Immunosuppression*, Immunology Today *14*, 296-302 (1993). Copyright © 1993 Elsevier Science Publishers Ltd., Oxford, U.K. Reprinted by permission of Elsevier Trends Journals.]

The phage display of peptides (Smith 1985, Parmley and Smith 1988) has been used for the construction of libraries of random peptides (Cwirla *et al.* 1990, Devlin *et al.* 1990, Scott and Smith 1990) and of protein domains (Bass *et al.* 1990, Markland *et al.* 1991) and may be applied to the mapping of discontinuous epitopes, probing the substrate specificity of proteases, and eliciting antibodies against epitopes (Hoess 1993). On the other hand, shape libraries may be of use in selecting molecules to bind a target on the basis of shape (Kenan *et al.* 1994).

Combinatorial libraries of synthetic peptides constitute a very powerful tool for the identification of lead compounds (Houghten and Dooley 1993). Let us, for example, assume that we would like to identify possible hexapeptides, $R_1R_2R_3R_4R_5R_6$ (where R_i denotes a natural amino acid), which might have a certain biological activity; the total number of hexapeptides is $64,000,000$ ($=20^6$). One could start by synthesizing 400 ($=20^2$) mixtures of hexapeptides, each mixture containing $160,000$ ($=20^4$) hexapeptides, all of which have the same R_1 and R_2. The appropriate R_1 and R_2 will be the ones corresponding to the mixture with the highest binding affinity. One would then synthesize 20 mixtures of hexapeptides, all of them with the same R_1 and R_2, with each mixture characterized by a different R_3 (hence the 20 mixtures) and containing $8,000$ ($=20^3$) hexapeptides. The appropriate R_3 would be those corresponding to mixtures with higher binding affinity than the original mixture which resulted in the selection of R_1 and R_2. Repetition of the procedure, until R_4, R_5, and R_6 are identified, would yield the optimum hexapeptides. [See the work of Houghten and Dooley (1993) for the specific details, different from the general scheme considered here.]

12.3.3 Site-directed Ligands

Once (a) lead compound(s) is(are) known, it is necessary to determine its binding geometry. As mentioned above, one can also proceed directly with this step and test different possible candidates, either peptides or peptidomimetics (Singh *et al.* 1991, Edmundson *et al.* 1993, Giannis and Kolter 1993, Eisen *et al.* 1994, Lam *et al.* 1994, Wlodawer 1994).

A visual docking, with user control, may be attempted first, followed by a proper docking simulation, using any of the procedures discussed in Section 11.4.2. Visualization should therefore be accomplished with an algorithm both capable of fast representation of time-dependent simulations and suitable for interactive manipulation; Heiden *et al.* (1993) have concluded that the *Marching Cube* algorithm of Lorensen and Clive (1987) may be the appropriate tool for this task. Another important point is that the structure of the target molecule may not be known, in which case it is necessary to find a *surrogate* protein that will permit the docking of the ligand and a study of the ligand-receptor interactions (Weinstein *et al.* 1985).

There is a number of software packages of possible use in this endeavor (Guida 1994), but their consideration is postponed to Section 12.3.5, as in many cases the study of the ligand-receptor interactions is performed in conjunction with the development of analogs, on the basis of various considerations.

12.3.4 Structure-activity Relationships

The specific ligand for the target molecule under consideration is characterized by structural factors, governing its action on the receptor as well as other features of its bioactivity, such as transport, metabolism, etc. (Peradejordi 1989). Only the former are considered here, in conjunction with the formation of the complex

$$L + R \rightleftarrows LR$$

characterized by the binding energy

$$\Delta E = E_{LR} - (E_L + E_R)$$

and association constant

$$K = \frac{[LR]}{[L][R]} = \frac{q_{LR}}{q_L q_R} e^{-\Delta E/k_B T}$$

where q_{LR}, q_L, q_R denote the corresponding partition functions (see Section 4.2.3). [The kinetics of the formation of the complex LR has been studied by Mahama and Linderman (1994) for the cellular response due to the activation of the so-called G-proteins.]

The association constant depends, through the binding energy and the partition functions, on the wave functions (see Chapter 3), which describe the electronic and nuclear structure of the three systems. Therefore, under the assumption that the bioactivity of the ligand is related to the formation of the complex LR, it may be concluded that it will be determined, ultimately, by the electronic and nuclear structure of the three systems. Within the context of a number of approximations, because of the difficulties in the evaluation of the partition functions, one may approximate the bioactivity as

$$\log A = a + b \, \Delta E$$

where a and b are constants to be determined, say, from the values of A and ΔE for a series of analogs.

The evaluation of relative binding energies of, say, a series of potential inhibitors might, in principle, be attempted by means of the free-energy perturbation (FEP) method (see Section 11.3.2). This approach, however, is not recommended (Aqvist *et al.* 1994) because the mutational path (for the transformation of one inhibitor into another) will involve, as a rule, changes in the molecular charge distribution and the creation/annihilation of atoms. In the semiempirical approach of Aqvist *et al.* (1994), the binding energy is approximated as

$$\Delta G = 1/2 <\Delta V_{e\ell}> + k <\Delta V_{vdw}>$$

in terms of Coulombic and van der Waals interaction energies, k denoting a parameter to be determined from the experimental information for known inhibitors. The mean values for the two contributions are obtained from the averages evaluated from *MD* simulations for the inhibitors, in solution and bound to the solvated protein, within a spherical boundary with a 16 Å radius; in the runs for the complexes, the protein atoms outside of the boundary are restrained to their crystallographic positions. Although the semiempirical parameter k will likely be force-field specific and reflect the nature of the active site of the protein, Aqvist *et al.* (1994) believe that this approach might be able to discriminate between a large number of inhibitors to a given target site, thus reducing the corresponding synthetic work.

The above equation, relating the bioactivity with ΔE, constitutes an example of a quantitative structure-activity relationship (*QSAR*), which could be used for the prediction of the activity of related analogs. *QSAR* have been in use since the early days of Quantum Chemistry (Fraga and Fraga 1976), when attempts were already made to correlate, say, the carcinogenic character of aromatic hydrocarbons with the values of effective charges, orbital energies, etc. A *QSAR* represents, essentially, an attempt at establishing a correlation between the bioactivities of a series of compounds and some selected characteristics, capable of being expressed numerically (see, e.g., Free and Wilson 1964, Hopfinger 1980, 1983, Hansch and Klein 1986, Cramer III *et al.* 1988, Marshall and Cramer III 1988). Such *QSAR* are developed for *similar* molecules, in order to rationalize their predictive value when applied to proposed candidates of unknown activity, and therefore an essential component consists of the definition of molecular similarity. A possible definition of similarity (Carbo *et al.* 1980, Carbo and Domingo 1987, Carbo and Calabuig 1992, Good *et al.* 1993) is based on

$$R_{AB} = \frac{\int P_A P_B d\tau}{\left(\int P_A^2 d\tau \right)^{1/2} \left(\int P_B^2 d\tau \right)^{1/2}}$$

where P_A, P_B denote the properties (electron density, electrostatic potential, shape, etc.) used for comparison and $d\tau$ represents the volume element. R_{AB} will have the value 1 when $A \equiv B$. [See also the work of Arteca and Mezey 1992 on shape similarities.] Because of the importance of electrostatic potentials in the study of molecular associations in general (see Sections 11.3.1 and 12.2.2), in complementarity-based simulations (see Section 11.4.2), and in the analysis of analogs, with common bioactivity, as considered here, it is appropriate to comment on their graphic representation because of the usefulness of a visual comparison. The problem in this connection (as for the representation of molecular surfaces; see Section 11.4.1) arises from the 3*D*-character of the electrostatic potential; the consequence is that the most-often used representations, whether isolines on parallel planes cutting through the molecules (see, e.g., Fraga and Fraga 1976) or color-coding on the van der Waals surface, are not very satisfactory. A more convenient representation is to be found in the topological feature maps (Gasteiger *et al.* 1994) obtained by application of the neural network of Kohonen (1982, 1989, 1990). The similarity in any molecular surface property may also be compared by the gnomic projection method of Dean and co-workers (Chau and Dean 1987, Dean and Chau 1987, Dean and Callow 1987, Dean *et al.* 1988, Borea *et al.* 1992), whereby one molecule is held fixed and the molecule to be compared is rotated by small angles.

The use of *QSAR* requires some caution. On one hand it must be emphasized that it is not proper to imply a relationship of causality in an empirical correlation which, at most, should be considered as a starting hypothesis for a detailed study of the problem (Peradejordi 1989). In addition, the comments (see Section 11.4.1) on complementarities and the conformational changes, which may occur upon binding, should be taken into consideration. And, finally, there are the special cases when covalent bonding is established (perhaps, even, with some groups leaving the reaction center).

12.3.5 Peptidomimetics

The preceding sections have examined briefly the three initial steps in a procedure for the *de novo* design of a new drug and we can now proceed to their integration. The basic idea, as already outlined in Section 12.3.1, is that if the structures of the target molecule and/or the specific ligand are known or may be established through computer simulations, it may be possible to use that information to design new therapeutic compounds (Corey *et al.* 1991, Owens *et al.* 1991, Saragoui *et al.* 1991, Jenks 1992), as illustrated in Fig. 12.2 (which complements and expands the general schematic representation of Fig. 12.1).

The simulation of the lead compound and of its interaction with the target molecule will provide static/interactive information regarding the basic requirements of possible pharmacophores (such as, in particular, the geometric arrangement of the functional groups essential for interaction with the receptor). That information may then be used in order to define criteria for a search for possible candidates in appropriate databases (Martin *et al.* 1988, Sheridan *et al.* 1989, Van Drie *et al.* 1989, Weininger and Weininger 1990, Martin 1991).

A number of software packages have been developed for the various (computational, statistical, databases, and graphics) components in this procedure. The ones most frequently used (without being exhaustive) are: ALADDIN (using geometric relationships to retrieve matching structures; Van Drie *et al.* 1989), CAVEAT (search of rigid structural templates; Bartlett *et al.* 1989, Lauri and

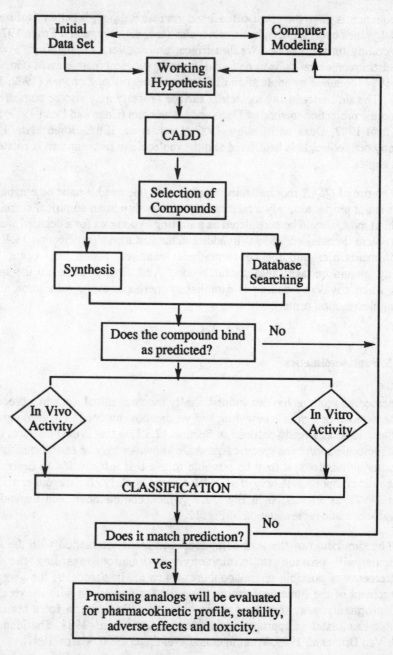

Figure 12.2. Interdisciplinary approach to computer-aided drug design (*CADD*) [From G.H. Loew, H.O. Villar, and I. Alkorta, *Strategies for Indirect Computer-Aided Drug Design*, Pharmaceutical Research *10*, 475-486 (1993). Copyright © 1993 Plenum Publishing Corporation, New York, U.S.A. Reprinted by permission of Plenum Publishing Corporation.]

Bartlett 1994), CAVITY (cavity search; Ho and Marshall 1990), CoMFA (field mapping; Cramer *et al.* 1988, Marshall and Cramer 1988), CONCORD (database; Pearlman 1987, 1993), DGEOM (distance geometry; program QCPE-590 of Blaney *et al.*), COSMIC/ASTRAL (integrated software framework for computational/graphics modeling; Vinter *et al.* 1987), DOCK (dock and score; Kuntz *et al.* 1982, DesJarlais *et al.* 1988, Kuntz 1992, Shoichet *et al.* 1991), FOUNDATION (geometry search; Ho and Marshall 1993), FRODO (graphic representation; Jones and Thirup 1986), GEMINI (packing arrangement; Singh *et al.* 1991), GOLPE (partial least squares analysis for statistical problems with large number of descriptors; Cruciani *et al.* 1993), GRID (exploration of binding site; Goodford 1989), HOOK (ligands with proper chemical/steric requirements for binding region; Eisen *et al.* 1994; see also Miranker and Karplus 1991), LUDI (*H*-bond and hydrophobic contact scanning; Böhm 1992*ab*), MACCS-3D (database search; Guner *et al.* 1991), MacroModel/BatchMin (conformational sampling; Mohamadi *et al.* 1990), MOLPAT (pharmacophore pattern search; Gund *et al.* 1974, Gund 1977), MD (graphic representation; Connolly 1983), POLDIAGNOSTICS (structural analysis; Baumann *et al.* 1989), PROTEUS (geometric modeling; Pabo and Suchanek 1986), and 3DSEARCH (Sheridan *et al.* 1989).

At that stage, when several candidates have been identified, one may proceed with a refinement of the prediction by means of quantum-chemical methods, even if it is not possible to operate at the desired level of sophistication because of lack of computational capabilities. As a rule, with chemical expertise, it is not difficult to design a new molecule in which given functional groups are positioned according to a given geometry. It may be necessary, however, to design a number of such molecules, with the final selection based on pharmacological, clinical, and economical considerations. The reader is referred to the work of Weaver (Weaver 1992, Bikker and Weaver 1992, 1993, Dakin and Weaver 1993, Bikker *et al.* 1994) for the general strategy and illustrative examples.

However, if an exhaustive search is contemplated, the picture changes dramatically, as described by Dean (1994). Assuming that the ligands may be constructed from n fragments and that each fragment may be connected in m_i ways, the possible number of combinations is

$$C = n! \prod_{i=1}^{n} m_i$$

This number further increases if branching and/or cyclization is considered and if different atoms may satisfy the valence-bond criteria at the various nodes.

Software packages are also available for these calculations but an appropriate background in quantum-mechanical treatments is really necessary, lest misleading conclusions are derived from the results obtained. Nevertheless, for completeness, we will mention the programs MNDO (Dewar and Thiel 1977), AM1 (Dewar *et al.*

234

1985), *MNDO-H* (Goldblum 1987), MNDO-PM3 (Stewart 1989), HONDO
(Dupuis *et al.* 1989, Dupuis and Maluendes 1991), GAUSSIAN (Frisch *et al.*
1992), etc., as well as other sophisticated programs included in MOTECC
(Clementi 1990).

Bibliography

References

E. Aarts and J. Korst. *Simulated Annealing and Boltzmann Machines*. John Wiley & Sons, New York (1989).

R.A. Abagyan and P. Argos. J. Mol. Biol. *225*, 519 (1992).

R.A. Abagyan and A.K. Mazur. J. Biomol. Struct. Dyn. *6*, 833 (1989).

R.A. Abagyan and M.M. Totrov. J. Mol. Biol. *235*, 983 (1994).

R.A, Abagyan, M.M. Totrov, and D. Kuznetsov. J. Comput. Chem. *15*, 488 (1994).

H. Abe, W. Braun, T. Noguti, and N. Go. Comp. Chem. *8*, 239 (1984).

E.E. Abola, F.C. Bernstein, S.H. Bryant, T.F. Koetzle, and J. Weng. *Protein Data Bank*, in *Crystallographic Databases - Information Content, Software Systems, Scientific Applications*, edited by F.H. Allen, G. Bergerhoff, and R. Sievers. Data Commission of the International Union of Crystallography, Bonn/Cambridge/Chester (1987).

F.F. Abraham. J. Chem. Phys. *61*, 1221 (1974).

F.F. Abraham, M. Mruzik, and G.M. Pound. Faraday Disc. 61, 34 (1976).

K.S. Ackerfeldt, R.M. Kim, D. Camac, J.T. Groves, J.D. Lear, and W.F. DeGrado. J. Am. Chem. Soc. *114*, 9656 (1992).

J.G. Adamson, N.E. Zhou, and R.S. Hodges. Current Opin. Struct. Biol. *4*, 428 (1993).

B. Alberts, D. Bray, J. Lewis, M. Raff, K. Roberts, and J. Watson. *Molecular Biology of the Cell*. Garland Publishing, Inc., New York (1989).

F.H. Allen, S.A. Bellard, M.D. Brice, B.A. Cartwright, A. Doubleday, H. Higgs, T. Hummelink, B.G. Hummelink-Peters, O. Kennard, W.D.S. Motherwell, J.R. Rogers, and D.G. Watson. Acta Cryst. *B35*, 2331 (1979).

M.P. Allen and D.J. Tildesley. *Computer Simulations of Liquids*. Clarendon Press, Oxford (1987).

N.L. Allinger. J. Am. Chem. Soc. *99*, 8127 (1977).

N.L. Allinger, Y.H. Yuh, and J.H. Lii. J. Am. Chem. Soc. *111*, 8551 (1989).

K.-H. Altmann, E. Altmann, and M. Mutter. Helv. Chim. Acta 75, 1198 (1992).

A. Amadei, A.B.M. Linssen, and H.J.C. Berendsen. Proteins Struct. Funct. and Genet. *17*, 412 (1993).

H.C. Andersen. J. Comput. Phys. *52*, 24 (1983).

S. Anderson, H.L. Anderson, and J.K.M. Sanders. Acc. Chem. Res. *26*, 469 (1993).

C.B. Anfinsen. Science *181*, 223 (1973).

C.B. Anfinsen, E. Haber, M. Sela, and F. White. Proc. Natl. Acad. Sci. USA *47*, 1309 (1961).

J. Aqvist, C. Medina, and J.-E. Samuelsson. Protein Eng. *7*, 385 (1994).

J.H. Arevalo, M.J. Taussig, and I.A. Wilson. Nature *365*, 859 (1993).

P. Argos, M. Vingron, and G. Vogt. Protein Eng. *4*, 375 (1991).

G.A. Arteca and P.G. Mezey. *Algebraic Approaches to the Shape Analysis of Biological Macromolecules*, in *Computational Chemistry. Structure, Interactions and Reactivity*, edited by S. Fraga. Elsevier, Amsterdam (1992).

M.Z. Atassi and T. Manshouri. Proc. Natl. Acad. Sci. USA *90*, 8282 (1993).

D.J. Bacon and W.F. Anderson. J. Mol. Biol. *191*, 153 (1986).

D.J. Bacon and J. Moult. J. Mol. Biol. *225*, 849 (1992).

A. Bairoch. Nucl. Ac. Res. *19*, 2241 (1991).

A. Bairoch. Nucl. Ac. Res. *21*, 3097 (1993).

A. Bairoch and B. Boeckmann. Nucl. Ac. Res. *21*, 3093 (1993).

J. Bajorath, R. Stenkamp, and A. Aruffo. Protein Sci. *2*, 1798 (1993).

E.N. Baker and R.E. Hubbard. Progr. Biophys. Mol. Biol. *44*, 97 (1984).

R.L. Baldwin. Trends Biochem. Sci. *14*, 291 (1989).

J.D. Baleja, M.W. Germann, J.H. van de Sande, and B.D. Sykes. J. Mol. Biol. *215*, 411 (1990).

J.B. Ball, P.R. Andrews, P.F. Alewood, and R.A. Hughes. FEBS Lett. *273*, 15 (1990).

B. Barbier and A. Black. J. Am. Chem. Soc. *110*, 6880 (1988).

B. Barbier and A. Black. J. Am. Chem. Soc. *114*, 3511 (1992).

M.A. Barcinski and A. Rosenthal. J. Exp. Med. *145*, 726 (1977).

M. Barinaga. Science *257*, 880 (1992).

J.A. Barker and R.O. Watts. Chem. Phys. Letters *3*, 144 (1969).

W.C. Barker, T.L. Hunt, and D.G. George. Protein Seq. Data Anal. *1*, 363 (1988).

W.C. Barker, D.G. George, H.-W. Mewes, F. Pfeiffer, and A. Tsugita. Nucl. Ac. Res. *21*, 3089 (1993).

D.J. Barlow, M.S. Edwards, and J.M. Thornton. Nature *322*, 747 (1986).

D. Bartel and J.W. Szostak. Science *261*, 1411 (1993).

P.A. Bartlett, G.T. Shea, S.J. Telfer, and S. Waterman. In *Molecular Recognition. Chemical and Biological Problems*, edited by S.M. Roberts. Royal Society of Chemistry, London (1989).

S. Bass, R. Greene, and J.A. Wells. Proteins *8*, 309 (1990).

R.M. Baum. Chem. & Eng. News 20 (1994).

G. Bauman, C. Frömmel, and C. Sander. Protein Eng. *2*, 329 (1989).

A. Baumgärtner. Annu. Rev. Phys. Chem. *35*, 419 (1984).

A. Baumgärtner. In *Applications of the Monte Carlo Method in Statistical Physics*, edited by K. Binder. Springer, Berlin (1987).

J.F. Bazan. Proc. Natl. Acad. Sci. USA *87*, 6934 (1990).

E. Beck and R. Berry. Biochem. *29*, 178 (1990).

C.R. Bedell, P.J. Goodford, F.E. Norrington, S. Wilkinson, and R. Wooton. Brit. J. Pharmac. *57*, 201 (1976).

C.J. Benham and M.S. Jafri. Protein Sci. *2*, 41 (1993).

S.A. Benner, D.L. Gerlof, and T.F. Jenny. Science *265*, 1642 (1994).

D. Benson, D.J. Lipman, and J. Ostell. Nucl. Ac. Res. *21*, 2693 (1993).

H.J.C. Berendsen, J.P.M. Postma, W.F. van Gunsteren, A. DiNola, and J.R. Haak. J. Chem. Phys. *81*, 3684 (1984).

P.H. Berens and K.R. Wilson. J. Chem. Phys. *74*, 4872 (1981).

B.A. Berg. Nature *361*, 708 (1993).

I. Berkover, G.K. Buckenmeyer, and J.A. Berzofsky. J. Immunol. *136*, 2498 (1986).

B. Berne and G.D. Harp. Adv. Chem. Phys. *17*, 63 (1970).

F.C. Bernstein, T.F. Koetzle, G.J.B. Williams, E.F. Meyer, Jr., M.D. Brice, J.R. Rodgers, O. Kennard, T. Shimanouchi, and M. Tasumi. J. Mol. Biol. *112*, 535 (1977).

J.A. Berzofsky, L.K. Richman, and D.J. Killion. Proc. Natl. Acad. Sci. USA *76*, 4046 (1979).

J.A. Berzofsky, K.B. Cease, J.L. Cornette, J.L. Spouge, H. Margalit, I.J. Berkover, M.F. Good, L.H. Miller, and C. DeLisi. Immunol. Rev. *98*, 9 (1987).

S.F. Betz, D.P. Raleigh, and W.F. DeGrado. Current Opin. Struct. Biol. *3*, 601 (1993).

D.L. Beveridge and F.M. Di Capua. *Free energy via molecular simulation: A primer*, in *Computer Simulation of Biomolecular Systems*, edited by W.F. van Gunsteren and P.K. Weiner. ESCOM Science Publishers, Leiden, The Netherlands (1989).

E.A. Bidacovich, S.G. Kalko, and R.E. Cachau. J. Mol. Struct. (Theochem) *210*, 455 (1990).

W.E. Biddison, P.E. Rao, M.A. Talle, G. Goldstein, and S. Shaw. J. Exp. Med. *156*, 1065 (1982).

J.J. Biesiadecki and R.D. Skeel. J. Comput. Phys. *109*, 318 (1993).

J.A. Bikker and D.F. Weaver. Can. J. Chem. *70*, 2449 (1992).

J.A. Bikker and D.F. Weaver. J. Mol. Struct. (Theochem) *281*, 173 (1993).

J.A. Bikker, J. Kubanek, and D.F. Weaver. Epilepsia *35*, 411 (1994).

M. Billeter, T.F. Havel, and I.D. Kuntz. Biopolymers *26*, 777 (1987).

K. Binder (editor). *Monte Carlo Methods in Statistical Physics*. Springer-Verlag, Berlin (1979).

K. Binder (editor). *Applications of the Monte Carlo Method in Statistical Physics*. Springer-Verlag, Berlin (1987).

V. Biou, J.F. Gibrat, J.M. Levin, B. Robson, and J. Garnier. Protein Eng. *2*, 185 (1988).

L. Birnbaumer, J. Abramowitz, and A. Brown. Biochim. Biophys. Acta *1031*, 163 (1990).

P.J. Bjorkman, M.A. Saper, B. Samraoui, W.S. Bennett, J.L. Strominger, and D.C. Wiley. Nature *329*, 506 (1987).

P.J. Bjorkman, M.A. Saper, B. Samraoui, W.S. Bennett, J.L. Strominger, and D.C. Wiley. Nature *329*, 512 (1987).

J.M. Blaney, G.M. Crippen, A. Dearing, and J.S. Dixon. *DGEOM*. Quantum Chemistry Program Exchange. Indiana University, Bloomington, IN, U.S.A. No. 590.

D.M. Blow, C.A. Wright, D. Kukla, A. Rühlmann, W. Steigemann, and R. Huber. J. Mol. Biol. *69*, 137 (1972).

T. Blundell, D. Carney, S. Gardner, F. Hayes, B. Howlin, T. Hubbard, J. Overington, D.A. Singh, B.L. Sibanda, and M. Sutcliffe. Eur. J. Biochem. *172*, 513 (1988).

J. Boberg, T. Salakoski, and M. Vihinen. Proteins Struct. Funct. and Genet. *14*, 265 (1992).

H.-J. Böhm. J. Comp.-Aided Mol. Design *6*, 61 (1992a).

H.-J. Böhm. J. Comp.-Aided Mol. Design *6*, 593 (1992b).

H. Bohr, J. Bohr, S. Brunak, R.M.J. Cotteril, B. Lautrop, L. Norkov, O.H. Olsen, and S.B. Peterson. FEBS Lett. *241*, 223 (1988).

H. Bohr, J. Bohr, S. Brunak, R.M.J. Cotteril, B. Lautrop, and S.B. Peterson. FEBS Lett. *261*, 43 (1990).

G. Bokoch, K. Bickford, and B. Bohl. J. Cell Biol. *106*, 1927 (1988).

G. Bolis and E. Clementi. J. Am. Chem. Soc. *99*, 5550 (1977).

A.M.J.J. Bonvin, R. Boelens, and R. Kaptein. J. Magn. Reson. *95*, 626 (1991).

P.A. Borea, P.M. Dean, I.L. Martin, and T.D.J. Perkins. Mol. Neuropharmac. *2*, 261 (1992).

T.V. Borchert, R.A. Abagyan, K.V.R. Kishan, J.Ph. Zeelen, and R.K. Wierenga. Structure *1*, 205 (1993).

B.A. Borgias and T.L. James. J. Magn. Reson. *79*, 493 (1988).

P. Bork. FEBS Lett. *257*, 191 (1989).

B. Borstnik, D. Pumpernik, D. Janezic, and A. Azman. *Molecular dynamics studies of molecular interactions*. In *Molecular Interactions*, edited by H. Ratajczak and W.J. Orville-Thomas. John Wiley & Sons Ltd., New York (1980).

P.E. Boscott, G.J. Barton, and W.G. Richards. Protein Eng. *6*, 261 (1993).

J.U. Bowie and D. Eisenberg. Current Opin. Struct Biol. *3*, 437 (1993).

J.U. Bowie, R. Luthy, and D. Eisenberg. Science *253*, 164 (1991).

D.B. Boyd. *Successes of Computer-Assisted Molecular Design*, in Reviews of Computational Chemistry, vol. 1, edited by K.B. Lipkowitz and D.B. Boyd. VCH Publishers, New York (1990).

R.J. Boyd and J.M. Ugalde. *Analysis of Wave Functions in Terms of One- and Two-Electron Density Functions*, in *Computational Chemistry. Structure, Interactions and Reactivity*, edited by S. Fraga. Elsevier Science Publishers, Amsterdam (1992).

M. Boyer, Z. Novak, E. Fraga, K. Oikawa, C.M. Kay, A. Fotedar, and B. Singh. Int. Immunol. *2*, 1221 (1990).

S.F. Boys and F. Bernardi. Molec. Phys. *19*, 553 (1970).

A. Brack and G. Spach. J. Am. Chem. Soc. *103*, 6319 (1981).

A. Brack and L.E. Orgel. Nature *256*, 383 (1975).

D.F. Bradley. J. Macromol. Sci.-Chem. *A4*, 741 (1970).

D.A. Brant and P.J. Flory. J. Am. Chem. Soc. *87*, 2791 (1965).

C.M. Breneman and K.B. Wiberg. J. Comput. Chem. *11*, 361 (1990).

T. Brodmeier and E. Pretsch. J. Comput. Chem. *15*, 588 (1994).

C.L. Brooks III. J. Phys. Chem. *90*, 6680 (1986).

C.L. Brooks III and D.A. Case. Chem Rev. *93*, 2487 (1993).

C.L. Brooks III and S.H. Fleischmann. J. Am. Chem. Soc. *112*, 3307 (1990).

C.L. Brooks III, M. Karplus, and B.M. Pettitt. Adv. Chem. Phys. *71*, 1 (1988).

C.L. Brooks III, M. Karplus, and B.M. Pettitt. *Proteins: A Theoretical Perspective of Dynamics, Structure, and Thermodynamics*. Wiley, New York (1988).

B.R. Brooks, R.E. Bruccoleri, B.D. Olafson, D.J. States, S. Swaminathan, and M. Karplus. J. Comput. Chem. *4*, 187 (1983).

W. Brostow, J.-P. Dussault, and B.L. Fox. J. Comput. Phys. *29*, 81 (1978).

F.F. Brown and P.J. Goodford. Brit. J. Pharmac. *60*, 337 (1997).

J.H. Brown, T. Jardetzky, M.A. Saper, B. Samraoui, P.J. Bjorkman, and D.C. Wiley. Nature *332*, 845 (1988).

R.E. Bruccoleri and M. Karplus. Macromolecules *18*, 2767 (1985).

R.E. Bruccoleri and M. Karplus. Biopolymers *26*, 137 (1987).

D. Brune and S. Kim. Proc. Natl. Acad. Sci. USA *91*, 2930 (1994).

A.T. Brünger, G.M. Clore, A.M. Gronenborn, and M. Karplus. Proc. Natl. Acad. Sci. USA *83*, 3801 (1986).

A. Brünger. *X-PLOR Version 1.5 Manual*. Howard Hughes Medical Institute and Department of Molecular Biophysics and Biochemistry, Yale University, New Haven (1987).

S.H. Bryant. Proteins Struct. Funct. and Genet. *5*, 233 (1989).

S.H. Bryant and C.E. Lawrence. Proteins Struct. Funct. and Genet. *16*, 92 (1993).

A.D. Buckingham. *Permanent and induced molecular moments and long-range intermolecular forces*, in *Intermolecular Forces*, edited by J.O. Hirschfelder. Adv. Chem. Phys., vol. 12. Interscience Publishers, New York (1967).

A.D. Buckingham, P.W. Fowler, and J.M. Hutson. Chem. Rev. *88*, 963 (1988).

D.D. Buechter and P. Schimmel. Crit. Rev. Biochem. Mol. Biol. *28*, 309 (1993).

H.B. Bull and K. Breese. Arch. Biochem. Biophys. *161*, 665 (1974).

A.W. Burgess and H.A. Scheraga. Proc. Nat. Acad. Sci. USA *72*, 1221 (1975).

B. Busetta, I.J. Tickle, and T.L. Blundell. J. Appl. Cryst. *16*, 432 (1983).

D. Byrne, J. Li, E. Platt, B. Robson, and P. Weiner. J. Computer-Aided Mol. Design *8*, 67 (1994).

R.E. Cachau, E.H. Serpersu, A.S. Mildvan, J.T. August, and L.M. Amzel. J. Mol. Recogn. *2*, 179 (1990).

A. Caflisch, P. Niederer, and M. Anliker. Proteins Struct. Funct. and Genet. *13*, 223 (1992).

A. Caflisch, P. Niederer, and M. Anliker. Proteins Struct. Funct. and Genet. *14*, 102 (1992).

R.S. Cahn, C.K. Ingold, and V. Prelog. Experientia *12*, 81 (1956).

R. Carbo and B. Calabuig. *Quantum Similarity: Definitions, Computational Details and Applications*, in *Computational Chemistry. Structure, Interactions and Reactivity*, edited by S. Fraga. Elsevier, Amsterdam (1992).

R. Carbo and L. Domingo. Int. J. Quantum Chem. *32*, 517 (1987).

R. Carbo and M. Klobukowski (editors). *Self-Consistent Field Theory. Theory and Applications.* Elsevier Science Publishers, Amsterdam (1991).

R. Carbo and J.M. Riera. *A General SCF Theory.* Springer-Verlag, Berlin (1978).

R. Carbo, L. Leyda, and M. Arnau. Int. J. Quantum Chem. *17*, 1185 (1980).

J.G. Carbonell and P. Langley. In *Encyclopedia of Artificial Intelligence.* Wiley-Interscience, New York (1987).

K.B. Cease, H. Margalit, J.L. Cornette, S.D. Putney, W.G. Robey, C. Ouyang, H.Z. Streicher, P.J. Fischinger, R.C. Gallo, C. DeLisi, and J.A. Berzofsky. Proc. Natl. Acad. Sci. USA *84*, 4249 (1987).

N.B. Centeno and J.J. Perez. Chem. Phys. Lett. *232*, 374 (1995).

H.S. Chan and K.A. Dill. Proc. Natl. Acad. Sci. USA *87*, 6388 (1990a).

H.S. Chan and K.A. Dill. J. Chem. Phys. *90*, 492 (1990b).

H.S. Chan and K.A. Dill. Ann. Rev. Biophys. Biophys. Chem. *20*, 447 (1991).

D. Chandler. *Introduction to Modern Statistical Mechanics.* Oxford University Press, Oxford (1987).

P.-L. Chau and P.M. Dean. J. Mol. Graph. *5*, 97 (1987).

Y. Chen, A. Brass, B.J. Pendleton, and B. Robson. Biopolymers *33*, 1307 (1993).

J. Cherfils, S. Duquerroy, and J. Janin. Proteins Struct. Funct. and Genet. *11*, 271 (1991).

C. Chothia. J. Mol. Biol. *105*, 1 (1976).

C. Chothia. Ann. Rev. Biochem. *53*, 537 (1984).

C. Chothia. Ann. Rev. Biochem. *59*, 1007 (1990).

C. Chothia. Nature *357*, 543 (1992).

C. Chothia and A. Finkelstein. Ann. Rev. Biochem. *59*, 1007 (1990).

C. Chothia and A.G. Murzin. Structure *1*, 217 (1993).

C. Chothia, M. Levitt, and D. Richardson. J. Mol. Biol. *145*, 215 (1981).

P.Y. Chou and G.D. Fasman. Biochem. *13*, 211 (1974).

P.Y. Chou and G.D. Fasman. Biochem. *13*, 222 (1974).

P.Y. Chou and G.D. Fasman. J. Mol. Biol. *115*, 135 (1977).

P.Y. Chou and G.D. Fasman. Adv. Enzymol. *47*, 45 (1978).

P.Y. Chou and G.D. Fasman. Ann. Rev. Biochem. *47*, 251 (1978).

P.Y. Chou and G.D. Fasman. Biophys. J. *26*, 367 (1979).

P. Cieplak, T.P. Lybrand, and P. Kollman. J. Chem. Phys. *86*, 6393 (1987).

D.A. Clark, G.J. Barton, and C.J. Rawlings. J. Mol. Graphics *8*, 94 (1990).

M. Clark, R.D. Cramer III, and N. Van Opdenbosch. J. Comput. Chem. *10*, 982 (1989).

E. Clementi. *Determination of Liquid Water Structure. Coordination Numbers for Ions and Solvation for Biological Molecules.* Springer-Verlag, Berlin (1976).

E. Clementi. *Computational Aspects for Large Chemical Systems.* Springer-Verlag, Berlin (1980).

E. Clementi and H. Popkie. J. Chem. Phys. *57*, 1077 (1972).

E. Clementi, F. Cavallone, and R. Scordamaglia. J. Am. Chem. Soc. *99*, 5531 (1977).

E. Clementi, G. Corongiu, M. Aida, U. Niesar, and G. Kneller. *Monte Carlo and Molecular Dynamics Simulations*. In *Modern Techniques in Computational Chemistry*, edited by E. Clementi. ESCOM, Leiden (1990).

B. Coghlan and S. Fraga. Comput. Phys. Commun. *36*, 391 (1985).

C. Cohen and D.A.D. Parry. TIBS *11*, 245 (1986).

F.E. Cohen, M.J.E. Sternberg, and W.R. Taylor. J. Mol. Biol. *148*, 253 (1981).

F.E. Cohen, R.M. Abarbanel, I.D. Kuntz, and R.J. Fletterick. Biochem. *22*, 4894 (1983).

F.E. Cohen, R.M. Abarbanel, I.D. Kuntz, and R.J. Fletterick. Biochem. *25*, 266 (1986).

F.E. Cohen, P.A. Kosen, I.D. Kuntz, L.B. Epstein, T.L. Ciardelly, and K.A. Smith. Science *234*, 349 (1986).

N. Colloc'h and J.-P. Mornon. J. Mol. Graphics *8*, 133 (1990).

N. Colloc'h, C. Etchebest, E. Thoreau, B. Henrissat, J.P. Mornon, U. Lessel, and D. Schomburg. Protein Eng. *7*, 1175 (1994).

V. Collura, J. Higo, and J. Garnier. Protein Sci. *2*, 1502 (1993).

E.U. Condon and G.H. Shortley. *The Theory of Atomic Spectra*. Cambridge University Press, Cambridge (1964).

M.L. Connolly. Quantum Chem. Program Exchange Bull. *1*, 75 (1981).

M.L. Connolly. Science *221*, 709 (1983).

M.L. Connolly. J. Appl. Crystallogr. *16*, 548 (1983).

M.L. Connolly. J. Appl. Crystallogr. *18*, 499 (1985).

M.L. Connolly. J. Mol. Graphics *4*, 3 (1986).

M.L. Connolly. Biopolymers *25*, 1229 (1986).

R.I. Corey, K.H. Altmann, and M. Mutter. Ciba Foundation Symp. *158*, 187 (1991).

J.L. Cornette, K.B. Cease, H. Margalit, J.L. Spouge, J.A. Berzofsky, and C. DeLisi. J. Mol. Biol. *195*, 659 (1987).

A.J. Corrigan and P.C. Huang. Comput. Programs Biomed. *15*, 163 (1982).

R.M.J. Cotterill and J.U. Madsen. In *Characterizing Complexity*, edited by H. Bohr. World Scientific, Singapore (1990).

R.M.J. Cotterill, E. Platt, and B. Robson. In *Computation on Biomolecular Structures: Achievements, Problems, and Perspectives,* edited by M. Soumpasis and T. Jovin. Springer, Berlin (1991).

D.G. Covell. J. Mol. Biol. *235*, 1032 (1994).

D.G. Covell and R.L. Jernigan. Biochem. *29*, 3287 (1990).

R.D. Cramer III, D.E. Patterson, and J.D. Bunce. J. Am. Chem. Soc. *110*, 5959 (1988).

R.D. Cramer III, J.D. Bunce, D.E. Patterson, and I.E. Frank. Quant. Struct.-Act. Relat. *7*, 18 (1988).

T.E. Creighton. *Proteins: Structural and Molecular Properties*. W.H. Freeman, New York (1983).

G.M. Crippen and T.F. Havel. *Distance Geometry and Molecular Conformation*. Wiley, New York (1988).

G.M. Crippen and V.N. Viswanadhan. Int. J. Peptide Protein Res. *24*, 279 (1984).

G. Cruciani, S. Clementi, and M. Baroni. *Variable selection in PLS analysis*, in *3D QSAR in Drug Design: Theory, Methods and Applications*, edited by H. Kubinyi. ESCOM, Leiden (1993).

D. Cvijovic and J. Klinowski. Science *267*, 664 (1995).

S.E. Cwirla, E.A. Peters, R.W. Barrett, and W.J. Dower. Proc. Natl. Acad. Sci. USA *87*, 6378 (1990).

G.P. Dado and S.H. Gellman. J. Am. Chem. Soc. *115*, 12609 (1993).

V. Daggett and M. Levitt. Annu. Rev. Biophys. Biomol. Struct. *22*, 353 (1993).

V. Daggett and M. Levitt. J. Mol. Biol. *232*, 600 (1993).

K.A. Dakin and D.F. Weaver. Seizure *2*, 21 (1993).

T. Dandekar and P. Argos. J. Mol. Biol. *236*, 844 (1994).

P. Dauber-Osguthorpe and D.J. Osguthorpe. J. Comput. Chem. *14*, 1259 (1993).

P. Dauber-Osguthorpe, V.A. Roberts, D.J. Osguthorpe, J. Wolff, M. Genest, and A.T. Hagler. Proteins Struct. Funct. and Genet. *4*, 31 (1988).

J. Dausset. Science *213*, 1469 (1981).

W.C. Davidon. A.E.C. Res. Develop. Report. ANL-5990 (1959).

W.C. Davidon. Comput. J. *10*, 406 (1968).

M.E. Davis and J.A. McCammon. J. Comput. Chem. *10*, 386 (1989).

M.E. Davis and J.A. McCammon. Chem. Rev. *90*, 509 (1990).

M.E. Davis, J.D. Madura, B.A. Luty, and J.A. McCammon. Comput. Phys. Commun. *62*, 187 (1990).

P.M. Dean. BioEssays *16*, 683 (1994).

P.M. Dean and P.-L. Chau. J. Mol. Graph. *5*, 152 (1987).

P.M. Dean and P. Callow. J. Mol. Graph. *5*, 159 (1987).

P.M. Dean, P. Callow, and P.-L. Chau. J. Mol. Graph. *6*, 28 (1988).

P. Debye and E. Hückel. Physik Z. *24*, 185 (1923).

W.F. DeGrado. Nature *365*, 488 (1993).

W.F. DeGrado and J.D. Lear. J. Am. Chem. Soc. *107*, 7684 (1985).

M. Delarue. Current Opin. Struct. Biol. *5*, 48 (1995).

G. Deleage, B. Tinland, and B. Roux. Anal. Biochem. *163*, 292 (1987).

C. DeLisi and J.A. Berzofsky. Proc. Natl. Acad. Sci. USA *82*, 7048 (1985).

R.L. DesJarlais, R.P. Sheridan, J.S. Dixon, I.D. Kuntz, and R. Venkatarghavan. J. Med. Chem. *29*, 2149 (1986).

R.L. DesJarlais, R.P. Sheridan, G.L. Seibel, J.S. Dixon, and I.D. Kuntz. J. Med. Chem. *31*, 722 (1988).

J.J. Devlin, L.C. Panganiban, and P.E. Devlin. Science *249*, 404 (1990).

M.J.S. Dewar and W. Thiel. J. Am. Chem. Soc. *99*, 4899 (1977).

M.J.S. Dewar, E.G. Zoebisch, E.F. Healy, and J.J.P. Stewart. J. Am. Chem Soc. *107*, 3902 (1985).

A. di Nola, D. Roccatano, and H.J.C. Berendsen. Proteins Struct. Funct. and Genet. *19*, 174 (1994).

H. Diaz, J.R. Espina, and J.W. Kelly. J. Am. Chem. Soc. *114*, 8316 (1992).

T.G. Dietterich and R.S. Michalski. In *Machine Learning: An Artificial Intelligence Approach*. Tioga Publishing Co., Palo Alto (1983).

K.A. Dill. Biochemistry *29*, 7133 (1990).

U. Dinur and A.T. Hagler. *New Approaches to Empirical Force Fields*, in *Reviews in Computational Chemistry*, Vol. 2, edited by K.B. Lipkowitz and D.B. Boyd. VCH Publishers, New York (1991).

L.C.W. Dixon and G.P. Szego (eds.). *Towards Global Optimization*. North-Holland and Elsevier Science Publishers, Amsterdam (1975).

C.M. Dobson. Curr. Biol. *4*, 636 (1994).

C.M. Dobson, P.A. Evans, and S.E. Radford. TIBS *19*, 31 (1994).

D.P. Dolata, A.R. Leach, and K. Prout. J. Comp.-Aided Mol. Design *1*, 73 (1987).

C.T. Dooley and R.A. Houghten. Life Sci. *52*, 1509 (1993).

R.F. Doolittle. Protein Sci. *1*, 191 (1992).

R.L. Dorit, L. Schoenbach, and W. Gilbert. Science *250*, 1377 (1990).

S. Duane, A.D. Kennedy, B.J. Pendleton, and D. Rowat. Phys. Lett *195*, 216 (1987).

M.J. Dudek and H.A. Scheraga. J. Comput. Chem. *11*, 121 (1990).

G. Duek and T. Scheuer. J. Comput. Phys. *90*, 161 (1990).

R.L. Dunbrack Jr. and M. Karplus. J. Mol. Biol. *230*, 543 (1993).

L.G. Dunfield, A.W. Burgess, and H.A. Scheraga. J. Phys. Chem. *82*, 2609 (1978).

M. Dupuis and S. Maluendes. *A General Atomic and Molecular Electronic Structure System*, in *Modern Techniques in Computational Chemistry*, edited by E. Clementi. ESCOM, Leiden (1991).

M. Dupuis, J.D. Watts, H.O. Villar, and G.J.B. Hurst. Comput. Phys. Commun. *52*, 415 (1989).

H.J. Dyson and P.E. Wright. Ann. Rev. Biophys. Biophys. Chem. *20*, 519 (1991).

H.J. Dyson, M. Rance, R.A. Houghten, R.A. Lerner, and P.E. Wright. J. Mol. Biol. *201*, 161 (1988).

S.E. Ealick and S.R. Armstrong. Current Opin. Struct. Biol. *3*, 861 (1993).

S.P. Edmondson. J. Magn. Reson. *98*, 283 (1992).

A.B. Edmundson, D.L. Harris, Z.-C. Fan, L.W. Guddat, B.T. Schley, B.L. Hanson, G. Tribbick, and H.M. Geysen. Proteins Struct. Funct. and Genet. *16*, 246 (1993).

J.T. Edsall and H.A. McKenzie. Adv. Biophys. *10*, 137 (1978).

J.T. Edsall and H.A. McKenzie. Adv. Biophys. *16*, 53 (1983).

A.V. Efimov. Current Opin. Struct. Biol. *3*, 379 (1993).

M. Eigen, W. Gardiner, P. Schuster, and R. Winkler-Oswaititsch. Scientific Amer. *244*, 88 (1981).

M.B. Eisen, D.C. Wiley, M. Karplus, and R.E. Hubbard. Proteins Struct. Funct. and Genet. *19*, 199 (1994).

D. Eisenberg and A.D. McLachlan. Nature *316*, 199 (1986).

F. Eisenhaber and P. Argos. J. Comput. Chem. *14*, 1272 (1993).

F. Eisenmenger, P. Argos, and R.A. Abagyan. J. Mol. Biol. *231*, 839 (1993).

R.A. Engh and R. Huber. Acta Cryst. *A47*, 392 (1991).

E.G. Engleman, C.J. Benike, C. Grumet, and R.L. Evans. J. Immunol. *127*, 2124 (1981).

D.L. Ermack and J.A. McCammon. J. Chem. Phys. *69*, 1352 (1978).

A.M. Eroshkin, P.A. Zhilkin, V.V. Shamin, S. Korolev, and B. Fedorov. Protein Eng. *6*, 997 (1993).

A. Eschenmoser and E. Loewenthal. Chem. Soc. Rev. *21*, 1 (1992).

G.D. Fasman (ed.) *Prediction of Protein Structure and the Principles of Protein Conformation*. Plenum Press, New York (1989).

J.-L. Fauchere and V. Pliska. Eur. J. Med. Chem. *18*, 369 (1983).

M. Feigel. Liebigs Ann. Chem. 459 (1989).

S. Feng, J.K. Chen, H. Yu, J.A. Simon, and S.L. Schreiber. Science *266*, 1241 (1994).

K.A. Fichthorn and W.H. Weinberg. J. Chem. Phys. *95*, 1090 (1991).

M.J. Field, P.A. Bash, and M. Karplus. J. Comput. Chem. *11*, 700 (1990).

J.S. Finer-Moore, A.A. Kossiakoff, J.F. Hurley, T. Earnest, and R.M. Stroud. Proteins Struct. Funct. and Genet. *12*, 203 (1992).

A.V. Finkelstein. Dokl. Akad. Nauk SSSR *223*, 744 (1975).

A.V. Finkelstein. Mol. Biol (USSR) *11*, 811 (1977).

A.V. Finkelstein and O.B. Ptitsyn. Progr. Biophys. Mol. Biol. *50*, 171 (1987).

C.F. Fischer. *The Hartree-Fock Method for Atoms*. John Wiley & Sons, New York (1977).

M. Fixman. J. Chem. Phys. *69*, 1527 (1978).

R. Fletcher and M.J.D. Powell. Comput. J. *6*, 163 (1963).

R. Fletcher and C.M. Reeves. Comput J. 7, 149 (1964).

P.J. Flory. *Principles of Polymer Chemistry*. Cornell University Press, Ithaca (1953).

P.J. Flory. *Statistical Mechanics of Chain Molecules*. Interscience Publishers, New York (1969).

P.J. Flory. *Statistical Mechanics of Chain Molecules*. Oxford University Press, Oxford (1989).

E. Fraga and S. Fraga. *Biomoléculas. Estudios Teóricos*. Editorial Alhambra, Madrid (1976).

S. Fraga. J. Comput. Chem. *3*, 329 (1982*a*).

S. Fraga. Can. J. Chem. *60*, 2606 (1982*b*).

S. Fraga. Comput. Phys. Commun. *29*, 351 (1983).

S. Fraga (editor). *Computational Chemistry. Structure, Interactions and Reactivity*. Elsevier Science Publishers, Amsterdam (1992).

S. Fraga and J. Muszynska. *Atoms in External Fields*. Elsevier Science Publishers, Amsterdam (1981).

S. Fraga and J.M.R. Parker. Amino Acids 7, 175 (1994).

S. Fraga and S.E. Thornton. Theor. Chim. Acta *85*, 61 (1993).

S. Fraga, J. Karwowski, and K.M.S. Saxena. *Handbook of Atomic Data*. Elsevier Science Publishers, Amsterdam (1976; second printing 1979).

S. Fraga, K.M.S. Saxena, and M. Torres. *Biomolecular Information Theory*. Elsevier Science Publishers, Amsterdam (1978).

S. Fraga, L. Seijo, M. Campillo, F. Moscardo, and B. Singh. J. Mol. Struct. (Theochem) *179*, 27 (1988).

S. Fraga, E. San Fabian, S. Thornton, and B. Singh. J. Mol. Recognition *3*, 65 (1990).

S. Fraga, S.E. Thornton, and B. Singh. J. Mol. Struct. (Theochem) *286*, 87 (1993).

H. Frauenfelder. *Structure and Motion in Membranes, Nucleic Acids and Proteins*. Adenine, New York (1985).

H. Frauenfelder, S.G. Sligar, and P.G. Wolynes. Science *254*, 1598 (1991).

S.M. Free and J.N. Wilson. J. Med. Chem. *7*, 395 (1964).

A.R. Friedman, V.A. Roberts, and J.A. Tainer. Proteins Struct. Funct. and Genet. *20*, 15 (1994).

M.J. Frisch, G.W. Trucks, M. Head-Gordon, P.M.W. Gill, M.W. Wong, J.B. Foresman, B.G. Johnson, H.B. Schlegel, M.A. Robb, E.S. Replogle, R. Gomperts, J.L. Andres, K. Raghavachari, J.S. Binkley, C. Gonzalez, R.L. Martin, D.J. Fox, D.J. Defrees, J. Baker, J.J.P. Stewart, and J.A. Pople. *Gaussian 92: Revision A*. Gaussian, Pittsburgh, PA (1992).

B. Furie, D.H. Bing, R.J. Feldmann, D.J. Robinson, J.P. Burnier, and B.C. Furie. J. Biol. Chem. *257*, 3875 (1982).

W.H. Gallagher and K.M. Crocker. Protein Sci. *3*, 1602 (1994).

J. Garnier, D.J. Osguthorpe, and B. Robson. J. Mol. Biol. *120*, 97 (1978).

J. Gasteiger, X. Li, C. Rudolph, J. Sadowski, and J. Zupan. J. Am. Chem. Soc. *116*, 4608 (1994).

C.W. Gear. Report ANL 7126. Argonne National Laboratory (1966).

C.W. Gear. Math. Comput. *21*, 146 (1967).

C.W. Gear. *Numerical Initial Value Problems in Ordinary Differential Equations*. Prentice-Hall, Englewood Cliffs, NJ (1971).

B.R. Gelin and M. Karplus. Proc. Nat. Acad. Sci. USA 72, 2002 (1975).

B.R. Gelin and M. Karplus. Biochem. *18*, 1256 (1979).

G. Geourjon and G. Deleage. Protein Eng. *7*, 157 (1994).

M.J. Gething and J. Sambrook. Nature *355*, 33 (1992).

M.R. Ghadiri and M.A. Case. Angew. Chem. Int. Ed. Engl. *9*, 1290 (1993).

M.R. Ghadiri and C. Choi. J. Am. Chem. Soc. *114*, 1630 (1992).

M.R. Ghadiri, C. Soares, and C. Choi. J. Am. Chem. Soc. *114*, 825 (1992).

C. Ghelis and J. Yon. *Protein Folding*. Academic Press, New York (1982).

A.K. Ghose and G.M. Crippen. J. Comput. Chem. *6*, 350 (1985).

A. Giannis and T. Kolter. Angew. Chem. Int. Ed. Engl. *32*, 1244 (1993).

K.D. Gibson and H.A. Scheraga. J. Comput. Chem. *11*, 468 (1990a).

K.D. Gibson and H.A. Scheraga. J. Comput. Chem. *11*, 487 (1990b).

T.J. Gibson, J.D. Thompson, and R.A. Abagyan. Protein Eng. *6*, 41 (1993).

M.K. Gilson and B. Honig. Nature *330*, 84 (1987).

M.K. Gilson and B. Honig. Proteins *3*, 32 (1988a).

M.K. Gilson and B. Honig. Proteins Struct. Funct. and Genet. *4*, 7 (1988b).

M.K. Gilson, A. Rashin, R. Fine, and B. Honig. J. Mol. Biol. *183*, 503 (1985).

M.K. Gilson, K.A. Sharp, and B. Honig. J. Comput. Chem. *9*, 327 (1987).

W. Glunt, T.L. Hayden, and M. Raydan. J. Comput. Chem. *14*, 114 (1993).

V. Glushko, P.J. Lawson, and F.R.N. Gurd. J. Biol. Chem. *247*, 3176 (1972).

N. Go and H.A. Scheraga. Macromolecules *3*, 178 (1970).

A. Godzik, A. Kolinski, and J. Skolnick. J. Mol. Biol. *227*, 227 (1992).

A. Godzik, A. Kolinski, and J. Skolnick. J. Computer-Aided Mol. Design *7*, 397 (1993).

D.E. Goldberg. *Genetic Algorithms in Search, Optimization and Machine Learning.* Addison-Wesley Publ., Reading, MA (1989).

A. Goldblum. J. Comput. Chem. *8*, 835 (1987).

A. Gonzalez-Lafont, J.M. Lluch, A. Oliva and J. Bertran. In *Modeling of Molecular Structures and Properties,* edited by J.L. Rivail. Elsevier Science Publishers, Amsterdam (1990), pp.165-171.

A.C. Good, S.-S. So, and W.G. Richards. J. Med. Chem. *36*, 433 (1993).

P.J. Goodford. J. Med. Chem. *28*, 849 (1985).

D.S. Goodsell and R.E. Dickerson. J. Med. Chem. *29*, 727 (1986).

D.S. Goodsell and A.J. Olson. Proteins *8*, 195 (1990).

D.S. Goodsell and A.J. Olson. Proteins Struct. Funct. and Genet. *8*, 195 (1990).

D.S. Goodsell and A.J. Olson. Trends Biol. Sci. *18*, 65 (1993).

H.L. Gordon and R.L. Somorjai. Proteins *14*, 249 (1992).

A.-D. Gorse and M. Pesquer. J. Comput. Chem. *15*, 1139 (1994).

P.M.D. Gray, N.W. Paton, G.J.L. Kemp, and J.E. Fothergill. Protein Eng. *3*, 235 (1990).

P. Green. Current Opin. Struct. Biol. *4*, 404 (1994).

J. Greer. J. Mol. Biol. *153*, 1027 (1981).

J. Greer. Proteins Stuct. Funct. and Genet. *7*, 317 (1990).

J. Greer and B.L. Bush. Proc. Natl. Acad. Sci. USA *75*, 303 (1978).

M. Gribskov, A.D. McLachlan, and D. Eisenberg. Proc. Natl. Acad. Sci. USA *84*, 4355 (1987).

N. Gronbech-Jensen and S. Doniach. J. Comput. Chem. *15*, 997 (1994).

A.M. Gronenborn and G.M. Clore. *Protein structure determination in solution by two-dimensional and three-dimensional NMR* in *Conformations and Forces in Protein Folding,* edited by B.T. Nall and K.A. Dill. American Association for the Advancement of Science, Washington (1990).

A. Grove, M. Mutter, J.E. Rivier, and M. Montal. J. Am. Chem. Soc. *115*, 1100 (1993).

H. Grubmüller, H. Heller, A. Windemuth, and K. Schulten. Mol. Simul. *6*, 121 (1991).

F. Guarnieri and W.C. Still. J. Comput. Chem. *11*, 1302 (1994).

C. Guerrier-Takada, K. Gardiner, T. Marsh, N. Pace, and S. Altman. Cell *35*, 849 (1983).

W.C. Guida. Current Opin. Struct. Biol. *4*, 777 (1994).

J.G. Guillet, M.Z. Lai, T.J. Briner, S. Buus, A. Sette, H.M. Grey, J.A. Smith, and M.L. Gefter. Science *235*, 865 (1987).

P. Gund. Prog. Mol. Subcell. Biol. *11*, 117 (1977).

P. Gund. *Retrospective and Prospective Successes of Molecular Modeling in the Pharmaceutical Industry.* Molecular Modeling Conference, East Brunswick, NJ, 2-4 October 1994.

P. Gund, W.T. Wipke, and R. Langridge. Comput. Chem. Res. Educ. Technol. *3*, 5 (1974).

O.F. Guner, D.W. Hughes, and L.M. Dumont. J. Chem. Int. Comput. Sci. *31*, 408 (1991).

M.A. Hadwiger. Protein Eng. *7*, 1283 (1994).

A.T. Hagler. Peptides Conformational Biol. Drug Res. 7, 213 (1985).

A.T. Hagler and J. Moult. Nature 272, 222 (1978).

A.T. Hagler and D. Osguthorpe. Biopolymers 19, 395 (1980).

A.T. Hagler, E. Huler, and S. Lifson. J. Am. Chem. Soc. 96, 5319 (1974).

A.T. Hagler, D.J. Osguthorpe, and B. Robson. Science 208, 599 (1980).

A.T. Hagler, D.J. Osguthorpe, P. Dauber-Osguthorpe, and J.C. Hempel. Science 227, 1309 (1985).

K.W. Hahn, W.A. Klis, and J.M. Stewart. Science 248, 1544 (1990).

C. Hansch and T.E. Klein. Acc. Chem. Res. 19, 392 (1986).

R.W. Harrison, I.V. Kourinov, and L.C. Andrews. Protein Eng. 7, 359 (1994).

S.C. Harrison and R. Durbin. Proc. Natl. Acad. Sci. USA 82, 4028 (1985).

T.N. Hart and R.J. Read. Proteins Struct. Funct. and Genet. 13, 206 (1992).

S.C. Harvey. Proteins Struct. Funct. and Genet. 5, 78 (1989).

T.F. Havel and M.E. Snow. J. Mol. Biol. 217, 1 (1991).

S. Hayward and J.F. Collins. Proteins Struct. Funct. and Genet. 14, 372 (1992).

B. Hazes and B.W. Dijkstra. Protein Eng. 2, 119 (1988).

D.P. Hearst and F.E. Cohen. Protein Eng. 7, 1411 (1994).

W. Heiden, T. Goetze, and J. Brickmann. J. Comput. Chem. 14, 246 (1993).

E. Helfand. J. Chem. Phys. 71, 5000 (1979).

M. Helmer-Citterich and A. Tramontano. J. Mol. Biol. 235, 1021 (1994).

S. Henikoff and J.G. Henikoff. Proteins Struct. Funct. and Genet. 17, 49 (1993).

J. Hermans. Current Opin. Struct. Biol. 3, 270 (1993).

J. Hermans, H.J.C. Berendsen, W.F. van Gunsteren, and J.P.M. Postma. Biopolymers 23, 1513 (1984).

J. Hermans, R.H. Yun, and A.G. Anderson. J. Comput. Chem. 13, 429 (1992).

G. Herzberg. Molecular Spectra and Molecular Structure. I. Spectra of Diatomic Molecules. D. Van Nostrand Company, Inc., Princeton (1957).

G. Herzberg. Molecular Spectra and Molecular Structure. II. Infrared and Raman Spectra of Polyatomic Molecules. D. Van Nostrand Company, Princeton (1960).

J. Higo and N. Go. J. Comput. Chem. 10, 376 (1989).

J. Higo, V. Collura, and J. Garnier. Biopolymers 32, 33 (1992).

J.S. Hill Gaston, P.F. Life, P.J. Jenner, M.J. Colston, and P.A. Bacon. J. Exp. Med. 171, 831 (1990).

D.A. Hinds and M. Levitt. J. Mol. Biol. 243, 668 (1994).

R. Hirschmann, K.C. Nicolaou, S. Pietranico, J. Salvino, E.M. Leahy, P.A. Sprengeler, G. Furst, and A.B. Smith. J. Am. Chem. Soc. 114, 9217 (1992).

C.M.W. Ho and G.R. Marshall. J. Comp.-Aided Mol. Design 4, 337 (1990).

C.M.W. Ho and G.R. Marshall. J. Comp.-Aided Mol. Design 7, 3 (1993).

U. Hobohm, M. Scharf, R. Schneider, and C. Sander. Protein Sci. 1, 409 (1992).

P. Hobza. Private communication (1985).

T.C. Hodgman. Comput. Appl. Biosci. 5, 1 (1989).

R.H. Hoess. Current Opin. Struct. Biol. 3, 572 (1993).

W.G.J. Hol, L.M. Halic, and C. Sander. Nature 294, 532 (1981).

J. Holland. Adaptation in Natural and Artificial Systems. The University of Michigan Press, Ann Arbor (1975).

L.H. Holley and M. Karplus. Proc. Natl. Acad. Sci. USA *86*, 152 (1989).

L. Holm and C. Sander. Proteins Struct. Funct. and Genet. *14*, 213 (1992).

L. Holm and C. Sander. Proteins Struct. Funct. and Genet. *19*, 256 (1994).

M. Holst, R.E. Kozack, F. Saied, and S. Subramanian. J. Biomol. Struct. Dynamics *11*, 1437 (1994).

B. Honig, A. Ray, and C. Levinthal. Proc. Natl. Acad. Sci. USA *73*, 1974 (1976).

B. Honig, K. Sharp, and A.-S. Yang. J. Phys. Chem. *97*, 1101 (1993).

J.J. Hopfield. Proc. Natl. Acad. Sci. USA *79*, 2554 (1982).

J.J. Hopfield and D. Tank. Biol. Cybernetics *52*, 141 (1985).

A.J. Hopfinger. J. Am. Chem. Soc. *102*, 7196 (1980).

A.J. Hopfinger. J. Med. Chem. *26*, 990 (1983).

T.P. Hopp and K.R. Woods. Proc. Natl. Acad. Sci. USA *78*, 3824 (1981).

T.P. Hopp. Progr. Clin. Biol. Res. *172B*, 367 (1985).

T.P. Hopp. Meth. Enzymol. *178*, 571 (1989).

N. Horton and M. Lewis. Protein Sci. *1*, 169 (1992).

R.A. Houghten and C.T. Dooley. Bioorg. Med. Chem. Lett. *3*, 405 (1993).

K. Huang. *Statistical Mechanics*. John Wiley & Sons, New York (1963).

M.C. Huang, J.M. Seyer, and A.H. Kang. J. Immunol. Meth. *129*, 77 (1990).

T.J.P. Hubbard and T.L. Blundell. Protein Eng. *1*, 159 (1987).

E.G. Hutchinson and J.M. Thornton. Protein Eng. *6*, 233 (1993).

E.G. Hutchinson and J.M. Thornton. Protein Sci. *3*, 2207 (1994).

M. Huysmans, J. Richelle, and S.J. Wodak. Proteins *11*, 59 (1991).

E. Iglesias, T.L. Sordo, and J.A. Sordo. Comput. Phys. Commun. *67*, 268 (1991).

E. Iglesias, T.L. Sordo, and J.A. Sordo. J. Mol. Struct. (Theochem) (in press).

E. Ising. Z. Phys. *31*, 253 (1925).

S.A. Islam and M.J.E. Sternberg. Protein Eng. *2*, 431 (1989).

Y. Isogai, G. Nemethy, and H.A. Scheraga. Proc. Natl. Acad. Sci. USA *74*, 414 (1977).

T. Ito, T. Fukushige, J. Makino, T. Ebisuzaki, S.K. Okumura, D. Sugimoto, H. Miyagawa, and K. Kitamura. Proteins Funct. Struct. and Genet. *20*, 139 (1994).

R. Jaenicke. Prog. Biophys. Molec. Biol. *49*, 117 (1987).

J. Janin. Nature *277*, 491 (1979).

J. Janin and C. Chothia. J. Mol. Biol. *100*, 197 (1976).

J. Janin and C. Chothia. J. Biol. Chem. *265*, 16027 (1990).

J. Janin and S. Wodak. Biopolymers *24*, 509 (1985).

J. Janin, S. Wodak, M. Levitt, and B. Maigret. J. Mol. Biol. *125*, 357 (1978).

W.P. Jencks. Adv. Enzymol. *43*, 219 (1975).

S. Jenks. J. Natl. Cancer Inst. *84*, 79 (1992).

T. Jiang and S.-H. Kim. J. Mol. Biol. *219*, 79 (1991).

S. Jindal, A.K. Dudani, B. Singh, C.B. Harley, and R.S. Gupta. Mol. Cell Biol. *9*, 2279 (1989).

K. Johnsson, R.K. Allemann, H. Widmer, and S.A. Benner. Nature *365*, 530 (1993).

K.S. Johnson and D.J. Chiswell. Current Opin. Struct. Biol. *3*, 564 (1993).

T.A. Jones and S. Thirup. EMBO J. *5*, 819 (1986).

D.T. Jones, W.R. Taylor, and J.M. Thornton. Nature *358*, 86 (1992).

W.L. Jorgensen. J. Am. Chem. Soc. *103*, 335 (1981).

W.L. Jorgensen. Acc. Chem. Res. *22*, 184 (1989).

W.L. Jorgensen. J. Am. Chem. Soc. *111*, 3770 (1989).

W.L. Jorgensen and C. Ravimohan. J. Chem. Phys. *83*, 3050 (1985).

W.L. Jorgensen and J. Tirado-Rives. J. Am. Chem. Soc. *110*, 1657 (1988).

W.L. Jorgensen, J.K. Buckner, S.E. Huston, and P.J. Rossky. J. Am. Chem. Soc. *109*, 1891 (1987).

W.L. Jorgensen, J.F. Blake, and J.K. Buckner. Chem. Phys. *129*, 193 (1989).

W. Kabsch and C. Sander. Biopolymers *22*, 2577 (1983).

W. Kabsch and C. Sander. FEBS Lett. *155*, 179 (1983).

W. Kabsch and C. Sander. Proc. Natl. Acad. Sci. USA *81*, 1075 (1984).

F. Kaden, I. Koch, and J. Selbig. J. Theor. Biol. *147*, 85 (1990).

E.T. Kaiser and F.J. Kezdy. Proc. Natl. Acad. Sci. USA *80*, 1137 (1983).

S.G. Kalko, R.E. Cachau, and A.M. Silva. Biophys. J. *63*, 1133 (1992).

P. Karathas, V.P. Sukhatme, L.A. Hezenberg, and J.R. Parnes. Proc. Natl. Acad. Sci. USA *81*, 7688 (1984).

M.E. Karpen, D.J. Tobias, and C.L. Brooks III. Biochem. *32*, 412 (1993).

M. Karplus and G.A. Petsko. Nature *347*, 631 (1990).

M. Karplus and J.A. McCammon. Ann. Rev. Biochem. *53*, 263 (1983).

M. Karplus and D.L. Weaver. Biopolymers *18*, 1421 (1979).

P.A. Karplus and G.E. Schulz. Naturwiss. *72*, 212 (1985).

P.A. Karplus and C. Faerman. Current Opin. Struct. Biol. *4*, 770 (1994).

J.F. Kaufman, C. Auffray, A.J. Korman, D.A. Schackelford, and J.L. Strominger. Cell *36*, 1 (1984).

S.H.E. Kaufmann. Immunol. Today *11*, 129 (1990).

P.T.P. Kaumaya, A.M. van Buskirk, E. Goldberg, and S.K. Pierce. J. Biol. Chem. *267*, 6338 (1992).

W. Kauzmann. Adv. Protein Chem. *3*, 1 (1959).

W.H. Kazmierski, H.I. Yamamura, and V.J. Hruby. J. Am. Chem. Soc. *113*, 2275 (1991).

M. Keller, E. Blochl, G. Wachtershauser, and K.O. Stetter. Nature *368*, 836 (1994).

T. Kenakin. *Pharmacologic Analysis of Drug-Receptor Interaction*. Raven Press, New York (1993).

D.J. Kenan, D.E. Tsai, and J.D. Keene. TIBS *19*, 57 (1994).

J.C. Kendrew, W. Klyne, S. Lifson, T. Miyazawa, G. Nemethy, D.C. Phillips, G.N. Ramachandran, and H.A. Scheraga. Biochemistry *9*, 3471 (1970).

K. Kikuchi, M. Yoshida, T. Maekawa, and H. Watanabe. Chem. Phys. Lett. *185*, 335 (1991).

K. Kikuchi, M. Yoshida, T. Maekawa, and H. Watanabe. Chem. Phys. Lett. *196*, 57 (1992).

T. Kikuchi, G. Nemethy, and H.A. Scheraga. J. Protein Chem. *7*, 473 (1988a).

T. Kikuchi, G. Nemethy, and H.A. Scheraga. J. Protein Chem. *7*, 491 (1988b).

P.S. Kim. Ann. Rev. Biochem. *59*, 631 (1990).

P.S. Kim and R.L. Baldwin. Ann. Rev. Biochem. *51*, 459 (1982).

E.S. Kimball and J.E. Coligan. Comp. Topics Molec. Immunol. *9*, 1 (1983).

R.D. King and M.J.E. Sternberg. J. Mol. Biol. *216*, 441 (1990).

R.D. King, S. Muggleton, R.A. Lewis, and M.J.E. Sternberg. Proc. Natl. Acad. Sci. USA *67*, 32 (1989).

J. King. Chem & Eng. News 32 (1989).

S. Kirkpatrick, C.D. Gelatt, and M.P. Vecchi. Science *220*, 671 (1983).

H. Kistenmacher, H. Popkie, and E. Clementi. J. Chem. Phys. *58*, 1683 (1973a).

H. Kistenmacher, H. Popkie, and E. Clementi. J. Chem. Phys. *58*, 5627 (1973b).

H. Kistenmacher, H. Popkie, and E. Clementi. J. Chem. Phys. *59*, 5842 (1973c).

H. Kistenmacher, H. Popkie, and E. Clementi. J. Chem. Phys. *60*, 4455 (1974).

D.B. Kitchen, L.H. Reed, and R.M. Levy. Biochemistry *31*, 10083 (1992).

I. Klapper, R. Hagstrom, R. Fine, K. Sharp, and B. Honig. Proteins *1*, 47 (1986).

J. Klein. Science *203*, 516 (1979).

J. Klein. *Natural History of the Major Histocompatibility Complex*. Wiley, New York (1986).

E.W. Knapp. J. Comput. Chem. *13*, 793 (1992).

E.W. Knapp and A. Irgens-Defregger. J. Comput. Chem. *14*, 19 (1993).

R.M.A. Knegtel, J.A.C. Rullmann, R. Boelens, and R. Kaptein. J. Mol. Biol. *235*, 318 (1994).

R.M.A. Knegtel, R. Boelens, and R. Kaptein. Protein Eng. *7*, 761 (1994).

D.G. Kneller, F.E. Cohen, and R. Langridge. J. Mol. Biol. *214*, 171 (1990).

P. Koehl and M. Delarue. Proteins Struct. Funct. and Genet. *20*, 264 (1994).

T. Kohonen. Biol. Cybern. *43*, 59 (1982).

T. Kohonen. *Self-Organization and Associative Memory*. Springer-Verlag, Berlin (1989).

T. Kohonen. Proc. IEEE *78*, 1460 (1990).

A. Kolinski and J. Skolnick. Proteins Struct. Funct. and Genet. *18*, 338 (1994).

A. Kolinski and J. Skolnick. Proteins Struct. Funct. and Genet. *18*, 353 (1994).

A. Kolinski, J. Skolnick, and R. Yaris. Biopolymers *26*, 937 (1987).

P. Kollman. Chem. Rev. *93*, 2395 (1993).

P.A. Kollman and K.A. Dill. J. Biomol. Struct. Dynam. *8*, 1103 (1991).

H. Kono and J. Doi. Proteins Struct. Funct. and Genet. *19*, 244 (1944).

A.P. Korn and D.R. Rose. Protein Eng. *7*, 961 (1994).

R.E. Kozack and S. Subramaniam. Protein Sci. *2*, 915 (1993).

P.J. Kraulis. J. Appl. Cryst. *24*, 946 (1990).

K. Kremer and K. Binder. Comp. Phys. Rev. *7*, 259 (1988).

A.M. Krensky, C.S. Reiss, J.W. Mier, J.L. Strominger, and S.J. Burakoff. Proc. Natl. Acad. Sci. USA *79*, 2365 (1982).

K. Kruger, P.J. Grabowski, A.J. Zaug, J. Sands, D.E. Gottschling, and T.R. Cech. Cell *31*, 147 (1982).

I.D. Kuntz. J. Am. Chem. Soc. *94*, 4009 (1972).

I.D. Kuntz. Science *257*, 1078 (1992).

I.D. Kuntz, J.M. Blaney, S.J. Oatley, R. Langridge, and T.E. Ferrin. J. Mol. Biol. *161*, 269 (1982).

J. Kyte and R.F. Doolittle. J. Mol. Biol. *157*, 105 (1982).

P.Y.S. Lam, P.K. Jadhav, C.J. Eyermann, C.N. Hodge, Y. Ru, L.T. Bacheler, J.L. Meek, M.J. Otto, M.M. Rayner, Y.N. Wong, C.-H. Chang, P.C. Weber, D.A. Jackson, T.R. Sharpe, and S. Erickson-Viitanen. Science 263, 380 (1994).

J.R. Lamb, J. Ivanyi, A.D.M. Rees, J.B. Rothbard, K. Howland, R.A. Young, and D.B. Young. EMBO J. 6, 1245 (1987).

M.H. Lambert and H.A. Scheraga. J. Comput. Chem. 10, 770 (1989a).

M.H. Lambert and H.A. Scheraga. J. Comput. Chem. 10, 798 (1989b).

M.H. Lambert and H.A. Scheraga. J. Comput. Chem. 10, 817 (1989c).

A.N. Lane. Biochim. Biophys. Acta 1049, 189 (1990).

M. Laskowski Jr., I. Kato, W. Ardelt, J. Cook, A. Denton, M.W. Empie, W.J. Kohr, S.J. Park, K. Parks, B.L. Schatzley, O.L. Scoenberger, M. Tashiro, G. Vichot, H.E.Whatley, A. Wieczorek, and M. Wieczorek. Biochemistry 26, 202 (1987).

R.A. Laskowski, D.S. Moss, and J.M. Thornton. J. Mol. Biol. 231, 1049 (1993).

R.H. Lathrop. Protein Eng. 7, 1059 (1994).

D. Lauffenburger and J. Linderman. Receptors: Models for Binding, Trafficking, and Signaling. Oxford University Press, New York (1993).

C.A. Laughton. Protein Eng. 7, 235 (1994).

G. Lauri and P.A. Bartlett. J. Comp.-Aided Mol. Design 8, 51 (1994).

A.R. Leach and I.D. Kuntz. J. Comput. Chem. 13, 730 (1992).

A.R. Leach and K. Prout. J. Comput. Chem. 11, 1193 (1990).

J.T.J. Lecomte and C.R. Mathews. Protein Eng. 6, 1 (1993).

B. Lee and F.M. Richards. J. Mol. Biol. 55, 379 (1971).

C. Lee and S. Subbiah. J. Mol. Biol. 217, 373 (1991).

R.H. Lee and G.D. Rose. Biopolymers 24, 1613 (1985).

S.E. Leicester, J.L. Finney, and R.P. Bywater. J. Mol. Graphics 6, 104 (1988).

S. Leikin and V.A. Parsegian. Proteins Struct. Funct. and Genet. 19, 73 (1994).

J.E. Lennard-Jones. Proc. Roy. Soc. London A106, 441 (1924).

A.M. Lesk and C. Chothia. J. Mol. Biol. 160, 325 (1982).

J.M. Levin, S. Pascarella, P. Argos and J. Garnier. Protein Eng. 6, 849 (1993).

C.L. Levinthal. J. Chim. Phys. 65, 44 (1968).

C.L. Levinthal. In Mossbauer Spectroscopy in Biological Systems, edited by P. Debrunner, J.C.M. Tsibris, and E. Munck. University of Illinois Press, Urbana (1969).

M. Levitt. J. Mol. Biol. 82, 393 (1974).

M. Levitt. J. Mol. Biol. 104, 59 (1976).

M. Levitt. Nature 294, 379 (1981).

M. Levitt. J. Mol. Biol. 168, 621 (1983).

M. Levitt and J. Greer. J. Mol. Biol. 114, 181 (1977).

M. Levitt and B.H. Park. Curr. Biol. 1, 223 (1993).

M. Levitt and A. Warshel. Nature 253, 694 (1975).

R.M. Levy and M. Karplus. Biopolymers 18, 2465 (1979).

B. Lewin. Genes. Oxford University Press, New York (1990).

M. Lewis and D.C. Rees. Science 230, 1163 (1985).

R.A. Lewis. Meth. Enzymol. 202, 126 (1991).

J. Li, E. Platt, B. Waskowycz, R.M.J. Cotterill, and B. Robson. Biophys. Chem. *43*, 221 (1992).

Z. Li and H.A. Scheraga. Proc. Natl. Acad. Sci. USA *84*, 6611 (1987).

Z. Li and H.A. Scheraga. J. Mol. Struct. (Theochem) *179*, 333 (1988).

M. Lieberman and T. Sasaki. J. Am. Chem. Soc. *113*, 1470 (1991).

S. Lifson. *Potential energy functions for structural molecular biology.* In *Structural Molecular Biology. Methods and Applications*, edited by D.B. Davies, W. Saenger, and S.S. Danyluk. Plenum Press, New York (1982).

J.-H. Lii and N.L. Allinger. J. Am. Chem. Soc. *111*, 8576 (1989).

V.I. Lim. J. Mol. Biol. *88*, 857 (1974).

V.I. Lim. J. Mol. Biol. *88*, 873 (1974).

K.B. Lipkowitz and D.B. Boyd. *Reviews in Computational Chemistry.* VCH Publishers, New York (1990; published annually).

K.B. Lipkowitz and R. Zegarra. J. Comput. Chem. *10*, 595 (1989).

E.R. Lippincott and R. Schroeder. J. Chem. Phys. *23*, 1099 (1955).

D.R. Littman, Y. Thomas, P.J. Maddon, L. Chess, and R. Axel. Cell *40*, 237 (1985).

A.M. Livingstone and C.G. Fathman. Ann. Rev. Immunol. *5*, 477 (1987).

J.M. Lluch, M. Moreno, and J. Bertran. In *Computational Chemistry. Structure, Interactions and Reactivity*, edited by S. Fraga. Elsevier Science Publishers, Amsterdam (1992).

G.H. Loew, H.O. Villar, and I. Alkorta. Pharmac. Res. *10*, 475 (1993).

W. Lorensen and H. Clive. Comp. Graph. *21*, 163 (1987).

V. Lounnas and M. Pettitt. Proteins Struct. Funct. and Genet. *18*, 133 (1994).

B. Lovejoy, S. Choe, D. Cascio, D. McRorie, W. DeGrado, and D. Eisenberg. Science *259*, 1288 (1993).

P.J. Maddon, D.R. Littman, M. Godfrey, D.E. Maddon, L. Chess, and R. Axel. Cell *42*, 93 (1985).

B. Madhusudan, R. Kodandapani, and M. Vijayan. Acta Cryst. *49*, 234 (1993).

P.A. Mahama and J.J. Linderman. Biophys. J. *67*, 1345 (1994).

G.I. Makhatadze and P.L. Privalov. J. Mol. Biol. *213*, 375 (1990).

P. Manavalan and P.K. Ponnuswamy. Nature *275*, 673 (1978).

F. Manca, J.A. Clarke, A. Miller, E.E. Sercarz, and N. Shastri. J. Immunol. *133*, 2075 (1984).

B. Mao. Biophys. J. *60*, 966 (1991).

B. Mao. Biochem. J. *288*, 109 (1992).

B. Mao and A.R. Friedman. Biophys. J. *58*, 803 (1990).

B. Mao, G.M. Maggiora, and K.C. Chou. Biopolymers *31*, 1077 (1991).

W. Markland, B.L. Roberts, M.J. Saxena, S.K. Guterman, and R.C. Ladner. Gene *109*, 13 (1991).

S. Marqusee, V.H. Robbins, and R.L. Baldwin. Proc. Natl. Acad. Sci. USA *86*, 5286 (1989).

P. Marrack and J. Kappler. Adv. Immunol. *38*, 1 (1986).

G. Marshall. Current Opin. Struct. Biol. *2*, 904 (1992).

G.R. Marshall. *Molecular Modeling in Drug Design*, in *Burger's Medicinal Chemistry and Drug Discovery*, edited by M.E. Wolff. Wiley-Interscience, New York (1994). 5th edition.

G.R. Marshall and R.D. Cramer III. Trends Pharm. Sci. *9*, 285 (1988).

Y.C. Martin. Methods Enzymol. *203*, 587 (1991).

Y.C. Martin, E.B. Danaher, C.S. May, and D. Weininger. J. Comp.-Aided Mol. Design *2*, 15 (1988).

C.R. Mathews. Current Opin. Struct. Biol. *1*, 28 (1991).

N. Matsushima, C.E. Creutz, and R.H. Kretsinger. Proteins Struct. Funct. and Genet. *7*, 125 (1990).

A.K. Mazur and R.A. Abagyan. J. Biomol. Struct. Dyn. *6*, 815 (1989).

J.A. McCammon, P.G. Wolynes, and M. Karplus. Biochemistry *18*, 927 (1979).

J.A. McCammon, S.H. Northrup, M. Karplus, and R.M. Levy. Biopolymers *19*, 2033 (1980).

D.B. McGarrah and R.S. Judson. J. Comput. Chem. *14*, 1385 (1993).

M.J. McGregor, S.A. Islam, and M.J.E. Sternberg. J. Mol. Biol. *198*, 295 (1987).

M.J. McGregor, T.P. Flores, and M.J.E. Sternberg. Protein Eng. *2*, 521 (1989).

M.J. McGregor, T.P. Flores, and M.J.E. Sternberg. Protein Eng. *3*, 459 (1990).

R.F. McGuire, F.A. Mommany, and H.A. Scheraga. J. Phys. Chem. *76*, 375 (1971).

D.R. McKelvey, C.L. Brooks, and M. Mokotoff. J. Protein Chem. *10*, 265 (1991).

D.A. McQuarrie. *Statistical Mechanics*. Harper & Row, New York (1976).

V. Mehra, D. Sweetser, and R.A. Young. Proc. Natl. Acad. Sci. USA *83*, 7013 (1986).

E.C. Meng, B.K. Shoichet, and I.D. Kuntz. J. Comput. Chem. *13*, 505 (1992).

E.C. Meng, D.A. Gschwend, J.M. Blaney, and I.D. Kuntz. Proteins Struct. Funct. and Genet. *17*, 266 (1993).

D.H. Menzel (editor). *Fundamental Formulas of Physics*. Dover Publications, Inc., New York (1960).

K.M. Merz Jr. J. Comput. Chem. *13*, 749 (1992).

N. Metropolis, A.W. Rosenbluth, M.N. Rosenbluth, A. Teller, and E. Teller. J. Chem. Phys. *21*, 1087 (1953).

S.C. Meuer, S.F. Schlossman, and E. Reinherz. Proc. Natl. Acad. Sci. USA *79*, 4395 (1982).

M. Migliore, G. Corongiu, E. Clementi, and G.C. Lie. J. Chem. Phys. *88*, 7766 (1988).

S.L. Miller. Science *117*, 528 (1953).

W.P. Minicozzi and D.F. Bradley. J. Comput. Phys. *4*, 118 (1969).

A. Miranker and M. Karplus. Proteins *11*, 29 (1991).

P.A. Mirau. J. Magn. Reson. *96*, 480 (1992).

T.M. Mitchell. Artif. Intell. *18*, 203 (1982).

S. Miyamoto and P.A. Kollman. J. Comput. Chem. *13*, 952 (1992).

S. Miyazawa and R.L. Jernigan. Macromolecules *18*, 534 (1985).

F. Mohamadi, N.G.J. Richards, W.C. Guida, R. Liskamp, M. Lipton, C. Caufield, G. Chang, T. Hendrickson, and W.C. Still. J. Comput. Chem. *11*, 440 (1990).

F.A. Momany and R. Rose. J. Comput. Chem. *13*, 888 (1992).

F.A. Momany, R.F. McGuire, A.W. Burgess, and H.A. Scheraga. J. Phys. Chem. *79*, 2361 (1975).

A. Mor, M. Amiche, and P. Nicolas. TIBS *17*, 481 (1992).

L.B. Morales, R. Garduño-Juarez, and D. Romero. J. Biomol. Struct. and Dynamics *9*, 951 (1992).

D. Moras. Trends Biochem. Sci. *17*, 159 (1992).

S.D. Morley, D.E. Jackson, M.R. Saunders, and J.G. Vinter. J. Comput. Chem. *13*, 693 (1992).

A.L. Morris, M.W. MacArthur, E.G. Huchinson, and J.M. Thornton. Proteins *12*, 345 (1992).

J. Moult and M.N.G. James. Proteins Struct. Funct. and Genet. *1*, 146 (1986).

J. Moult and R. Unger. Biochem. *30*, 3816 (1991).

W.G. Moulton and R.A. Kromhout. J. Chem. Phys. *25*, 34 (1956).

S. Muggleton. New Gen. Computing *8*, 295 (1991).

S. Muggleton and C. Feng. In Proceedings of the First Conference on Algorithmic Learning Theory, edited by S. Arikawa, S. Goto, S. Ohsuga, and T. Yokomori. Japanese Society for Artificial Intelligence, Tokyo (1990).

S. Muggleton, R.D. King, and M.J.E. Sternberg. Protein Eng. *5*, 647 (1992).

H. Müller-Krumbhaar and K. Binder. J. Stat. Phys. *8*, 1 (1973).

N. Muller. TIBS *17*, 459 (1992).

R.S. Mulliken. J. Chem. Phys. *23*, 1833 (1955a).

R.S. Mulliken. J. Chem. Phys. *23*, 1841 (1955b).

R.S. Mulliken. J. Chem. Phys. *23*, 2338 (1955c).

R.S. Mulliken. J. Chem. Phys. *23*, 2343 (1995d).

M.E. Munk, B. Schoel, S. Modrow, R.W. Karr, R.A. Young, and S.H.E. Kaufmann. J. Immunol. *143*, 2844 (1989).

K.P. Murphy and S.J. Gill. J. Mol. Biol. *222*, 699 (1991).

A.G. Murzin and A.V. Finkelstein. J. Mol. Biol. *204*, 749 (1988).

M. Mutter and G. Tuchscherer. Macromol. Chem. Rapid Commun. *9*, 437 (1988).

M. Mutter and S. Vulleumies. Angew. Chem. *28*, 535 (1989).

M. Mutter, R. Gassmann, U. Buttkus, and K.-H. Altmann. Angew. Chem. Int. Ed. Engl. *30*, 1514 (1991).

S. Nakagawa, H.-A. Yu, M. Karplus, and H. Umeyama. Proteins Struct. Funct. and Genet. *16*, 172 (1993).

H. Nakamura, K. Katayanagi, K. Morikawa, and M. Ikehara. Nucl. Ac. Res. *19*, 1817 (1991).

T. Nakazawa, H. Kawai, Y. Okamoto, and M. Fukugita. Protein Eng. *5*, 495 (1992).

P. Nambi, A. Wierzbicki, and S.A. Allison. J. Phys. Chem. *95*, 9595 (1991).

M.A. Navia and D.A. Pettie. Immunol. Today *14*, 296 (1993).

A. Nayeem, J. Vila, and H.A. Scheraga. J. Comput. Chem. *12*, 594 (1991).

G. Nemethy and H.A. Scheraga. FASEB J. *4*, 3189 (1990).

G. Nemethy, M.S. Pottle, and H.A. Scheraga. J. Phys. Chem. *87*, 1883 (1983).

W. Nerdal, D.R. Hare, and B.R. Reid. Biochem. *28*, 10008 (1989).

R. Neubig, R. Gantzos, and R. Brazier. Mol. Pharmacol. *28*, 475 (1985).

J.T. Ngo and J. Marks. Protein Eng. *5*, 313 (1992).

T. Niermann and K. Kirschner. Protein Eng. *4*, 137 (1990).

M. Nilges and A.T. Brünger. Protein Eng. *4*, 649 (1991).

M. Nilges, G.M. Clore, and A.M. Gronenborn. FEB *229*, 317 (1988).

M. Nilges, J. Habazettl, A.T. Brunger, and T.A. Holak. J. Mol. Biol. *219*, 499 (1991).

O. Nilsson. J. Mol. Graphics *8*, 192 (1990).

O. Nilsson and O. Tapia. J. Mol. Struct. (Theochem) *256*, 295 (1992).

T. Noguti and N. Go. J. Phys. Soc. Japan *52*, 3283 (1983).

T. Noguti and N. Go. Biopolymers *24*, 527 (1985).

H.F. Noller. J. Bacteriology *175*, 5297 (1993).

A. Nordsieck. Math. Comput. *16*, 22 (1962).

S.H. Northrup and J.A. McCammon. Biopolymers *19*, 1001 (1990).

S.H. Northrup, S.A. Allison, and J.A. McCammon. J. Chem. Phys. *80*, 1517 (1984).

S.H. Northrup, J.O. Boles, and J.C.L. Reynolds. Science *241*, 67 (1988).

J. Novotny and K. Sharp. Progr. Biophys. Mol. Biol. *58*, 203 (1992).

H. Oberoi and N.M. Allewell. Biophys. J. *65*, 48 (1993).

S. Oikawa, M. Tsuda, H. Kato, and T. Urabe. Acta Cryst. *B41*, 437 (1985).

M. Oka, Y. Baba, A. Kagemoto, and A. Nakajima. Polymer J. *22*, 135 (1990*a*).

M. Oka, Y. Baba, A. Kagemoto, and A. Nakajima. Polymer J. *22*, 416 (1990*b*).

M. Oka, Y. Baba, A. Kagemoto, and A. Nakajima. Polymer J. *22*, 555 (1990*c*).

M. Oka, Y. Baba, A. Kagemoto, and A. Nakajima. Pept. Chem. 319 (1991).

M. Oka, R. Nishikawa, Y. Baba, and A. Kagemoto. Pept. Chem. 253 (1992).

H.B. Oldham and J.C. Myland. *Fundamentals of Electrochemical Science*. Academic Press, San Diego (1994).

T. Ooi, R.A. Scott, G. Vanderkooi, and H.A. Scheraga. J. Chem. Phys. *46*, 4410 (1967).

C. Orengo. In *Patterns in Protein Sequence and Structure*, edited by W.R. Taylor. Springer-Verlag, New York (1992).

C.A. Orengo. Current Opin. Struct. Biol. *4*, 429 (1994).

C.A. Orengo and J.M. Thornton. Structure *1*, 105 (1993).

C.A. Orengo, N.P. Brown, and W.R. Taylor. Proteins Struct. Funct. and Genet. *14*, 139 (1992).

C.A. Orengo, T.P. Flores, D.T. Jones, W.R. Taylor, and J.M. Thornton. Current Biol. *3*, 131 (1993).

C.A. Orengo, T.P. Flores, W.R. Taylor, and J.M. Thornton. Protein Eng. *6*, 485 (1993).

C.A. Orengo, D.T. Jones, and J.M. Thornton. Nature *372*, 631 (1994).

L. Orgel. Scientific Amer. *271*, 77 (1994).

J. Oro. Ann. N.Y. Acad. Sci. *108*, 464 (1963).

J. Oro. Biochem. Biophys. Res. Commun. *2*, 407 (1960).

D. Osguthorpe. The Biochemist *11*, 4 (1989).

G. Ourisson and Y. Nakatani. Chemistry and Biology *1*, 11 (1994).

R.A. Owens, P.D. Gesellchen, B.J. Houchins, and R.D. Dimarchi. Biochem. Biophys. Res. Commun. *181*, 402 (1991).

C.O. Pabo and E.G. Suchanek. Biochem. *25*, 5987 (1986).

256

K.A. Palmer and H.A. Scheraga. J. Comput. Chem. *12*, 505 (1991).

S. Park and K.W. Miller. Computing Practices *31*, 1192 (1988).

J.M.R. Parker. Unpublished results (1994).

J.M.R. Parker and R.S. Hodges. Peptide Res. *4*, 347 (1991a).

J.M.R. Parker and R.S. Hodges. Peptide Res. *4*, 355 (1991b).

J.M.R. Parker and R.S. Hodges. *HPLC hydrophobicity parameters: Prediction of surface and interior regions in proteins.* In *High Performance Liquid Chromatography of Peptides and Proteins: Separation, Analysis and Conformation*, edited by C.T. Mant and R.S. Hodges. CRC Press, Boca Raton (1991c).

J.M.R. Parker and R.S. Hodges. J. Comput.-Aided Mol. Design *8*, 193 (1994).

J.M.R. Parker, D. Guo, and R.S. Hodges. Biochem. *25*, 5425 (1986).

S.E. Parmley and G.P. Smith. Gene *73*, 305 (1988).

K. Parsaye, M. Chignell, S. Koshafian, and H. Wong. *Intelligent Databases. Object-oriented, Deductive Hypermedia Technologies.* Wiley & Sons, New York (1989).

S. Pascarella and P. Argos. Protein Eng. *5*, 121 (1992).

S. Pascarella, A. Colosimo, and F. Bossa. *Computational analysis of protein sequencing data* in *Laboratory Methodology in Biochemistry*, edited by C. Fini, A. Floridi, V.N. Finelly, and B. Wittman-Liebold. CRC Press, Boca Raton, Florida (1990).

N. Pattabiraman, M. Levitt, T.E. Ferrin, and R. Langridge. J. Comput. Chem. *6*, 432 (1985).

C.H. Paul. J. Mol. Biol. *155*, 53 (1982).

M.R. Pear, S.H. Northrup, J.A. McCammon, M. Karplus, and R.M. Levy. Biopolymers *20*, 626 (1981).

L.H. Pearl and A. Honegger. J. Mol. Graphics *1*, 9 (1983).

D.A. Pearlman. J. Biomol. *NMR 4*, 1 (1994).

D.A. Pearlman and P.A. Kollman. J. Mol. Biol. *220*, 457 (1991).

R.S. Pearlman. Chem. Design Auto. News *2*, 1 (1987).

R.S. Pearlman. Chem. Design Auto. News *8*, 3 (1993).

R.S. Pearlman. In *3D QSAR in Drug Design: Theory, Methods and Applications*, edited by H. Kubinyi. ESCOM Scientific Publishers, Leiden (1993).

F. Peradejordi. *Farmacologia molecular cuántica.* In *Química Teórica*, vol. II, edited by S. Fraga. Consejo Superior de Investigaciones Científicas, Madrid (1989).

A. Perczel, M. Hollosi, P. Sandor, and G.D. Fasman. Int. J. Peptide Protein Res. *41*, 223 (1993).

J.J. Perez, H.O. Villar, and G.A. Arteca. J. Phys. Chem. *98*, 2318 (1994).

K.S. Pitzer. Adv. Chem. Phys. *2*, 59 (1959).

D. Plochocka, P. Zielenkiewicz, and A. Rabczenko. Protein Eng. *2*, 115 (1988).

J.W. Ponder and F.M. Richards. J. Mol. Biol. *193*, 775 (1987).

S. Pongor, V. Skerl, M. Cserzo, Z. Hatsagi, G. Simon, and V. Bevilacqua. Nucl. Ac. Res. *21*, 3111 (1993).

S. Pongor, V. Skerl, M. Cserzo, Z. Hatsagi, G. Simon, and V. Bevilacqua. Protein Eng. *6*, 391 (1993).

H. Popkie, H. Kistenmacher, and E. Clementi. J. Chem. Phys. *59*, 1325 (1973).

M.J.D. Powell. Comput. J. *7*, 155 (1964).

M.J.D. Powell. *Minimization of Functions of Several Variables*, in *Numerical Analysis. An Introduction*, edited by J. Walsh. Academic Press, New York (1966).

W.H. Press, S.A. Teukolsky, W.T. Vetterling, and B.P. Flannery. *Numerical Recipes in Fortran. The Art of Scientific Computing*. Cambridge University Press, Cambridge (1986; second edition 1992).

L.G. Presta and G.D. Rose. Science *240*, 1632 (1988).

P.L. Privalov. Ann. Rev. Biophys. Biophys. Chem. *18*, 47 (1989).

P.L. Privalov and G.I. Makhatadze. J. Mol. Biol. *213*, 385 (1990).

P.L. Privalov and G.I. Makhatadze. J. Mol. Biol. *232*, 660 (1993).

O.B. Ptitsyn. Dokl. Acad. Nauk. *210*, 1212 (1973).

O.B. Ptitsyn. J. Protein Chem. *6*, 272 (1987).

O.B. Ptitsyn and A.V. Finkelstein. Biopolym. *22*, 15 (1983).

O.B. Ptitsyn and A.V. Finkelstein. Protein Eng. *2*, 443 (1989).

O.B. Ptitsyn and A.A. Rashin. Biophys. Chem. *3*, 1 (1975).

O.B. Ptitsyn and G.V. Semisotnov. In *Conformations and Forces in Protein Folding*, edited by B.T. Nall and K.A. Dill. American Association for the Advancement of Science, Washington (1991).

N. Qian and T.J. Sejnowski. J. Mol. Biol. *119*, 537 (1988).

Z. Qiang, K. Rosenfeld, S. Vajda, and C. DeLisi. J. Comput. Chem. *14*, 556 (1993).

A. Radzicka and R. Wolfenden. Biochemistry *27*, 1664 (1988).

G.N. Ramachandran. C. Ramakrishnan, and V.J. Sasisekharan. J. Mol. Biol. *7*, 95 (1963).

G.N. Ramachandran and V. Sasisekharan. Adv. Protein Chem. *23*, 283 (1968).

G. Ranghino and E. Clementi. Gazz. Chim. Ital. *108*, 157 (1978).

B.G. Rao and U.C. Singh. J. Am. Chem. Soc. *111*, 3125 (1989).

A.A. Rashin and B. Honig. J. Mol. Biol. *173*, 515 (1984).

A.E. Reed, R.B. Weinstock, and F. Weinhold. J. Chem. Phys. *83*, 735 (1985).

A.E. Reed, L.A. Curtiss, and F. Weinhold. Chem. Rev. *88*, 899 (1988).

F.M. Richards. Annu. Rev. Biophys. Bioeng. *6*, 151 (1977).

F.M. Richards and C.E. Kundrot. Proteins Struct. Funct. and Genet. *3*, 71 (1988).

J.S. Richardson. Adv. Protein Chem. *34*, 167 (1981).

R.J. Roberts and D. Macelis. Nucl. Ac. Res. *21*, 3125 (1993).

J. Robertus. Struct. Biol. *1*, 352 (1994).

B. Robson and D.J. Osguthorpe. J. Mol. Biol. *132*, 19 (1979).

B. Robson and E. Platt. J. Mol. Biol. *188*, 259 (1986).

N.K. Rogers, G.R. Moore, and M.J.E. Sternberg. J. Mol. Biol. *182*, 613 (1985).

S. Romano and E. Clementi. Gazz. Chim. Ital. *108*, 315 (1978).

S. Romano and E. Clementi. Int. J. Quantum Chem. *17*, 1007 (1980).

M.J. Rooman, J. Rodriguez, and S.J. Wodak. J. Mol. Biol. *213*, 327 (1990).

M.J. Rooman, J.P.A. Kochar, and S.J. Wodak. J. Mol. Biol. *221*, 961 (1991).

M.J. Rooman, J.P.A. Kocher, and S.J. Wodak. Biochemistry *31*, 10226 (1992).

C.C.J. Roothaan. Rev. Modern Phys. *23*, 69 (1951).

G.D. Rose and R. Wolfenden. Ann. Rev. Biophys. Biomol. Struct. *22*, 381 (1993).

G.D. Rose, A.R. Geselowitz, G.J. Lesser, R.H. Lee, and M.H. Zehfus. Science *229*, 834 (1985).

G.D. Rose, L.M. Gierasch, and J.A. Smith. Adv. Protein Chem. *37*, 1 (1985).

B. Rost and C. Sander. J. Mol. Biol. *232*, 584 (1993).

B. Rost and C. Sander. Protein Eng. *6*, 831 (1993).

B. Rost, R. Schneider, and C. Sander. TIBS *18*, 120 (1993).

B. Rost, C. Sander, and R. Schneider. J. Mol. Biol. *233*, 13 (1994).

I. Roterman, K.D. Gibson, and H.A. Scheraga. J. Biomol. Struct. Dynam. *7*, 391 (1989).

J.B. Rothbard. Ann. Inst. Past. *137E*, 518 (1986).

J.B. Rothbard, R.I. Lechler, K. Howland, V. Bal, D.D. Eckels, R. Sekaly, E.O. Long, W.R. Taylor, and J.R. Lamb. Cell *52*, 515 (1988).

M. Rubio, F. Torrens, and J. Sanchez-Marin. J. Comput. Chem. *14*, 647 (1993).

A. Ruhlmann, D. Kukla, P. Schwager, K. Bartels, and R. Huber. J. Mol. Biol. *77*, 417 (1973).

D.E. Rumelhart and J.E. McClelland (eds.). *Parallel Distributed Processing: Explorations in the Microstructure of Cognition. Vol. 1. Foundations.* MIT Press, Cambridge, MA (1986).

D.E. Rumelhart, G.E. Hinton, and R.J. Williams. Nature *323*, 533 (1986).

R. Rupp, H. Acha-Orbea, H. Hengartner, R. Zinkernagel, and R. Joho. Nature *315*, 425 (1985).

R.B. Russell and G.J. Barton. Proteins Struct. Funct. and Genet. *14*, 309 (1992).

J.-P. Ryckaert, G. Ciccotti, and H.J.C. Berendsen. J. Comput. Phys. *23*, 327 (1977).

M. Saito. Mol. Simulation *8*, 321 (1992).

M.E. Saks, J.R. Sampson, and J.N. Abelson. Science *263*, 191 (1994).

F.R. Salemme. J. Mol. Biol. *102*, 563 (1976).

F.R. Salemme. Progr. Biophys. Mol. Biol. *42*, 95 (1983).

A. Sali and T.L. Blundell. J. Mol. Biol. *212*, 403 (1990).

A. Sali and T.L. Blundell. J. Mol. Biol. *234*, 779 (1993).

A. Sali and J.P. Overington. Protein Sci. *3*, 1582 (1994).

A. Sali, J.P. Overington, M.S. Johnson, and T.L. Blundell. Trends Biochem. Sci. *15*, 235 (1990).

A. Sali, E. Shakhnovich, and M. Karplus. J. Mol. Biol. *235*, 1614 (1994).

C. Sander and R. Schneider. Proteins Struct. Funct. and Genet. *9*, 56 (1991).

C. Sander and R. Schneider. Nucl. Ac. Res. *21*, 3105 (1993).

M.A.S. Saqi, A. Bates, and M.J.E. Sternberg. Protein Eng. *5*, 305 (1992).

H.V. Saragoui, D. Fitzpatrick, A. Raktabotr, H. Nakaniski, M. Kahn, and M.I. Greene. Science *253*, 792 (1991).

T. Sasaki and E.T. Kaiser. J. Am. Chem. Soc. *111*, 380 (1989).

P.S. Schimmel and B. Henderson. Proc. Natl. Acad. Sci. USA *91*, 11283 (1994).

A. Schrake and J.A. Rupley. J. Mol. Biol. *79*, 351 (1973).

H. Schrauber, F. Eisenhaber, and P. Argos. J. Mol. Biol. *230*, 592 (1993).

E. Schrödinger. Ann. Physik *79*, 361 (1926a).

E. Schrödinger. Ann. Physik *79*, 480 (1926b).

E. Schrödinger. Ann. Physik *79*, 734 (1926c).

E. Schrödinger. Ann. Physik *80*, 437 (1926d)

E. Schrödinger. Ann. Physik *81*, 109 (1926e)

R. Schroeder and E.R. Lippincott. J. Phys. Chem. *61*, 921 (1957).

G.D. Schuler, S.F. Altschul, and D.J. Lipman. Proteins Struct. Funct. and Genet. *9*, 180 (1991).

G.E. Schulz. Ann. Rev. Biophys. Chem. *17*, 1 (1988).

G.E. Schulz and R.H. Schirmer. *Principles of Protein Structure*. Springer-Verlag, Berlin (1979).

S. Schulze-Kremer. In *Symbols versus Neurons? Advanced Applications in Artificial Intelligence*, edited by J. Stender and T. Addis. IOS Press, Amsterdam (1990).

S. Schulze-Kremer and R.D. King. Protein Eng. *5*, 377 (1992).

R.H. Schwartz. Ann. Rev. Immunol. *3*, 237 (1985).

R.H. Schwartz, B.S. Box, E. Fraga, C. Chen, and B. Singh. J. Immunol. *135*, 2598 (1985).

R. Scordamaglia, F. Cavallone, and E. Clementi. J. Am. Chem. Soc. *99*, 5545 (1977).

J.K. Scott and G.P. Smith. Science *249*, 386 (1990).

R.A. Scott and H.A. Scheraga. J. Chem. Phys. *45*, 2091 (1966).

B.R. Seavey, E.A. Farr, W.M. Westler, and J.L. Markely. J. Biomol. *NMR 1*, 217 (1991).

L. Seijo, M. Klobukowski, B.K. Mitchell, and S. Fraga. J. Biol. Phys. *14*, 107 (1986).

Y. Seto, Y. Ikeuchi, and M. Kanehisa. Proteins Struct. Funct. and Genet. *8*, 341 (1990).

A. Sette, S. Buus, S. Colon, C. Miles, and H.M. Grey. J. Immunol. *141*, 45 (1988).

U. Sezerman, S. Vajda, J. Cornette, and C. DeLisi. Protein Sci. *2*, 1827 (1993).

E.I. Shakhnovich and E.M. Gutin. J. Chem. Phys. *93*, 5967 (1990).

E.I. Shakhnovich and E.M. Gutin. J. Theoret. Biol. *149*, 537 (1991).

P.S. Shenking and D.Q. McDonald. J. Comput. Chem. *15*, 899 (1994).

R.P. Sheridan, R. Nilakantan, A.I. Rusinko, N. Bauman, K.S. Haraki, and R. Venkataraghavan. J. Chem. Inf. Comput. Sci. *29*, 255 (1989).

R.P. Sheridan, A. Rushinko III, R. Nilakantan, and R. Venkataraghavan. Proc. Natl. Acad. Sci. USA *86*, 8165 (1989).

R. Shimonkevitz, S. Colon, J.W. Kappler, P. Marrack, and H.M. Grey. J. Immunol. *133*, 2067 (1984).

B.K. Shoichet and I.D. Kuntz. J. Mol. Biol. *221*, 327 (1991).

A. Sikorski and J. Skolnick. J. Mol. Biol. *215*, 183 (1990).

A.M. Silva, R.E. Cachau, and D.J. Goldstein. Biophys. J. *52*, 595 (1987).

K. Singer, J.V.L. Singer, and A.J. Taylor. Mol. Phys. *37*, 1239 (1979).

J. Singh, J. Saldanha, and J.M. Thornton. Protein Eng. *4*, 251 (1991).

B. Singh, K.C. Lee, and E. Fraga. Pept. Res. *2*, 120 (1989).

M.J. Sippl. J. Mol. Biol. *213*, 859 (1990).

M.J. Sippl. J. Computer-Aided Mol. Design *7*, 473 (1993).

J. Skolnick and A. Kolinski. Ann. Rev. Phys. Chem. *40*, 207 (1989).

J. Skolnick and A. Kolinski. J. Mol. Biol. *221*, 499 (1991).

J. Skolnick, A. Kolinski, and R. Yaris. Proc. Natl. Acad. Sci. USA *85*, 5057 (1988).

J.C. Slater and J.G. Kirkwood. Phys. Rev. *37*, 682 (1931).

P.E. Smith, R.M. Brunne, A.E. Mark, and W.F. van Gunsteren. J. Phys. Chem. *97*, 2009 (1993).

H.O. Smith, T.M. Annau, and S. Chandrasegaran. Proc. Natl. Acad. Sci. USA *87*, 826 (1990).

G.P. Smith. Science *228*, 1315 (1985).

V.H. Smith. Phys. Scr. *15*, 147 (1977).

S.F. Sneddon and C.L. Brooks III. In *Molecular Structures in Biology*, edited by K. Prout. Oxford University Press, Oxford (1991).

M.E. Snow. J. Comput. Chem. *13*, 579 (1992).

M.E. Snow. Proteins Struct. Funct. and Genet. *15*, 183 (1993).

M.E. Snow and L.M. Amzel. Proteins *1*, 267 (1986).

J. Sodek, R.S. Hodges, L.B. Smillie, and L. Jurasek. Proc. Natl. Acad. Sci. USA *69*, 3800 (1972).

T.L. Sordo, J.A. Sordo, and R. Florez. J. Comput. Chem. *11*, 291 (1990).

J.A. Sordo, M. Probst, G. Corongiu, S. Chin, and E. Clementi. J. Am. Chem. Soc. *105*, 1702 (1987).

J.A. Sordo, M. Probst, S. Chin, G. Corongiu, and E. Clementi. *Non-empirical pair potentials for the interaction between amino acids*. In *Structure and Dynamics of Nucleic Acids, Proteins, and Membranes*, edited by E. Clementi and S. Chin. Plenum, New York (1986).

T.R. Sosnick, L. Mayne, R. Hiller, and S.W. Englander. Struct. Biol. *1*, 149 (1994).

R. Sowdhamini, N. Srinivasan, B. Shoichet, D.V. Santi, C. Ramakrishnan, and P. Balaram. Protein Eng. *3*, 95 (1989).

P.J. Steinbach and B.R. Brooks. J. Comput. Chem. *15*, 667 (1994).

T.A. Steitz. Current Opin. Struct. Biol. *1*, 139 (1991).

M.J.E. Sternberg and J.S. Chickos. Protein Eng. *7*, 149 (1994).

M.J.E. Sternberg and S.A. Islam. Protein Eng. *4*, 125 (1990).

G.W. Stewart III. J. Assoc. Comput. Math. *14*, 72 (1967).

J.J.P. Stewart. J. Comput. Chem. *10*, 209 (1989).

D. Stickle and R. Barber. Mol. Pharmacol. *36*, 437 (1989).

A. Stoltzfus, D. Spencer, M. Zuker, J.M. Logsdon Jr, and R.F. Doolittle. Science *265*, 202 (1994).

T.P. Straatsma and J.A. McCammon. Annu. Rev. Phys. Chem. *43*, 407 (1992).

J.E. Straub and D. Thirumalai. Proc. Natl. Acad. Sci. USA *90*, 809 (1993).

W.B. Streett, D.J. Tildesley, and G. Saville. Mol. Phys. *35*, 639 (1978).

V.P. Sukhatme, K.C. Sizer, A.C. Vollmer, T. Hunkapiller, and J.R. Parness. Cell, *40*, 591 (1985).

M.C. Surles, J.S. Richardson, D.C. Richardson, and F.B. Brooks Jr. Protein Science *3*, 198 (1994).

R. Susnow, N. Senko, and T. Ocain. J. Comput. Chem. *15*, 1074 (1994).

S.L. Swain. Immunol. Rev. *74*, 129 (1983).

A.G. Sykes. Chem. Soc. Rev. *14*, 283 (1985).

A.G. Sykes. Adv. Inorg. Chem. *36*, 377 (1991).

J.A. Tainer, E.D. Getzoff, H. Alexander, R.A. Houghten, A.J. Olson, and R.A. Lerner. Nature *312*, 127 (1984).

J.P. Tam. Proc. Natl. Acad. Sci. USA *85*, 5409 (1988).

S. Tanaka and H.A. Scheraga. Proc. Nat. Acad. Sci. USA *72*, 3802 (1975).

S. Tanaka and H.A. Scheraga. Macromolecules *9*, 945 (1976).

O. Tapia, R. Cardenas, J. Andres, J. Krechl, M. Campillo, and F. Colonna-Cesari. Int. J. Quantum Chem. *39*, 767 (1991).

C. Taylor. Biochem. J. *272*, 1 (1990).

W.R. Taylor. J. Theor. Biol. *119*, 205 (1986).

W.R. Taylor. *Protein structure prediction.* In *Nucleic Acid and Protein Sequence Analysis. A Practical approach,* edited by M.J. Bishop and C.J. Rawlings. IRL Press, Oxford (1987).

W.R. Taylor. Protein Eng. *2*, 77 (1988).

W.R. Taylor. Protein Eng. *6*, 593 (1993).

W.R. Taylor and D.T. Jones. Current Opin. Struct. Biol. *1*, 327 (1991).

W.R. Taylor and C.A. Orengo. J. Mol. Biol. *208*, 1 (1989).

W.R. Taylor, J.M. Thornton, and W.G. Turnell. J. Mol. Graphics *1*, 30 (1983).

M.W. Teeter. Proc. Natl. Acad. Sci. USA *81*, 6014 (1984).

M.M. Teeter. Annu. Rev. Biophys. Biophys. Chem. *20*, 577 (1991).

N. Thanki, J.M. Thornton, and J.M. Goodfellow. J. Mol. Biol. *202*, 637 (1988).

V. Thery, D. Rinaldi, J.-L. Rivail, B. Maigret, and G.G. Ferenczy. J. Comput. Chem. *15*, 269 (1994).

J.M. Thornton. Nature *335*, 10 (1988).

J.M. Thornton and B.L. Sibanda. J. Mol. Biol. *167*, 443 (1983).

J.M. Thornton, M.S. Edwards, W.R. Taylor, and D.J. Barlow. EMBO J. *5*, 409 (1986).

S. Thornton and S. Fraga. Can. J. Chem. *69*, 1636 (1991).

S. Thornton, E. San Fabian, S. Fraga, J.M.R. Parker, and R.S. Hodges. J. Mol. Struct. (Theochem) *226*, 87 (1991).

S. Thornton, E. San Fabian, S. Fraga, J.M.R. Parker, and R.S. Hodges. J. Mol. Struct. (Theochem) *232*, 321 (1991).

N. Tomioka, A. Itai, and Y. Iitaka. J. Comp.-Aided Mol. Design *1*, 197 (1987).

C. Toniolo, M. Crisma, F. Formaggio, and G. Caruicchioni. Biopolymers *33*, 1061 (1993).

A.E. Torda and W.F. van Gunsteren. J. Comput.Chem. *15*, 1331 (1994).

A.E. Torda, R.M. Scheek, and W.F. van Gunsteren. Chem. Phys. Lett. *157*, 289 (1989).

A.E. Torda, R.M. Scheek, and W.F. van Gunsteren. J. Mol. Biol. *214*, 223 (1990).

F. Torrens, A.M. Sanchez de Meras, and J. Sanchez Marin. J. Mol. Struct. (Theochem) *166*, 135 (1988).

F. Torrens, E. Orti, and J. Sanchez-Marin. J. Mol. Graph. *9*, 254 (1991).

M. Totrov and R. Abagyan. J. Comput. Chem. *15*, 1105 (1994).

A.R.M. Townsend, J. Rothbard, F.M. Gotch, C. Bahadur, D. Wraith, and A.J. McMichael. Cell *44*, 959 (1986).

G. Tuchscherer and M. Mutter. J. Pept. Sci. *1*, 3 (1995).

G. Tuchscherer, C. Servis, G. Corradin, U. Blum, J. Rivier, and M. Mutter. Protein Sci. *1*, 1377 (1992).

G. Tuchscherer, B. Dörner, U. Sila, B. Kamber, and M. Mutter. Tetrahedron *49*, 3559 (1993).

E.C. Tyler, M.R. Horton, and P.R. Karuse. Comp. Biomed. Res. *24*, 72 (1991).

R. Unger and J.L. Sussman. J. Comp.-Aided Mol. Design *7*, 457 (1993).

R. Unger, D. Harel, S. Wherland, and J.L. Susman. Proteins *5*, 355 (1989).

R. Unger, D. Harel, S. Wherland, and J.L. Sussman. Biopolymers *30*, 499 (1990).

A. Valencia Herrera, L. Menendez Arias, and L. Serrano Pubul. Anales de Quimica *84*, 223 (1988).

J.H. Van Drie, D. Weininger, and Y.C. Martin. J. Comp.-Aided Mol. Design *3*, 225 (1989).

W. van Eden, J.E.R. Thole, R. van der Zee, A. Noordzij, J.D.A. van Embden, E.J. Hensen, and I.R. Cohen. Nature *331*, 171 (1988).

C.W.G. van Gelder, F.J.J. Leusen, J.A.M. Leunissen, and J.H. Noordik. Proteins Struct. Funct. and Genet. *18*, 174 (1994).

W.F. van Gunsteren. Mol. Phys. *40*, 1015 (1980).

W.F. van Gunsteren and H.J.C. Berendsen. Mol. Phys. *34*, 1311 (1977).

W.F. van Gunsteren and H.J.C. Berendsen. J. Mol. Biol. *176*, 559 (1984).

W.F. van Gunsteren and H.J.C. Berendsen. Molec. Simul. *1*, 173 (1988).

W.F. van Gunsteren and M. Karplus. Macromolecules *15*, 1528 (1982).

W.F. van Gunsteren and A.E. Mark. Eur. J. Biochem. *204*, 947 (1992).

W.F. van Gunsteren, H.J.C. Berendsen, J. Hermans, W.G.J. Hol, and J.P.M. Postma. Proc. Natl. Acad. Sci. USA *80*, 4315 (1983).

W.F. van Gunsteren, P. Gross, A.E. Torda, H.J.C. Berendsen, and R.C. Schack. *On deriving spatial protein structure from NMR or X-ray diffraction data* in *Protein Conformation*. Ciba Foundation Symposium 161. Wiley Interscience, Chichester, England (1991).

M.H.V. van Regenmortel. Immunol. Today *10*, 266 (1989).

R.C. van Schaik, W.F. van Gunsteren, and H.J.C. Berendsen. J. Comput. Aided Mol. Design *6*, 97 (1992).

G. Vasmatsis. New approaches to constrained minimization in protein structure determination. Ph.D. Thesis. Boston University, Boston (1992).

M. Vasquez and H.A. Scheraga. Biopolymers *24*, 1437 (1985).

M. Vasquez, G. Nemethy, and H.A. Scheraga. Macromolecules *16*, 1043 (1983).

M. Vasquez, G. Nemethy, and H.A. Scheraga. Chem. Rev. *94*, 2183 (1994).

M.C. Vega, C. Aleman, E. Giralt, and J.J. Perez. J. Biomol. Struct. and Dynamics *10*, 1 (1992).

M. Vihinen, E. Torkkila, and P. Riikonen. Proteins Struct. Funct. and Genet. *19*, 141 (1994).

J. Vila, R.L. Williams, M. Vasquez, and H.A. Scheraga. Proteins Struct. Funct. and Genet. *10*, 199 (1991).

J.G. Vinter, A. Davis, and M.R. Saunders. J. Comput.-Aided Mol. Design *1*, 31 (1987).

M.V. Volkenstein. *Configurational Statistics of Polymer Chains*. Interscience Publishers, New York (1963).

G.F. Voronoi. J. Reine Ang. Math. *134*, 198 (1908).

G. Vriend and C. Sander. Proteins Struct. Funct. and Genet. *11*, 52 (1991).

G. Vriend, C. Sander, and P.F.W. Stouten. Protein Eng. *7*, 23 (1994).

H. Wako. J. Protein Chem. *8*, 733 (1989).

A. Wallqvist and M. Ullner. Proteins Struct. Funct. and Genet. *18*, 267 (1994).

P.H. Walls and M.J.E. Sternberg. J. Mol. Biol. *228*, 277 (1992).

M.C. Wang and G.E. Uhlenbeck. Rev. Mod. Phys. *17*, 323 (1945).

D.J. Ward, A.M. Brass, J. Li, E. Platt, Y. Chen, and B. Robson. *Theoretical approaches to peptide drug design*. In *Peptide Pharmaceuticals. Approaches to the Design of Novel Drugs*, edited by D.J. Ward. Elsevier Science Publishers, Amsterdam (1991).

P.K. Warme, F.A. Momany, S.V. Rumball, R.W. Tuttle, and H.A. Scheraga. Biochemistry *13*, 768 (1974).

J. Warwicker. J. Theoret. Biol. *121*, 199 (1986).

J. Warwicker. J. Mol. Biol. *206*, 381 (1989).

J. Warwicker and H.C. Watson. J. Mol. Biol. *157*, 671 (1982).

D.V. Waterhous and W.C. Johnson Jr. Biochem. *33*, 2121 (1994).

T.H. Watts, J. Gariepy, G.K. Schoolnik, and H.M. McConnell. Proc. Natl. Acad. Sci. USA *82*, 5480 (1985).

D.F. Weaver. Seizure *1*, 223 (1992).

P.K. Weiner and P.A. Kollman. J. Comput. Chem. *2*, 287 (1981).

F. Weinhold and J.E. Carpenter. *The Structure of Small Molecules and Ions*. Plenum, New York (1988).

D. Weininger and J.L. Weininger. *Chemical Structures and Computers*, in *Comprehensive Medicinal Chemistry*. Vol. 4. *Quantitative Drug Design*, edited by C. Hansch, P.G. Sammes, J.B. Taylor, and C.A. Ramdsen. Pergamon, Elmsford, NY (1990).

H. Weinstein, M.N. Liebman, and C.A. Venanzi. *Theoretical Principles of Drug Action: The Use of Enzymes to Model Receptor Recognition and Activity,* in *New Methods in Drug Research*, edited by A. Makriyannis. J.R. Prous (1985). Vol. 1.

J.S. Weissman and P.S. Kim. Science *253*, 1386 (1991).

L. Wesson and D. Eisenberg. Protein Sci. *1*, 227 (1992).

E. Westhof, D. Altschuh, D. Moras, A.C. Bloomer, A. Mondragon, A. Klug, and M.H.V. van Regenmortel. Nature *311*, 123 (1984).

D.B. Wetlaufer and E. Ristow. Annu. Rev. Biochem. *42*, 139 (1973).

K.B. Wiberg. J. Comput. Chem. *2*, 53 (1981).

D.R. Wilde, P. Marrack, J. Kappler, D.P. Dyalinas, and F.W. Fitch. J. Immunol. *131*, 2178 (1983).

R.L. Williams, J. Vila, G. Perrot, and H.A. Scheraga. Proteins Struct. Funct. and Genet. *14*, 110 (1992).

M.A. Williams, J.M. Goodfellow, and J.M. Thornton. Protein Sci. *3*, 1224 (1994).

C.M. Wilmot and J.M. Thornton. J. Mol. Biol. *203*, 221 (1988).

C.M. Wilmot and J.M. Thornton. Protein Eng. *3*, 479 (1990).

C. Wilson and S. Doniach. Proteins Struct. Funct. and Genet. *6*, 193 (1989).

E.B. Wilson, J.C. Decius, and P.C. Cross. *Molecular Vibrations*. McGraw-Hill, New York (1955).

T. Wilson and A. Klausner. Biotechnol. *2*, 511 (1984).

D.S. Wishart, B.D. Sykes, and F.M. Richards. J. Mol. Biol. *222*, 311 (1991).

D.S. Wishart, B.D. Sykes, and F.M. Richards. Biochem. *31*, 1647 (1992).

D.S. Wishart, R.F. Boyko, L. Willard, F.M. Richards, and B.D. Sykes. CABIOS *10*, 121 (1994).

A. Wlodawer. Pharmacotherapy *14*, 104 (1994).

S.J. Wodak and J. Janin. J. Mol. Biol. *124*, 323 (1978).

S. Wolfe, D.F. Weaver, and K. Yang. Can. J. Chem. *66*, 2687 (1988).

S. Wolfe, S. Bruder, D.F. Weaver, and K. Yang. Can. J. Phys. *66*, 2703 (1988).

R. Wolfenden. Biochemistry *17*, 201 (1978).

R. Wolfenden, P.M. Cullis, and C.C.F. Southgate. Science *206*, 575 (1979).

R. Wolfenden, L. Anderson, P.M. Cullis, and C.C.F. Southgate. Biochem. *20*, 849 (1981).

J.T. Wong. Proc. Nat. Acad. Sci. USA *72*, 1909 (1975).

C.N. Woolfson, A. Cooper, M.M. Harding, D.H. Williams, and P.A. Evans. J. Mol. Biol. *229*, 502 (1993).

C.S. Wright, A.R. Alden, and J. Kraut. Nature *221*, 253 (1969).

T.T. Wu and E.A. Kabat. J. Mol. Biol. *75*, 13 (1973).

K. Wüthrich. *Six years of protein structure determination by NMR spectroscopy: what have we learned?* in *Protein Conformation*. Ciba Foundation Symposium 161. Wiley Interscience, Chichester, England (1991).

R.Y. Yada, R.L. Jackman, and S. Nakai. Int. J. Peptide Protein Res. *31*, 98 (1988).

H. Yamamoto. Private communication (1995).

T. Yamato, J. Higo, Y. Seno, and N. Go. Proteins Struct. Funct. and Genet. *16*, 327 (1993).

J. Yao, R.A. Greenkorn, and K.C. Chao. Mol. Phys. *46*, 587 (1982).

M. Ycas. J. Protein Chem. *9*, 177 (1990).

P. Yip and D.A. Case. J. Magn. Reson. *83*, 643 (1989).

M.D. Yoder, N.T. Keen, and F. Jurnak. Science *260*, 1503 (1993a).

M.D. Yoder, S.E. Lietze, and F. Jurnak. Structure *1*, 241 (1993b).

B.J. Yoon and A.M. Lenhoff. J. Comput. Chem. *11*, 1080 (1990).

B.J. Yoon and A.M. Lenhoff. J. Phys. Chem. *96*, 3130 (1992).

S. Yoshioki. J. Comput. Chem. *15*, 684 (1994).

D.B. Young, J. Ivanyi, J.H. Cox, and J.R. Lamb. Immunol. Today *8*, 215 (1987).

K. Yue and K.A. Dill. Proc. Natl. Acad. Sci. USA *92*, 146 (1995).

K. Yue, K.M. Fiebig, P.D. Thomas, H.S. Chan, and E.I. Shakhnovich. Proc. Natl. Acad. Sci. USA *92*, 325 (1995).

S. Yue. Protein Eng. *4*, 177 (1990).

W.I. Zangwill. Comput. J. *10*, 293 (1967).

R.J. Zauhar and R.S. Morgan. J. Mol. Biol. *186*, 815 (1985).

R.J. Zauhar and R.S. Morgan. J. Comput. Chem. *9*, 171 (1988).

R.J. Zauhar and R.S. Morgan. J. Comput. Chem. *11*, 603 (1990).

L. Zhang and J. Hermans. Proteins Struct. Funct. and Genet. *16*, 384 (1993).

H.-X. Zhou. Biophys. J. *64*, 1711 (1993a).

H.-X. Zhou. Biophys. J. *65*, 955 (1993b).

N.E. Zhou, C.M. Kay, and R.S. Hodges. J. Mol. Biol. *237*, 500 (1994).

B.-Y. Zhu, N.E. Zhou, P.D. Semchuk, C.M. Kay, and R.S. Hodges. Int. J. Peptide Protein Res. *40*, 171 (1992).

Z.-Y. Zhu, A. Sali, and T.L. Blundell. Protein Eng. *5*, 43 (1992).

B.H. Zimm and J.K. Bragg. J. Chem. Phys. *31*, 526 (1958).

K. Zimmerman. J. Comput. Chem. *12*, 310 (1991).

S.S. Zimmerman, M.S. Pottle, G. Nemethy, and H.A. Scheraga. Macromolecules *10*, 1 (1977).

S.S. Zimmerman. *Theoretical methods in the analysis of peptide conformation.* In *The Peptides*, edited by V.J. Hruby. Academic Press, New York (1985).

R.N. Zuckermann. Current Opin. Struct. Biol. *3*, 580 (1993).

J.E. Zull and S.K. Smith. TIBS *15*, 257 (1990).

R. Zwanzig, A. Szaba, and B. Bagchi. Proc. Natl. Acad. Sci. USA *89*, 20 (1992).

R.W. Zwanzig. J. Chem. Phys. *22*, 1420 (1954).

R.W. Zwanzig. Ann. Rev. Phys. Chem. *16*, 67 (1965).

Reference Texts

The following texts may be found to be useful in connection with diverse subjects. They are listed in chronological order, with brief comments on possible information of interest.

P.J. Flory. *Statistical Mechanics of Chain Molecules.* Interscience Publishers, New York (1969)
A detailed mathematical treatment of Statistical Mechanics

S. Fraga, K.M.S. Saxena, and M. Torres. *Biomolecular Information Theory.* Elsevier Science Publishers, Amsterdam (1978)
The origins of amino acids and peptides, conformational energy maps of amino acids, minimization techniques

G.E. Schulz and R.H. Schirmer. *Principles of Protein Structure.* Springer-Verlag, Berlin (1979)
Early review of non-covalent forces in proteins, patterns of folding, secondary structure prediction, thermodynamics, protein evolution, protein function

C. Ghelis and J. Yon. *Protein Folding.* Academic Press, New York (1982)
Early review of theoretical and experimental approaches to protein folding

B. Robson and J. Garnier. *Introduction to Proteins and Protein Engineering.* Elsevier Science Publishers, Amsterdam (1986)
Case study of lysozyme and metalloproteins, calculations of secondary and tertiary structures, design of novel peptides and proteins, *GOR* prediction method in BASIC. An extensive list of conformation energy calculation references

M.J. Bishop and C.J. Rawlings (editors). *Nucleic Acid and Protein Sequence Analysis: A Practical Approach*. IRL Press, Oxford (1987)
Sequence alignment procedures

H. Gould and J. Tobochnik. *An Introduction to Computer Simulation Methods: Applications to Physical Systems*. Addison-Wesley Publishing Co., Reading (1988)
Part 1: Simulations with deterministic processes. Part 2: Simulations with random processes

G.D. Fasman (editor). *Prediction of Protein Structure and the Principles of Protein Conformation*. Plenum Press, New York (1989).
Principles and patterns of protein conformation, theoretical studies of protein stability, stabilization energies of protein conformations, prediction of secondary structure (*CF, GOR*), prediction of tertiary structure, prediction of packing, hydrophobicity profiles, structure prediction of membrane proteins

H. Neurath (editor). *Perspectives in Biochemistry*. Vol. 2 American Chemical Society, Washington (1989)
Dominant forces in protein folding

J.B. Hook and G. Poste (editors). *Protein Design and the Development of New Therapeutics and Vaccines*. Plenum Press, New York (1990)
Dynamic processes in proteins by X-ray diffraction, mass-spectrometric determination of peptide/protein structure, knowledge-based modeling and design, computer simulations of site-specific mutations, chemical approaches to protein engineering, targeted drug delivery, directed mutagenesis of proteins

K.B. Lipkowitz and D.B. Boyd (editors). *Reviews in Computational Chemistry*. VCH Publishers, New York (1990; published annually)
Volume II (1991): Survey of conformational search methods, parameterization in molecular mechanics, semiempirical molecular orbital methods, computational chemistry literature, software for molecular modeling
Volume III (1992): optimization methods, prediction of structure of oligopeptides, modeling with *NMR*-defined constraints

M.P. Allen and D.J. Tildesley. *Computer Simulations of Liquids*. Oxford University Press, New York (1991)
Good review of Statistical Mechanics, molecular dynamics, Monte Carlo methods, non-equilibrium molecular dynamics, Bownian dynamics; several applications

C. Branden and J. Tooze. *Introduction to Protein Structure*. Garland Publishing Inc., New York (1991)
Excellent description of protein motifs, alpha, alpha/beta, and parallel beta and *DNA* structures, enzymes, viruses, membrane proteins, prediction and design of proteins

D.J. Ward (editor). *Peptide Pharmaceuticals*: *Approaches to the Design of Novel Drugs*. Elsevier Science Publishers, Amsterdam (1991)

Chemical and theoretical approaches to peptide drug design

T.E. Creighton (editor). *Protein Folding*. W.H. Freeman and Company, New York (1992)
Folded and unfolded proteins, physical basis for protein stability, theoretical studies of thermodynamics and dynamics, kinetics of unfolding and refolding, and the molten globule

M. Gribskov and J. Devereux (editors). *Sequence Analysis Primer*. W.H. Freeman and Company, New York (1992)
Identification of simple sites and transcription signals and coding region identification in *DNA* sequences; detailed description of protein similarity and homology algorithms, detailed example of *DNA* and protein sequence analysis, partial list of software suppliers, list of *FTP* sites and bulletin boards available at the time

A.R. Rees, M.J. Sternberg, and R. Wetzel (editors). *Protein Engineering: A Practical Approach*. IRL Press, Oxford (1992)
Protein crystallography, *NMR*, sequence databases, design of protein structures, protein stability, modification and expression of proteins in prokaryotic, yeast, and mammalian systems, phage display

W.R. Taylor (editor). *Patterns in Protein Sequence and Structure*. Springer-Verlag, Berlin (1992)
Several reviews on pattern matching, sequence alignment, and structure comparison

G. Weber. *Protein Interactions*. Chapman and Hall, New York (1992)
Protein-ligand and protein-protein interactions

T.E. Creighton. *Proteins: Structures and Molecular Properties*. 2nd edition. W.H. Freeman and Company, New York (1993)
Chemical properties, biosynthesis and evolutionary origin of proteins, physical interactions and conformation properties of proteins, interaction with other molecules

J. Zupan and J. Gasteiger. *Neural Networks for Chemists*. VCH Publishers, New York (1993)
Basic concepts of neural networks, detailed description of one-layer and multilayer networks, examples of *QSAR* and secondary structure predictions

Chemical Reviews *93* (7) (1993)
Molecular mechanics and modeling review, electrostatic interactions, free energy calculations, transition state modeling, simulation of peptide conformational dynamics and thermodynamics, simulation of *H*-bond dynamics

Appendix 1. Constants and Units

The constants and units of interest for this work are:

A1.1 Constants

Avogadro number:	$N_A = 6.022045 \cdot 10^{23}$ mol^{-1}
Planck constant:	$h/2\pi = 1.0545887 \cdot 10^{-27}$ erg s
Boltzmann constant:	$k_B = 1.380662 \cdot 10^{-16}$ erg K^{-1}

[Taken from E.R. Cohen and B.N. Taylor, *Fundamental Constants. 1986 Adjustments*, Europhysics News *18*, 65 (1987). See also E.R. Cohen and B.N. Taylor, Chem. Ref. Data *17*, 1795 (1988).]

A1.2 Units

In the system of atomic units, the electron mass (at rest), the charge of the electron, and Planck constant ($h/2\pi$) are assigned the value 1. In this system, distances are expressed in *bohrs* and energies in *Hartrees*, with the following conversion factors:

Distance:	1 bohr = 052917706$\cdot 10^{-8}$ cm
	1 Å = 10^{-8} cm
Energy:	1 hartree = $4.359814445 \cdot 10^{-11}$ erg
	= 27.211608 eV
	= 219474.6354 cm^{-1}
	1 erg = 10^{-7} J
	1 kcal = 4.184 kJ

The *SI* prefixes and symbols for submultiples are:

deci (*d*) 10^{-1}	centi (*c*) 10^{-2}	milli (*m*) 10^{-3}
micro (*μ*) 10^{-6}	nano (*n*) 10^{-9}	pico (*p*) 10^{-12}
femto (*f*) 10^{-15}	atto (*a*) 10^{-18}	

Appendix 2. Amino Acid Data

Each amino acid is identified by its full name and its 3- and 1-letter symbols. The information collected for each amino acid is as follows: atom symbol, atomic class (see Table 3.1), effective charge, and x-, y-, and z-coordinates. Two sets of data are given for arginine, aspartic acid, cysteine, glutamic acid, and lysine: neutral and protonated (on the side chain) forms for arginine and lysine, neutral and ionized (on the side chain) forms for aspartic and glutamic acids, and neutral and dehydrogenated forms of cysteine.

The effective charges have been obtained, by renormalization (in order to reproduce the total charge of the residue) of the average effective charges in Table 3.1.

The coordinates were taken from the work of Clementi and co-workers (1977) and transformed so that the backbone amino-group and the C-alpha and C' atoms lie on the xy-plane, with one of the amino hydrogens at the centre of coordinates and the N atom on the x-axis. Regarding the amino group, the transformation was needed in order to change the hybridization of the N atom from sp^3 to sp^2. The other changes were introduced for simplicity in the handling of the residues during the formation of the peptide bond.

The atoms are given in the same order for all the residues: one amino Hydrogen, amino Nitrogen, C-alpha, C', Hydrogen attached to the C-alpha carbon, atoms of the side chain, atoms of the carboxylic group, and the other amino Hydrogen.

The original coordinates are from the following works:

Ala: M.S. Lehmann, T.F. Koetzle, and W.C. Hamilton. J. Am. Chem. Soc. 94, 2657 (1972)

Arg: M.S. Lehmann, J.J. Verbist, W.C. Hamilton, and T.F. Koetzle. J. Chem.Soc., Perkin Trans. 2, 133 (1973)

Asn: M. Ramanadham, S.K. Kikka, and R. Chidambaram. Acta Crystallogr., Sect. B 28,3000 (1972)

Asp: S.T. Rao. Acta Crystallogr., Sect. B 29, 1718 (1973)

Cys: D.D. Jones, L. Bernal, M.N. Frey, and T.F. Koetzle. Acta Crystallogr., Sect. B 30, 1220 (1974)

Glu: A. Sequeira, H. Rajagopal, and R. Chidambaram. Acta Crystallogr., Sect. B 28, 2514 (1972)

Gln: T.F. Koetzle, M.N. Frey, M.S. Lehmann, and W.C. Hamilton. Acta Crystallogr., Sect. B 29, 2571 (1973)

Gly: P.-G. Jonsson and A. Kvick. Acta Crystallogr., Sect. B, 28, 1827 (1972)

His: P. Eddington and M.H. Harding. Acta Crystallogr., Sect. B 30, 204 (1974)

Ile: K. Torii and Y. Iitaka. Acta Crystallogr., Sect. B 27, 2237 (1971)

Leu: L. Golic and W.C. Hamilton. Acta Crystallogr., Sect. B 28, 1265 (1972)

Lys: T.F. Koetzle, M.S. Lehmann, J.J. Verbist, and W.C. Hamilton. Acta Crystallogr., Sect. B 28, 3207 (1972)

Met: A.M. Mathieson. Acta Crystallogr. 5, 332 (1952)

Phe: G.V. Gurskaya. Sov. Phys.-Crystallogr. 9, 6 (1965)

Pro: Y. Mitsui, M. Tsuboi, and Y. Iitaka. Acta Crystallogr., Sect. B 25, 2182 (1969)

Ser: M.N. Frey, M.S. Lehmann, T.F. Koetzle, and W.C. Hamilton. Acta Crystallogr., Sect. B 29, 876 (1973)

Thr: D.P. Shoemaker, J. Donohue, V. Shoemaker, and R.B. Coray. J. Am. Chem. Soc. 72, 2328 (1950)

Trp: R.A. Pasternak. Acta Crystallogr. 9, 341 (1950)

Tyr: M.N. Frey, T.F. Koetzle, M.S. Lehmann, and W.C. Hamilton. J. Chem. Phys. 58, 2547 (1973)

Val: K. Torii and Y Iitaka. Acta Crystallogr., Sect. B 26 (1317 (1970)

Alanine - ala - A

H	1	0.26476	1.51500	-0.87469	0.00000
N	11	-0.55658	1.01000	0.00000	0.00000
C	8	-0.13563	1.75338	1.28757	0.00000
C	5	0.50862	1.36169	2.11981	1.22389
H	2	0.20305	2.81550	1.04139	0.07239
C	6	-0.61083	1.47082	2.04524	-1.29112
H	3	0.20405	0.40857	2.21402	-1.40246
H	3	0.20405	1.96760	3.00655	-1.28465
H	3	0.20405	1.81629	1.47804	-2.14388

O	10	-0.41090	0.30668	1.83908	1.81588
H	4	0.40212	1.83771	3.52010	2.31850
O	9	-0.54151	2.19098	3.08168	1.55330
H	1	0.26476	0.00000	0.00000	0.00000

Arginine - arg - R
neutral

H	1	0.25866	1.51500	0.87469	0.00000
N	11	-0.56929	1.01000	0.00000	0.00000
C	8	-0.13873	1.74714	-1.27676	0.00000
C	5	0.49690	3.14179	-1.10297	0.58168
H	2	0.19837	1.24351	-2.01960	0.63036
C	7	-0.39357	1.85015	-1.85158	-1.41454
N	11	-0.56929	3.96954	-1.02290	-4.44968
N	11	-0.56929	5.53508	-2.89977	-5.20001
N	12	-0.32575	5.42639	-0.78385	-6.35806
C	7	-0.39357	2.64562	-0.99917	-2.39376
C	7	-0.39357	3.30054	-1.83338	-3.44124
C	5	0.49690	4.97362	-1.55634	-5.37248
H	2	0.19837	2.26681	-2.85202	-1.32473
H	2	0.19837	0.84487	-1.99611	-1.77079
H	2	0.19837	1.99240	-0.29501	-2.87712
H	2	0.19837	3.35943	-0.40935	-1.91070
H	2	0.19837	4.04042	-2.42954	-2.93673
H	2	0.19837	2.64020	-2.53168	-3.86727
H	1	0.25866	3.52249	-0.15062	-4.74440
H	1	0.25866	5.22635	-3.53935	-4.48491
H	1	0.25866	6.26633	-3.21879	-5.83842
H	1	0.25866	6.06331	-1.18770	-7.03558
O	10	-0.42029	3.59759	0.03419	0.77045
H	4	0.39285	4.64433	-1.97179	1.19226
O	9	-0.55387	3.78999	-2.21164	0.85350
H	1	0.25866	0.00000	0.00000	0.00000

protonated

H	1	0.25866	1.51500	0.87469	0.00000
N	11	-0.56929	1.01000	0.00000	0.00000
C	8	-0.13873	1.74714	-1.27676	0.00000
C	5	0.49690	3.14179	-1.10297	0.58168
H	2	0.19837	1.24351	-2.01960	0.63036
C	7	-0.39357	1.85015	-1.85158	-1.41454
N	11	-0.38396	3.96954	-1.02290	-4.44968
N	11	-0.38396	5.53508	-2.89977	-5.20001
N	12	-0.14041	5.42639	-0.78385	-6.35806
C	7	-0.39357	2.64562	-0.99917	-2.39376
C	7	-0.39357	3.30054	-1.83338	-3.44124
C	5	0.68224	4.97362	-1.55634	-5.37248
H	2	0.19837	2.26681	-2.85202	-1.32473

H	2	0.19837	0.84487	-1.99611	-1.77079
H	2	0.19837	1.99240	-0.29501	-2.87712
H	2	0.19837	3.35943	-0.40935	-1.91070
H	2	0.19837	4.04042	-2.42954	-2.93673
H	2	0.19837	2.64020	-2.53168	-3.86727
H	1	0.25866	3.52249	-0.15062	-4.74440
H	1	0.25866	5.22635	-3.53935	-4.48491
H	1	0.25866	6.26633	-3.21879	-5.83842
H	1	0.25866	6.06331	-1.18770	-7.03558
H	1	0.25866	5.11532	0.11630	-6.68482
O	10	-0.42029	3.59759	0.03419	0.77045
H	4	0.39285	4.64433	-1.97179	1.19226
O	9	-0.55387	3.78999	-2.21164	0.85350
H	1	0.25866	0.00000	0.00000	0.00000

Asparagine - asn - N

H	1	0.26414	1.51500	0.87469	0.00000
N	11	-0.55787	1.01000	0.00000	0.00000
C	8	-0.13594	1.74818	-1.27857	0.00000
C	5	0.50743	1.91373	-1.82504	-1.41179
H	2	0.20258	2.76606	-1.08394	0.39955
C	7	-0.38567	1.13817	-2.25243	0.99732
O	10	-0.41185	-0.64039	-2.55485	-0.61063
N	13	-0.63440	-0.81101	-3.65958	1.36443
C	5	0.50743	-0.16825	-2.83267	0.51202
H	1	0.26414	-1.69362	-4.03087	1.11650
H	1	0.26414	-0.37414	-3.89913	2.23267
H	2	0.20258	1.85222	-3.09690	1.13232
H	2	0.20258	1.01672	-1.85451	1.98499
O	10	-0.41185	1.47123	-1.15926	-2.36417
H	4	0.40118	2.60680	-3.20342	-2.41369
O	9	-0.54276	2.50628	-2.93645	-1.44376
H	1	0.26414	0.00000	0.00000	0.00000

Aspartic acid - asp - D
neutral

H	1	0.26573	1.51500	-0.87469	0.00000
N	11	-0.55456	1.01000	0.00000	0.00000
C	8	-0.13514	1.76815	1.31316	0.00000
C	5	0.51048	3.28187	1.04934	-0.05916
H	2	0.20379	1.52767	1.75645	-0.84803
C	7	-0.38339	1.38794	2.18817	1.18169
O	10	-0.40941	-0.84747	2.16098	0.33729
O	9	-0.53954	-0.40625	3.39472	2.13138
C	5	0.51048	-0.06698	2.57226	1.17707
H	2	0.20379	1.90496	2.98584	1.16938
H	2	0.20379	1.57072	1.74145	2.01586
H	4	0.40359	-1.32190	3.66812	2.08837

O	10	-0.40941	4.02414	2.04270	-0.21283
H	4	0.40359	4.61523	-0.21830	-0.02292
O	9	-0.53954	3.66853	-0.19898	0.05685
H	1	0.26573	0.00000	0.00000	0.00000

ionized

H	1	0.26573	1.51500	-0.87469	0.00000
N	11	-0.55456	1.01000	0.00000	0.00000
C	8	-0.13514	1.76815	1.31316	0.00000
C	5	0.51048	3.28187	1.04934	-0.05916
H	2	0.20379	1.52767	1.75645	-0.84803
C	7	-0.38339	1.38794	2.18817	1.18169
O	10	-0.77268	-0.84747	2.16098	0.33729
O	9	-0.77268	-0.40625	3.39472	2.13138
C	5	0.51048	-0.06698	2.57226	1.17707
H	2	0.20379	1.90496	2.98584	1.16938
H	2	0.20379	1.57072	1.74145	2.01586
O	10	-0.40941	4.02414	2.04270	-0.21283
H	4	0.40359	4.61523	-0.21830	-0.02292
O	9	-0.53954	3.66853	-0.19898	0.05685
H	1	0.26573	0.00000	0.00000	0.00000

Cysteine - cys - C

neutral

H	1	0.25981	1.51500	0.87469	0.00000
N	11	-0.56690	1.01000	0.00000	0.00000
C	8	-0.13814	1.75222	-1.28556	0.00000
C	5	0.49911	3.23572	-0.97188	-0.10643
H	2	0.19925	1.59473	-1.76010	0.97631
C	21	-0.51471	1.27482	-2.22237	-1.09674
S	22	0.12014	1.65413	-1.71550	-2.78228
H	2	0.19925	1.75706	-3.20265	-0.95835
H	2	0.19925	0.18694	-2.40153	-0.99894
H	23	0.05860	2.70656	-2.48493	-2.43210
O	10	-0.41852	3.67921	0.14199	-0.02164
H	4	0.39460	4.91488	-1.85035	-0.32283
O	9	-0.55155	3.94097	-2.05580	-0.29798
H	1	0.25981	0.00000	0.00000	0.00000

dehydrogenated

H	1	0.25981	1.51500	0.87469	0.00000
N	11	-0.56690	1.01000	0.00000	0.00000
C	8	-0.13814	1.75222	-1.28556	0.00000
C	5	0.49911	3.23572	-0.97188	-0.10643
H	2	0.19925	1.59473	-1.76010	0.97631
C	21	-0.33597	1.27482	-2.22237	-1.09674
S	22	0.00000	1.65413	-1.71550	-2.78228
H	2	0.19925	1.75706	-3.20265	-0.95835

H	2	0.19925	0.18694	-2.40153	-0.99894
O	10	-0.41852	3.67921	0.14199	-0.02164
H	4	0.39460	4.91488	-1.85035	-0.32283
O	9	-0.55155	3.94097	-2.05580	-0.29798
H	1	0.25981	0.00000	0.00000	0.00000

Glutamic acid - glu - E
neutral

H	1	0.26478	1.51500	0.87469	0.00000
N	11	-0.55655	1.01000	0.00000	0.00000
C	8	-0.13562	1.75100	-1.28344	0.00000
C	5	0.50865	3.21509	-1.01034	0.37110
H	2	0.20306	1.31778	-1.92545	0.75945
C	7	-0.38476	1.73899	-1.97502	-1.37414
O	10	-0.41088	1.57619	-3.70730	-3.50898
O	9	-0.54148	-0.64079	-3.57908	-3.66345
C	7	-0.38476	0.38926	-2.52310	-1.83026
C	5	0.50865	0.50349	-3.30699	-3.07458
H	4	0.40214	-0.56691	-4.14376	-4.46287
H	2	0.20306	2.44801	-2.79079	-1.34300
H	2	0.20306	2.09855	-1.28234	-2.10176
H	2	0.20306	-0.01441	-3.16546	-1.03311
H	2	0.20306	-0.37208	-1.72319	-1.97354
O	10	-0.41088	3.70641	0.10435	0.28653
H	4	0.40214	4.83370	-1.89195	0.82931
O	9	-0.54148	3.83518	-2.07696	0.76993
H	1	0.26478	0.00000	0.00000	0.00000

ionized

H	1	0.26478	1.51500	0.87469	0.00000
N	11	-0.55655	1.01000	0.00000	0.00000
C	8	-0.13562	1.75100	-1.28344	0.00000
C	5	0.50865	3.21509	-1.01034	0.37110
H	2	0.20306	1.31778	-1.92545	0.75945
C	7	-0.38476	1.73899	-1.97502	-1.37414
O	10	-0.77511	1.57619	-3.70730	-3.50898
O	9	-0.77511	-0.64079	-3.57908	-3.66345
C	7	-0.38476	0.38926	-2.52310	-1.83026
C	5	0.50865	0.50349	-3.30699	-3.07458
H	2	0.20306	2.44801	-2.79079	-1.34300
H	2	0.20306	2.09855	-1.28234	-2.10176
H	2	0.20306	-0.01441	-3.16546	-1.03311
H	2	0.20306	-0.37208	-1.72319	-1.97354
O	10	-0.41088	3.70641	0.10435	0.28653
H	4	0.40214	4.83370	-1.89195	0.82931
O	9	-0.54148	3.83518	-2.07696	0.76993
H	1	0.26478	0.00000	0.00000	0.00000

Glutamine - gln - Q

H	1	0.26340	1.51500	-0.87469	0.00000
N	11	-0.55942	1.01000	0.00000	0.00000
C	8	-0.13632	1.75782	1.29527	0.00000
C	5	0.50600	1.16855	2.34594	1.02043
H	2	0.20200	2.77907	1.06714	0.31418
C	7	-0.38675	1.77529	1.91854	-1.39271
C	7	-0.38675	0.40992	2.35759	-1.88746
C	5	0.50600	0.46440	3.06832	-3.21764
N	13	-0.63616	-0.71127	3.23695	-3.81904
O	10	-0.41300	1.52044	3.46870	-3.69979
H	2	0.20200	2.44946	2.78812	-1.35190
H	2	0.20200	2.21273	1.20054	-2.09652
H	2	0.20200	-0.28356	1.51812	-1.96448
H	2	0.20200	-0.03108	3.05848	-1.18736
H	1	0.26340	-1.57033	2.95701	-3.37005
H	1	0.26340	-0.78012	3.74864	-4.67611
O	10	-0.41300	0.00195	2.02760	1.45086
H	4	0.40005	1.50932	3.82037	1.97696
O	9	-0.54427	1.91748	3.38635	1.29826
H	1	0.26340	0.00000	0.00000	0.00000

Glycine - gly - G

H	1	0.24939	1.51500	0.87469	0.00000
N	11	-0.58859	1.01000	0.00000	0.00000
C	8	-0.14343	1.71748	-1.22539	0.00000
C	5	0.47910	1.22636	-2.22281	1.04499
H	2	0.19126	1.51302	-1.62592	-0.99281
H	2	0.19126	2.79289	-1.07774	0.08953
O	10	-0.43453	1.95776	-3.21149	1.27655
H	4	0.37878	-0.10058	-2.68102	2.23437
O	9	-0.57265	0.07653	-1.97480	1.62490
H	1	0.24939	0.00000	0.00000	0.00000

Histidine - his - H

H	1	0.27626	1.51500	0.87469	0.00000
N	11	-0.53264	1.01000	0.00000	0.00000
C	8	-0.12980	1.75653	-1.29303	0.00000
C	5	0.53070	3.23626	-1.03497	-0.36444
H	2	0.21186	1.69673	-1.65435	0.91931
C	7	-0.36823	1.11481	-2.26356	-1.00418
N	12	-0.30478	-1.24674	-1.52084	-0.59246
N	15	-0.45476	-2.27499	-3.44566	-0.22269
C	14	-0.07499	-0.32874	-2.55112	-0.69002
C	14	-0.07499	-2.39681	-2.11359	-0.30353
C	14	-0.07499	-0.95403	-3.74012	-0.46418

H	16	0.26275	-2.93661	-4.08235	0.00180
H	2	0.21186	1.63953	-3.07525	-0.96851
H	2	0.21186	1.22339	-1.97149	-1.94163
H	16	0.26275	-3.20018	-1.62986	-0.18279
H	16	0.26275	-0.57222	-4.64092	-0.46841
O	10	-0.39323	3.48712	-0.01336	-1.03602
H	4	0.41958	4.96115	-1.66566	-0.24512
O	9	-0.51822	4.10160	-1.92083	0.06897
H	1	0.27626	0.00000	0.00000	0.00000

Isoleucine - ile - I

H	1	0.27071	1.51500	0.87469	0.00000
N	11	-0.54419	1.01000	0.00000	0.00000
C	8	-0.13261	1.76470	-1.30718	0.00000
C	5	0.52005	1.33876	-2.10380	1.24773
H	2	0.20761	1.50131	-1.86785	-0.88484
C	7	-0.37621	3.28003	-1.05136	-0.00108
C	7	-0.37621	3.72020	-0.23782	-1.26783
C	6	-0.59723	5.23468	-0.00729	-1.35675
C	6	-0.59723	4.03514	-2.37988	0.08190
H	3	0.20863	5.46161	0.55619	-2.24974
H	3	0.20863	5.74228	-0.95991	-1.39383
H	3	0.20863	5.56651	0.54340	-0.48941
H	3	0.20863	5.09802	-2.19020	0.08067
H	3	0.20863	3.77795	-2.99408	-0.76837
H	3	0.20863	3.76178	-2.89359	0.99261
H	2	0.20761	3.22739	0.72301	-1.24659
H	2	0.20761	3.40254	-0.77638	-2.14834
H	2	0.20761	3.52468	-0.46957	0.87580
O	10	-0.40175	1.12798	-3.34887	1.11120
H	4	0.41116	1.03132	-2.06335	3.06160
O	9	-0.52945	1.22215	-1.43833	2.37228
H	1	0.27071	0.00000	0.00000	0.00000

Leucine - leu - L

H	1	0.27071	1.51500	-0.87469	0.00000
N	11	-0.54419	1.01000	0.00000	0.00000
C	8	-0.13261	1.75590	1.29194	0.00000
C	5	0.52005	3.22815	0.99547	-0.00438
H	2	0.20761	1.55334	1.78444	-0.80192
C	7	-0.37621	1.28091	2.11064	1.20571
C	7	-0.37621	1.80042	3.53434	1.32366
C	6	-0.59723	1.28306	4.13330	2.62945
C	6	-0.59723	1.34389	4.38257	0.14316
H	2	0.20761	0.31562	2.15198	1.16809
H	2	0.20761	1.55466	1.63919	2.01064

H	2	0.20761	2.76680	3.52316	1.33484
H	3	0.20863	1.56337	3.57118	3.40033
H	3	0.20863	1.58455	5.07242	2.72595
H	3	0.20863	0.25930	4.15366	2.64724
H	3	0.20863	1.64799	3.98315	-0.71697
H	3	0.20863	1.65530	5.32078	0.24131
H	3	0.20863	0.31918	4.42693	0.08707
O	10	-0.40175	3.63535	-0.11405	0.33262
H	4	0.41116	5.02577	1.94961	-0.23918
O	9	-0.52945	3.97362	1.96782	-0.37606
H	1	0.27071	0.00000	0.00000	0.00000

Lysine - lys - K
neutral

H	1	0.25359	1.51500	0.87469	0.00000
N	11	-0.57984	1.01000	0.00000	0.00000
C	8	-0.14130	1.75350	-1.28778	0.00000
C	5	0.48717	1.60979	-1.95407	-1.37551
H	2	0.19449	2.80479	-1.06638	0.17907
C	7	-0.40086	1.23488	-2.23621	1.07556
N	11	-0.57984	0.20055	-3.10026	5.87028
C	7	-0.40086	1.26626	-1.65444	2.48990
C	7	-0.40086	0.82848	-2.68244	3.52260
C	8	-0.14130	0.64573	-2.07138	4.89591
H	2	0.19449	0.20717	-2.52367	0.84326
H	2	0.19449	1.87336	-3.12714	1.04162
H	2	0.19449	2.27914	-1.29460	2.73180
H	2	0.19449	0.61045	-0.78976	2.57011
H	2	0.19449	-0.11740	-3.12541	3.21831
H	2	0.19449	1.57251	-3.48202	3.57950
H	2	0.19449	1.56662	-1.63709	5.26323
H	2	0.19449	-0.13595	-1.31226	4.86838
H	1	0.25359	-0.60523	-3.59941	5.50526
H	1	0.25359	0.97880	-3.73237	6.07893
O	10	-0.42808	0.69754	-1.55843	-2.12094
H	4	0.38516	3.07455	-2.96628	-0.91148
O	9	-0.56414	2.47812	-2.89929	-1.64781
H	1	0.25359	0.00000	0.00000	0.00000

protonated

H	1	0.25359	1.51500	0.87469	0.00000
N	11	-0.57984	1.01000	0.00000	0.00000
C	8	-0.14130	1.75350	-1.28778	0.00000
C	5	0.48717	1.60979	-1.95407	-1.37551
H	2	0.19449	2.80479	-1.06638	0.17907
C	7	-0.40086	1.23488	-2.23621	1.07556
N	11	0.16657	0.20055	-3.10026	5.87028

C	7	-0.40086	1.26626	-1.65444	2.48990
C	7	-0.40086	0.82848	-2.68244	3.52260
C	8	-0.14130	0.64573	-2.07138	4.89591
H	2	0.19449	0.20717	-2.52367	0.84326
H	2	0.19449	1.87336	-3.12714	1.04162
H	2	0.19449	2.27914	-1.29460	2.73180
H	2	0.19449	0.61045	-0.78976	2.57011
H	2	0.19449	-0.11740	-3.12541	3.21831
H	2	0.19449	1.57251	-3.48202	3.57950
H	2	0.19449	1.56662	-1.63709	5.26323
H	2	0.19449	-0.13595	-1.31226	4.86838
H	1	0.25359	-0.60523	-3.59941	5.50526
H	1	0.25359	0.97880	-3.73237	6.07893
H	1	0.25359	0.16620	-2.36792	6.56483
O	10	-0.42808	0.69754	-1.55843	-2.12094
H	4	0.38516	3.07455	-2.96628	-0.91148
O	9	-0.56414	2.47812	-2.89929	-1.64781
H	1	0.25359	0.00000	0.00000	0.00000

Methionine - met - M

H	1	0.25836	1.51500	-0.87469	0.00000
N	11	-0.56991	1.01000	0.00000	0.00000
C	8	-0.13888	1.76039	1.29972	0.00000
C	5	0.49632	1.40089	2.11211	1.24518
H	2	0.19814	2.82393	1.11329	-0.02467
C	7	-0.39400	1.32178	2.15365	-1.27962
S	22	0.11947	1.19305	2.55453	-3.92355
C	21	-0.51745	1.63826	1.44263	-2.58489
C	21	-0.51745	1.33198	1.45241	-5.31354
H	2	0.19814	1.06922	0.52802	-2.65657
H	2	0.19814	2.69083	1.20541	-2.63199
H	2	0.19814	1.84184	3.10032	-1.26251
H	2	0.19814	0.25795	2.33558	-1.23001
H	3	0.19911	1.63111	0.47350	-4.96865
H	3	0.19911	2.07095	1.83298	-6.00297
H	3	0.19911	0.37698	1.38150	-5.81320
O	10	-0.42075	0.29094	2.01466	1.72145
H	4	0.39239	1.95807	3.29766	2.53781
O	9	-0.55448	2.32086	2.89449	1.75736
H	1	0.25836	0.00000	0.00000	0.00000

Phenylalanine - phe - F

H	1	0.25851	1.51500	0.87469	0.00000
N	11	-0.56959	1.01000	0.00000	0.00000
C	8	-0.13880	1.74784	-1.27797	0.00000
C	5	0.49662	1.32369	-1.98707	-1.24854

H	2	0.19826	2.81362	-1.10056	-0.00734
C	7	-0.39378	1.56281	-2.06649	1.32028
C	19	-0.03290	0.07555	-2.48013	1.59367
C	17	-0.22414	-0.45418	-3.62481	1.02109
C	17	-0.22414	-1.76976	-3.97855	1.27071
C	17	-0.22414	-2.55537	-3.18791	2.09356
C	17	-0.22414	-2.02530	-2.04356	2.66589
C	17	-0.22414	-0.71012	-1.68992	2.41595
H	2	0.19826	2.16624	-2.96048	1.27518
H	2	0.19826	1.90509	-1.45135	2.13933
H	16	0.24588	0.15836	-4.24139	0.38013
H	16	0.24588	-2.18322	-4.87116	0.82457
H	16	0.24588	-3.58128	-3.46461	2.28934
H	16	0.24588	-2.63786	-1.42585	3.30646
H	16	0.24588	-0.29688	-0.79763	2.86331
O	10	-0.42051	0.54015	-1.62632	-2.03081
H	4	0.39263	1.70834	-3.54735	-2.14370
O	9	-0.55417	1.99992	-3.14452	-1.34531
H	1	0.25851	0.00000	0.00000	0.00000

Proline - pro - P

N	11	-0.53582	1.07987	0.00000	0.00000
C	8	-0.13057	1.60269	-1.32818	-0.36477
C	5	0.52776	1.64033	-2.25636	0.83857
C	7	-0.37043	2.99220	-1.10197	-0.99544
C	7	-0.37043	2.76612	0.20232	-1.71289
C	7	-0.37043	1.61780	0.92995	-1.07062
H	2	0.21069	1.95642	1.85584	-0.62979
H	2	0.21069	0.85258	1.14630	-1.80110
H	2	0.21069	3.65837	0.80814	-1.65503
H	2	0.21069	2.53883	0.01061	-2.75100
H	2	0.21069	3.25437	-1.89460	-1.68008
H	2	0.21069	3.76357	-1.01673	-0.24418
H	2	0.21069	0.95241	-1.75994	-1.11129
O	10	-0.39558	1.06982	-1.99083	1.90555
H	4	0.41725	2.32926	-3.86044	1.41889
O	9	-0.52132	2.33346	-3.34874	0.61895
H	1	0.27473	0.00000	0.00000	0.00000

Serine - ser - S

H	1	0.24911	1.51500	0.87469	0.00000
N	11	-0.58917	1.01000	0.00000	0.00000
C	8	-0.14357	1.75346	-1.28771	0.00000
C	5	0.47856	3.24751	-1.01742	0.19626
H	2	0.19105	1.39146	-1.86420	0.86491
C	18	0.00749	1.49444	-2.08543	-1.26446

O	9	-0.57322	2.09362	-1.47774	-2.39150
H	2	0.19105	1.94288	-3.07607	-1.14023
H	2	0.19105	0.41817	-2.22153	-1.41283
H	4	0.37835	1.46943	-0.82744	-2.77992
O	10	-0.43496	3.61411	0.16825	0.32675
H	4	0.37835	4.92041	-1.76665	0.35793
O	9	-0.57322	4.03208	-2.06874	0.21155
H	1	0.24911	0.00000	0.00000	0.00000

Threonine - thr - T

H	1	0.26140	1.51500	0.87469	0.00000
N	11	-0.56358	1.01000	0.00000	0.00000
C	8	-0.13733	1.75564	-1.29149	0.00000
C	5	0.50216	1.40391	-2.10337	-1.23252
H	2	0.20047	2.73254	-0.98767	-0.19168
C	18	0.00786	1.44562	-2.01882	1.32275
O	9	-0.54832	1.78759	-1.07093	2.32867
C	6	-0.61852	2.25207	-3.27930	1.49021
H	2	0.20047	0.45260	-2.40122	1.49201
H	3	0.20145	1.99710	-3.74796	2.42986
H	3	0.20145	3.30443	-3.03648	1.48267
H	3	0.20145	2.03329	-3.95717	0.67842
H	4	0.39701	1.83038	-1.51578	3.51652
O	10	-0.41607	0.31571	-1.86604	-1.76757
H	4	0.39701	1.94223	-3.44977	-2.36547
O	9	-0.54832	2.29580	-2.98654	-1.61517
H	1	0.26140	0.00000	0.00000	0.00000

Tryptophan - trp - W

H	1	0.25833	1.51500	0.87469	0.00000
N	11	-0.56998	1.01000	0.00000	0.00000
C	8	-0.13889	1.73044	-1.24784	0.00000
C	5	0.49626	1.52177	-2.07571	-1.33242
H	2	0.19812	2.70226	-0.89038	0.05072
C	7	-0.39404	1.35823	-2.04576	1.24912
N	15	-0.48664	1.98653	-0.06282	4.46414
C	17	-0.22429	3.22797	-3.20344	3.64013
C	17	-0.22429	3.96813	-3.44107	4.81709
C	17	-0.22429	4.00422	-2.58218	5.90568
C	17	-0.22429	3.37724	-1.37224	5.88941
C	20	0.18161	2.65279	-1.11972	4.72110
C	19	-0.03292	2.55928	-2.00062	3.60500
C	14	-0.08025	1.74653	-1.40677	2.63421
C	14	-0.08025	1.35348	-0.24321	3.17090
H	16	0.24570	3.17210	-3.85524	2.83112
H	16	0.24570	4.49490	-4.30767	4.95222

H	16	0.24570	4.50279	-2.78859	6.79462
H	16	0.24570	3.38405	-0.67403	6.74944
H	16	0.24570	0.84935	0.57363	2.78770
H	16	0.24570	1.80800	0.70712	5.01572
H	2	0.19812	1.82590	-2.92974	1.11171
H	2	0.19812	0.29377	-2.13323	1.20965
O	10	-0.42079	0.48954	-1.91134	-2.04705
H	4	0.39235	2.09314	-3.45131	-2.41315
O	9	-0.55454	2.44034	-2.94220	-1.69008
H	1	0.25833	0.00000	0.00000	0.00000

Tyrosine - tyr - Y

H	1	0.25629	1.51500	-0.87469	0.00000
N	11	-0.57423	1.01000	0.00000	0.00000
C	8	-0.13993	1.75404	1.28872	0.00000
C	5	0.49234	1.33298	2.12688	1.20831
H	2	0.19655	2.81839	1.04836	0.08194
C	7	-0.39699	1.51802	2.05488	-1.31343
O	9	-0.55868	-3.67519	4.21169	-2.10507
C	19	-0.03317	0.11495	2.59094	-1.48184
C	17	-0.22596	-0.90679	1.80476	-2.02322
C	17	-0.22596	-0.18569	3.90672	-1.13246
C	18	0.00771	-2.18771	2.31674	-2.21569
C	17	-0.22596	-1.45786	4.43658	-1.32709
C	18	0.00771	-2.45198	3.64357	-1.88355
H	2	0.19655	2.22318	2.88396	-1.33486
H	2	0.19655	1.77930	1.39832	-2.14921
H	16	0.24376	-0.69663	0.78549	-2.33247
H	16	0.24376	0.59609	4.54382	-0.72210
H	16	0.24376	-2.96497	1.69349	-2.64178
H	16	0.24376	-1.67478	5.47064	-1.07621
H	4	0.38925	-4.28837	3.57542	-2.53139
O	10	-0.42394	0.31019	1.76781	1.81419
H	4	0.38925	1.71378	3.58737	2.26053
O	9	-0.55868	2.09085	3.15651	1.50196
H	1	0.25629	0.00000	0.00000	0.00000

Valine - val - V

H	1	0.27235	1.51500	0.87469	0.00000
N	11	-0.54077	1.01000	0.00000	0.00000
C	8	-0.13178	1.75765	-1.29497	0.00000
C	5	0.52321	3.24265	-1.02427	0.15950
H	2	0.20887	1.43526	-1.85410	0.86599
C	7	-0.37385	1.40908	-2.13563	-1.25014
C	6	-0.59348	1.59999	-3.63628	-0.99642
C	6	-0.59348	2.24707	-1.65196	-2.48371

H	3	0.20990	1.99349	-2.24829	-3.34805
H	3	0.20990	3.30033	-1.76072	-2.26791
H	3	0.20990	2.02643	-0.61375	-2.68468
H	3	0.20990	2.10276	-4.52672	-1.34444
H	3	0.20990	0.53051	-3.78639	-1.04121
H	3	0.20990	1.89254	-3.43323	0.02304
H	2	0.20887	0.36590	-1.97032	-1.47684
O	10	-0.39923	3.64933	0.13258	-0.06001
H	4	0.41365	4.90309	-1.72186	0.53702
O	9	-0.52612	4.00425	-2.03002	0.51953
H	1	0.27235	0.00000	0.00000	0.00000

The average values for the main chain bond lengths and bond angles are given below [R.A. Laskowski, D.S. Moss, and J.M. Thornton, J. Mol. Biol. *231*, 1049 (1993)]:

bond	length (Å)	atoms	angle (degrees)
C–N	1.32	C–N–C_α	121
C–O	1.24	C_α–C–N	116
C_α–C	1.52	C_α–C–O	120
C_α–C_β	1.53	C_β–C_α–C	110
N–C_α	1.46	N–C_α–C	111
		N–C_α–C_β	110
		O–C–N	123

Lecture Notes in Chemistry

For information about Vols. 1–26
please contact your bookseller or Springer-Verlag

Editorial Policy

This series aims to report new developments in chemical research and teaching - quickly, informally and at a high level. The type of material considered for publication includes:

1. Preliminary drafts of original papers and monographs

2. Lectures on a new field, or presenting a new angle on a classical field

3. Seminar work-outs

4. Reports of meetings, provided they are
   a) of exceptional interest and

   b) devoted to a single topic.

Texts which are out of print but still in demand may also be considered if they fall within these categories.

The timeliness of a manuscript is more important than its form, which may be unfinished or tentative. Thus, in some instances, proofs may be merely outlined and results presented which have been or will later be published elsewhere. If possible, a subject index should be included. Publication of Lecture Notes is intended as a service to the international chemical community, in that a commercial publisher, Springer-Verlag, can offer a wider distribution to documents which would otherwise have a restricted readership. Once published and copyrighted, they can be documented in the scientific literature.

Manuscripts

Manuscripts should comprise not less than 100 and preferably not more than 500 pages. They are reproduced by a photographic process and therefore must be typed with extreme care. Symbols not on the typewriter should be inserted by hand in indelible black ink. Corrections to the typescript should be made by pasting the amended text over the old one, or by obliterating errors with white correcting fluid. Authors receive 50 free copies and are free to use the material in other publications. The typescript is reduced slightly in size during reproduction; best results will not be obtained unless the text on any one page is kept within the overall limit of 18 x 26.5 cm (7 x $10^1/_2$ inches). The publishers will be pleased to supply on request special stationary with the typing area outlined.

Manuscripts should be sent to one of the editors or directly to Springer-Verlag, Heidelberg.